CHUR: C00359500161
TITLE: Adv. in insect phys.
CLASS: 595.701
VOL : 16 YR: 1982
COPY : 1 LOC : E
7-day loan

AF616323

Advances in

Insect Physiology

Volume 16

Advances in Insect Physiology

edited by

M. J. BERRIDGE
J. E. TREHERNE
and V. B. WIGGLESWORTH

Department of Zoology, The University
Cambridge, England

Volume 16

1982

ACADEMIC PRESS

A Subsidiary of Harcourt Brace Jovanovich, Publishers

London New York
Paris San Diego San Francisco São Paulo
Sydney Tokyo Toronto

ACADEMIC PRESS INC. (LONDON) LTD
24/28 Oval Road
London NW1 7DX

United States Edition published by
ACADEMIC PRESS INC.
111 Fifth Avenue
New York, New York 10003

British Library Cataloguing in Publication Data

Advances in insect physiology.
Vol. 16
1. Insects – Physiology
595.7′01′05 QL495

ISBN 0–12–024216–8
ISSN 0065–2806

Typeset by Bath Typesetting Ltd., Bath
and printed in Great Britain by John Wright & Sons (Printing) Ltd, Bristol

Contributors

J. Beetsma

Department of Entomology, Agricultural University, Binnenhaven 7, Wageningen, The Netherlands

E. A. Bernays

The Centre for Overseas Pest Research, College House, Wrights Lane, London W8 5SJ, UK

R. F. Chapman

The Centre for Overseas Pest Research, College House, Wrights Lane, London W8 5SJ, UK

A. J. Howells

Department of Biochemistry, Faculty of Science, Australian National University, Canberra, ACT 2600, Australia

N. A. Pyliotis

Anatomical Pathology Department, Prince Henry's Hospital, Melbourne, Victoria 3004, Australia

S. J. Simpson

The Centre for Overseas Pest Research, College House, Wrights Lane, London W8 5SJ, UK

K. M. Summers

Genetics Laboratory, Department of Biochemistry, University of Oxford, South Parks Road, Oxford OX1 3QU, UK

J. de Wilde

Department of Entomology, Agricultural University, Binnenhaven 7, Wageningen, The Netherlands

P. G. Willmer

Department of Zoology, University of Cambridge, Downing Street, Cambridge CB2 3EJ, UK

Contents

Microclimate and the Environmental Physiology of Insects

P. G. Willmer

Department of Zoology, Cambridge, UK

1 Introduction

During the last few decades of intensive entomological research a number of review articles have considered the relationship between climate and the distribution, abundance or development of insects (Uvarov, 1931; Buxton, 1932a; Kühnelt, 1934; Gunn, 1942; Ludwig, 1945; Andrewartha and Birch 1954; Wellington, 1957; Klomp, 1962). Such studies have usually been from the standpoint of the ecologist, and so have highlighted the grosser effects of weather patterns on insect populations (Andrewartha and Birch 1954; Lowry, 1967; Henson, 1968). Some of the authors have also noted the primacy of very local conditions in determining the location and activity of such small animals (Grimm, 1937; Mazek-Fialla, 1941; Smith, 1954; Cloudsley-Thompson, 1962a), and the inappropriateness of the conventional meteorological approaches (Wellington 1957; Wellington *et al.*, 1966; Lowry, 1967); Cloudsley-Thompson (1962a) in particular has produced a wealth of examples of climatic effects on insect distribution on a very small scale. There is now, however, scope for an extension of such approaches. Firstly, the science of microclimatology has benefited from much improved techniques, with more appropriate instrumentation becoming available (Cloudsley-Thompson, 1962a, 1967; Monteith, 1972; Rosenberg, 1974; Schwerdtpfeger, 1976; Unwin, 1980), so that it is now possible to analyse the physical environment on a scale (both spatial and temporal) relevant to its insect inhabitants. Secondly, while it has often been recognised that the insects' responses to climate must be mediated through physiological mechanisms, it has only recently become feasible to examine some of the changing physiological parameters under field conditions, thereby allowing direct assessment of the importance of ever changing surroundings.

The major climatic parameters, whether on a macro- or microscale, are the temperature of the environment and its water content. The direct effects of climatic change should thus be sought on the physiology of thermal balance and of water balance respectively. Any insect must, if averaged through time, maintain its energy (heat) and its water contents constant. Both rates of metabolism and rates of water loss are partly functions of internal physiology; but both can be directly modified by the environment. Hence the environmental variables (which are principally temperature, humidity, radiation and wind) and the physiological parameters are all interdependent, and the transductions of energy and moisture across the interface of the insect cuticle can never be reduced to a simple model. Thus new approaches and techniques to assess all these variables acting in concert form a crucial aspect of this area of environmental physiology, and are central to this review. In the light of these developments, considerations of the whereabouts and extent of microclimatic gradients, of their location and use by insects, and of their particular physiological effects are also presented. However, it must also be stressed that climate has many less "direct" effects on insects which may nevertheless be vital in the longer term and which are themselves mediated physiologically. These include effects on the timing and control of reproduction and development for the insect itself, on parallel components of the life cycles of predators, pathogens and parasites, and on food sources, whether plant or animal; thus rendering the whole area of interactions between insects and weather exceedingly complex.

2 Microclimate

2.1 THE PROBLEMS OF MICROCLIMATE

Micrometeorology is concerned with the atmospheric processes occurring at or just above the earth's surface (Sutton, 1953; Geiger, 1965). Gross climatic changes are determined by large-scale energy redistributions consequent upon differential solar heating (Monteith, 1960; Geiger, 1965; Rosenberg, 1974). But on a scale relevant to individual animals, energy may be redistributed much more rapidly and drastically, with sharp vertical gradients of temperature and humidity arising close to heated surfaces. Horizontal movements of air may be very limited close to such surfaces, with static "boundary layers" comparable to those occurring at the interfaces of any flow system. Hence the vertical gradients created by varied heating of, and evaporation from, different substrates may be maintained with minimal mixing for considerable periods, though the extent and even the direction of gradients will change gradually through the diurnal cycle. Both the steepness and the stability of such gradients are therefore of some importance, and will vary with the nature of the substrate, its proximity to other surfaces having different properties, the solar radiation inputs and the wind strength. Hence the question of scale arises to pose some inevitable problems for the biological investigator: how close together spatially, and how frequently, must measurements be taken to give a meaningful picture of an animal's physical environment? Clearly the answer will depend not only on the factors listed above, but also on the size of the animal; for studies of large mammals, the instrumentation in a conventionally-sited Stevenson screen may be adequate, but for the entomologist readings as finely-spaced as is practically possible may be essential.

At the same time, it is unfortunate that many of the techniques available for climatic measurements have only a moderate response time, or require a minimum volume of air for sampling well above that which is spatially desirable. Any measurement schedule is thus likely to be a compromise between conflicting requirements when small arthropods are involved (c.f. Cloudsley-Thompson, 1962a; Monteith, 1973). And the techniques themselves often involve an "uncertainty principle", modifying just that parameter which they purport to measure, so that it may be necessary to forego some degree of accuracy in favour of realism (MacFadyen, 1967). All these problems of technique and instrumentation are dealt with more fully in a recent review by Unwin (1980), which gives valuable guidance on the selection of appropriate methodology. In the end the "scale" of microclimate is very much a matter to be defined by the experimenter, bearing all these factors in mind.

Studies of an insect's environmental physiology require assessment of the conditions in which a given species lives, so that besides microclimatic data one also needs a good idea of the proportions of time spent in different types of activity and in different places, both by day and by night and through a season (c.f. Bursell, 1974b). Once this information is available it is sometimes possible to relate bioclimatic measurements to standard meteorological data, deriving "average" microclimate from general air temperatures and humidities (e.g. Haufe and Burgess, 1956); but, particularly where radiation plays an important role in determining the habitat temperatures (Wellington, 1950; Henson, 1958; Shepherd, 1958) this is a dangerous exercise and may prove unsatisfactory. Thus it is always particularly crucial to specify the techniques and timing of measurements, and to avoid averaging or extrapolating from inadequate data.

The concept of microclimate, and also some of the associated problems, become particularly relevant to the arthropod biologist if the "processes at the earth's surfaces" are taken to include those at the surfaces of plants and animals themselves. For while the physical environment influences the growth and form of living material, so the living surfaces are also contributing to, or even creating, unique microenvironments around themselves, both by their own gains and losses of energy and of water and by interactive shading effects on neighbouring physical substrates. Because of the added complexity which this interdependence brings to the problem, it is necessary to study insect physiology in relation to specific microhabitats; the mere accumulation of climatic data, coupled with extrapolations from physiological studies in static laboratory simulations, will never provide adequate analyses. The section which follows therefore considers both biotic and abiotic situations in which microclimatic gradients of relevance to insects can occur, as an introduction to the scope of this area of research.

2.2 MICROCLIMATIC CONDITIONS

In terms of physiological effect, the most important parameters of climate to a small terrestrial animal will be the temperature and moisture content (whether saturation deficit or relative humidity) of the air. These two factors are continuously modified by the other two major climatic variables, solar radiation and wind. Insects are especially vulnerable to all of these features of the weather because of their small size and proportionately large surface area; the compensation for this is their ability to exploit much more finely graded habitats, escaping from harsh ambient conditions into more favourable microniches.

This section considers examples of the gradients of temperature and RH which can occur under different conditions, and of their stability in the discrete microhabitats used by insects; it draws largely on detailed studies published since an earlier review by Cloudsley-Thompson (1962a). This should provide some indication of the extreme variability of conditions available to a mobile insect within relatively small areas.

2.2.1 *Abiotic situations*

(*a*) *Soil and rock* Many insects, both as larvae and adults, inhabit cracks or holes in soil, sand or rock. In temperate or hot climates the surface of such substrates may suffer extreme fluctuations in temperature or humidity, and can constitute one of the least hospitable of all environments. Surface temperatures may reach 80–90°C in deserts, 40°C in excess of air temperature (Berry and Cloudsley-Thompson, 1960), and quite inconsistent with normal physiological functioning. Yet many soils have a considerable insulating effect, so that cracks within a few cm of the surface can provide an equable habitat even in the hottest deserts (Madge, 1965; Edney, 1960; Cloudsley-Thompson, 1962b). Solar radiation is the prime determinant of soil temperature (Cloudsley-Thompson, 1962a), which will in turn affect evaporation and hence humidity conditions within air spaces; and received radiation will itself depend on such factors as soil slope and colour, heat capacity and drainage (c.f. Rosenberg, 1974), so that conditions beneath the surface cannot be simply predicted. However, many investigations of subsoil conditions have been published, and valuable information is contained in reviews by Kristensen (1959), Geiger (1965) and Rosenberg (1974).

Taking the extreme case of deserts, diurnal temperature variations in the upper soil layers may be as great as the annual range (e.g. Sinclair, 1922; Hamilton, 1971), whereas at 20–100 cm depth temperatures may be almost constant. Similarly humidities may vary from almost 0% at the surface to a reasonable 50% at only 50 cm depth (Pierre, 1958; see also Hadley, 1970; Edney *et al.*, 1974). Hence it is essential to know the depth at which an insect lives, or the time spent at different depths, whether in natural crevices or in a burrow: even in deserts, fossorial insects have a ready choice of a range of microclimates without expending much energy in locating them. Above the surface insects may still have some scope for choice, since temperature gradients can be very steep over the first few mm, and many desert beetles have very long legs to take advantage of this fact (Cloudsley-Thompson, 1956; Hamilton, 1971). The environmental physiology of desert arthropods has received a good deal of recent attention, much of it reviewed by Edney (1967a, 1974) and Cloudsley-Thompson (1975); it is clear that the majority of the animals concerned do spend at least part of their time within the soil, and tend to be crepuscular or even nocturnal. Some further examples of the conditions which they thus take advantage of are shown in Fig. 1.

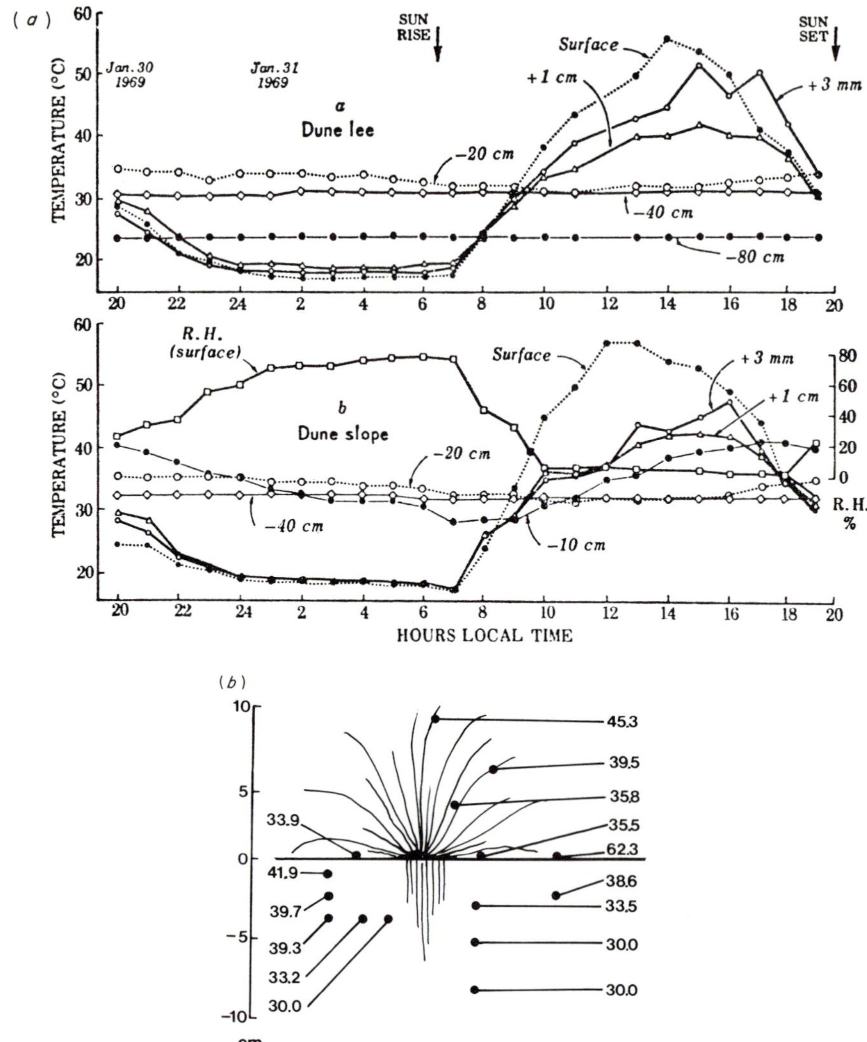

Fig. 1 Microclimatic conditions in deserts. (*a*) Temporal and spatial variation of conditions above and below the surface of the Namib desert during summer. From Holm and Edney, 1973. (*b*) Conditions at midday in the Namib desert. Figures on the right show the air and soil temperatures (°C), while those on the left show the values of T_b for white beetles (*Onymacris langi*) found at various sites. (After Hamilton, 1975)

In nondesert conditions, the microclimate of burrows in soil has been described rather less often, though studies of bees (Michener *et al.*, 1958) and of spiders (Humphreys, 1975) are available, the latter showing as much as a

40% difference between soil surface humidity and that only 10 cm down. An early study by Mail (1932) showed that the depth at which beetles hibernated depended upon soil temperature, which was in turn a function of insulation and soil texture. Recent studies of a sand-dwelling wasp (Willmer, unpublished) provide more detailed profiles of burrow conditions; examples of temperature and humidity gradients are shown in Fig. 2, and indicate almost constant hygrothermal conditions only 10 cm beneath the surface, through a summer day. During wet periods, lower soil layers are likely to be even more moist than this, and for some soil dwelling arthropods this can produce problems of having excess water (Galbreath, 1975).

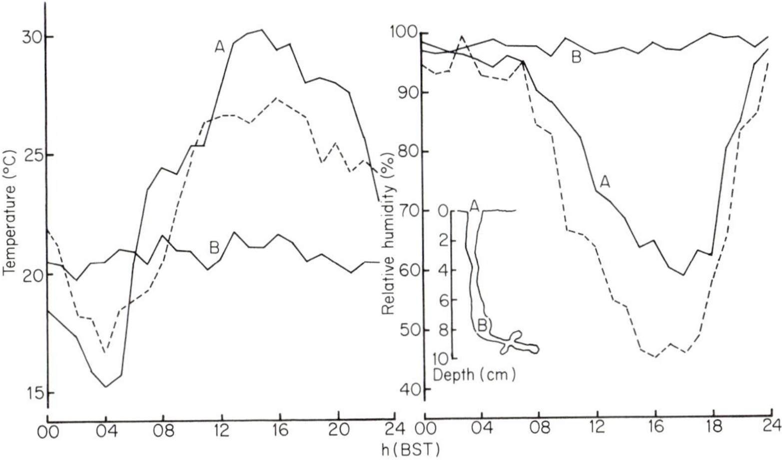

Fig. 2 Conditions of temperature and humidity at different depths within the burrow of a soil-nesting wasp (*Cerceris arenaria*), through a summer day. Dashed lines show ambient conditions. (Willmer, unpublished)

In cold climates there may be more difficulties to overcome (Corbet, 1972). However, conditions within the soil can be considerably modified and improved (from the insect's point of view) by the insulation from a layer of ice or snow, since these surfaces have a very high thermal reflectivity (Monteith, 1973), so that soil may remain habitable for a cold-hardy insect at most depths (Mail, 1932; Wellington, 1950; Downes, 1965). Some arctic insects may even be "forced" to burrow under snow to avoid freezing (Somme and Ostbye, 1969). During the brief arctic summers, though, insolation is sufficient to produce very large temperature excesses at ground level, which both plants and insects exploit (Corbet, 1972).

(*b*) *Caves* A second major site of "usable" abiotic microclimates are those which occur in caves, and the environmental physiology of troglodytes has again become a target for arthropod biologists (Bull and Mitchell, 1972;

Howarth, 1980). Within deep caves, conditions of stable temperature and constant very high humidity can be found at all seasons in both tropical and temperate zones (Howarth, 1980) and even in deserts (Buxton, 1932b). Recently physiologists have begun to consider the problems of water balance and osmotic physiology in such conditions, but there is considerable scope for the application of direct measurements in these situations.

(*c*) *Over water* A third type of abiotic microclimate can be found in association with water. Over standing pools, lakes and seas and to a lesser extent over running water and on marine shores boundary layers of relatively cool and humid air may be found on even the hottest days (Geiger, 1965; Monteith, 1973). These can be potentially important to the insects, which constitute a major part of the surface fauna because of the potential their cuticle has to provide hydrophobic areas. Examples of the microclimatic conditions around ponds, streams and seas on a fairly coarse scale are discussed by Geiger (1965). On a more appropriate scale for insects there has been little work, but the studies of Willmer (1982, and unpublished) of insects resting on and around lily pads include information on the hygrothermal conditions. Profiles of temperature and RH are shown in Fig. 3; these show the favourability of conditions within the first few mm, as experienced by the gerrids, gyrinids and flies which can make use of this zone.

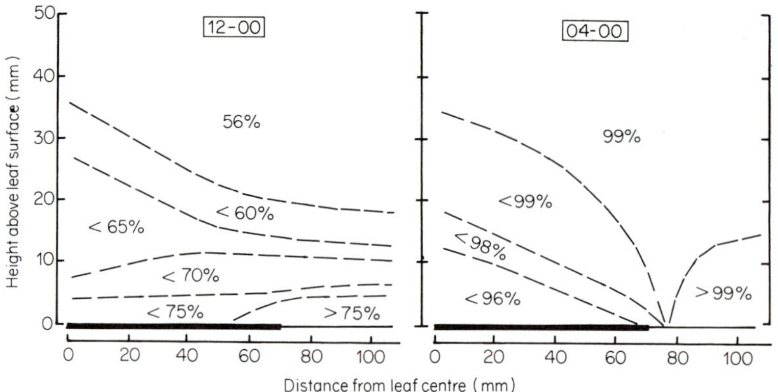

Fig. 3 Humidity profiles above floating lily leaves (thickened baseline) and open water, measured by the potassium acetate droplet technique (see text). The lily leaf provides a zone of relative hygrothermal stability and is exploited by a wide range of insects as a resting site. (From Willmer, 1982)

2.2.2 *Biotic situations*

(*a*) *Plants* The presence of plants ameliorates microclimatic conditions to a marked degree, and a considerable literature attests to the effects of herbage on soil temperatures and humidities (Shanks, 1956; Kristensen, 1959; MacHattie and McCormack, 1961; Mattsson, 1965; Geiger, 1965; Unwin, 1978). Apart from the indirect effects on soil microclimate, plants create their own unique microhabitats, which many insects exploit. These range from gall-makers and leaf-miners enclosed within the plant, through those insects living on leaves or flowers, to less specialised forms living beneath plant canopies or in the resultant humus and debris. The potential interactions between plant and insect are clearly enormous.

Habitats within plant tissues. Insects living actually within plant tissues become largely dependent upon the regulatory mechanisms of the plant itself. They will in general be in thermal equilibrium with the neighbouring tissues, at a temperature determined by the plant's transpiration, and their water balance will be crucially dependent upon the concentrations of the plant sap. There has been remarkably little work on the physiology of plant "parasites" however, the difficulties of sampling plant fluids being a notorious deterrent. Perhaps the most instructive examples are those concerning sap-feeding insects, especially the Homoptera, which if not strictly within the plant are at least functionally plugged into it. Here the osmotic and water balance problems may be extreme, with an excess of water, and perhaps of some organic solutes (Cheung and Marshall, 1973; Marshall and Cheung, 1975; Downing, 1980); some of the resultant physiological mechanisms are discussed by Cheung and Marshall (1973) and Marshall and Cheung (1974). The effects of water stress in the host plant are known to be manifested in the insects feeding on it (Kennedy and Booth 1959; Downing, 1980), and there is evidence that temperature tolerances of psyllids are modified by the water balance and evaporative rates of their food plants (Hoffman *et al.*, 1975). Osmotic balance may be less of a problem for an insect actually within a gall or leaf-mine, when plant cells rather than the flowing sap form the main dietary intake. And in all these cases, temperature maintenance at least is likely to be relatively simple, with much control delegated to the plant. The total physiological stresses may therefore be reduced in such microhabitats, and the predictability of the environment should also reduce the complexity of homeostatic mechanisms required.

Habitats on leaves and flowers. The microhabitats associated with leaf surfaces and flowers have not received a great deal of attention, yet those studies which are available are particularly instructive. Some microclimatic analyses of leaves (Ramsay *et al.*, 1938; Haarløv *et al.*, 1952; Raschke, 1956; Norgaard, 1956; Henson, 1958; Lewis, 1962; Flitters, 1968; Douwes, 1976; Burrage, 1976) have shown the higher humidities to be expected close to the transpiring surface, and particularly on the lower side; and in some cases these surfaces are also measurably cooler. Gradients may also be considerably steeper at one surface than at the other; some examples are shown in

Fig. 4. These effects have clear significance for the insect occupants of leaves. The flea beetles studied by Tahvanainen (1972) distributed themselves in accordance with microclimate, and desert caterpillars also moved around their foodplants to utilise favourable microclimates and stabilise their body temperature (Casey, 1976a). Cicadas (Heath and Wilkin, 1970) and psyllids (Hoffmann *et al.*, 1975) also profit from cooler moister air on the shaded sides of plant stems. Caterpillars of *Pieris brassicae* showed changes of water balance which could be correlated with their position within the microclimatic frameworks around cabbage leaves (Willmer, 1980a). In the case of insects which produce leaf rolls, these effects may be even more marked; nettle leaf rolls made by *Pleuroptya* larvae maintain an environment always in excess of 95% RH, and the larval water balance benefits accordingly (Willmer, 1980a). Similarly the leaf rolls made by *Choristoneura* show profound microclimatic effects, hanging vertically to produce a "chimney effect" and thus giving a steady internal temperature 8°C above ambient when insolated (Henson, 1958).

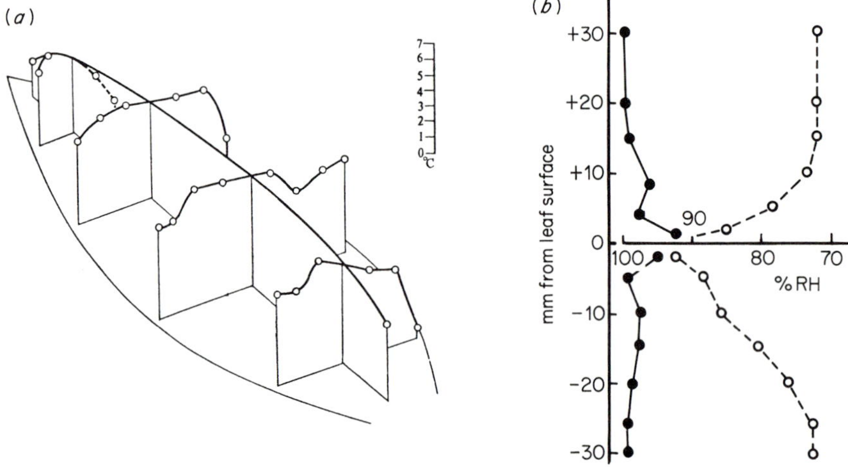

Fig. 4 Profiles of microclimate around leaf surfaces. (*a*) The temperature excesses recorded above a *Canna* leaf (from Raschke 1956). (*b*). Humidity profiles above and below leaves of *Brassica oleracea* (cabbage), at 14.00 (○) and 22.00 (●) in summer. (Willmer, unpublished)

Flowers may provide yet more specialised microhabitats, since their colouration and form can affect the temperature (Hocking and Sharplin, 1965; Kevan, 1975) and humidity (Corbet *et al.*, 1979) within them (see also Büdel, 1956; Baker and Hurd, 1968). Thermal balance in arctic insects can be materially aided by basking within flowers (Hocking and Sharplin, 1965), which may even act as parabolic reflectors (Kevan, 1975); the efficacy of flower basking has also been shown, though less dramatically, for temperate insects and flowers (Lack, 1976; Dafni *et al.*, 1981). Humidity may be very high within corollas, especially when these are of elongate form where the trapped air may equilibrate with watery nectar (Corbet *et al.*, 1979); in the extreme case of closed buds, the RH may be around 99% at all times (Willmer 1980a and unpublished observations). Figure 5 shows some examples of the range of conditions to be expected, and further examples of profiles are given in Corbet and Willmer (1981).

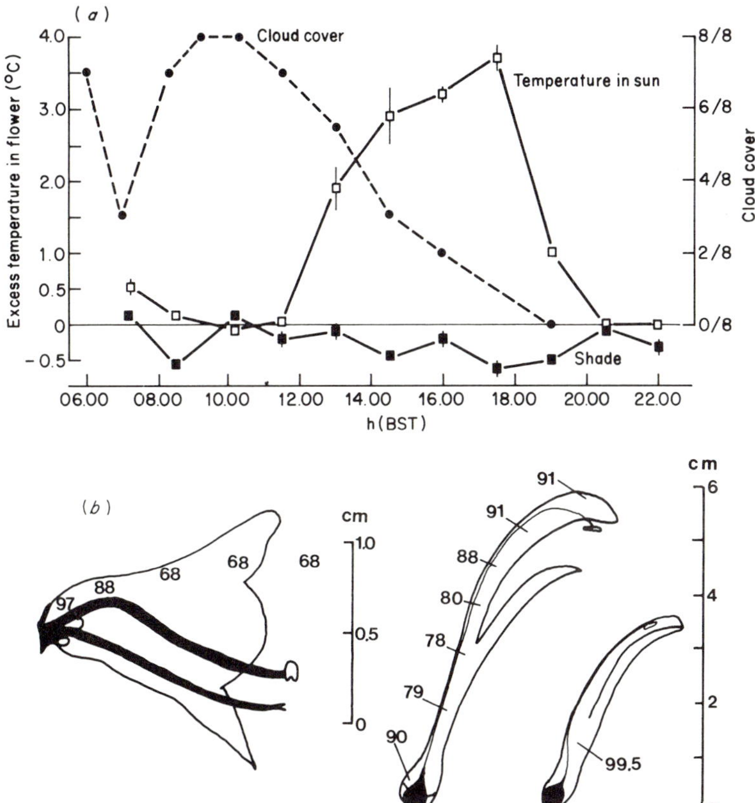

Fig. 5 Temperature (*a*) and relative humidity (*b*) conditions in flowers. (*a*) shows the changes in temperature excess recorded in *Crataegus* flowers in sun (□) or shade (■) through a day (from Corbet, Unwin and Prŷs-Jones, 1979). (*b*) shows humidity profiles (as % RH) measured using the droplet technique in flowers of *Echium vulgare* (left) and of *Justicia aurea* (right; mature flower and bud). (Willmer, unpublished)

On a broader scale the conditions within and below stands of vegetation may be much more favourable to insects than those over bare ground, and there is scope for considerable vertical stratification, particularly within uniform crops (Broadbent, 1950; Willmer unpublished). Both temperature and humidity will depend upon incoming radiation, on longwave radiation losses, and on the wind through its effects on turbulence and convective heat and moisture exchanges. Hence the height, architecture, reflectivity and density of herbage will modify any microclimatic profiles. There may be hot layers immediately above the ground, with cooler damper air trapped beneath the leaf canopy (Waterhouse, 1950; Willmer, unpublished observations), and these gradients may invert on a calm night (Waterhouse, 1950). Such effects can be related to the distributions or feeding patterns of many insects, including sawfly larvae (Waterhouse, 1955), caterpillars (Casey, 1976a) and ants and bees (Willmer and Corbet, 1981).

Debris and dead vegetation. Further favourable microclimates occur in dead or decaying vegetation, and humus supports a rich invertebrate fauna diagnostic of the equable conditions. In a temperate pine woodland site, conditions under 2–3 cm of needles and debris showed only a 3°C and 17% RH range over a 24 hour cycle in which ambient conditions varied by 13°C and 45% RH (Willmer, unpublished). Extreme microclimatic effects in such sites have also been reported by Stoutjesdijk, (1977). Some observations are also available for dead wood: boring insects inhabit galleries of remarkably constant temperature (Haarløv and Petersen, 1952), and bark beetles lay their eggs within logs to avoid both overheating in highly insolated sectors and excessive water in the shaded damp areas (Schimitschek, 1931). Nesting carpenter bees may choose to make their holes on the more insolated side of available wood for similar reasons (Corbet and Willmer, 1980).

(*b*) *Animals* Endoparasites. For any arthropod living within a larger host, and particularly within a homeothermic vertebrate, the problem of maintaining a favourable microhabitat is largely delegated to the physiological mechanisms of the host. In such situations the parasite need expend little or no energy in managing its own environment. For parasites of other arthropods the advantages are less profound, though there has been little direct work to demonstrate this (see Fisher, 1971). The osmotic environment may stay reasonably constant, at least while the host remains functional and is still much larger than its attacker; though Corbet (1968) showed quite marked changes in the osmotic pressure of haemolymph from *Ephestia* when parasitised by *Nemeritis*, and further demonstrated that the parasite could not feed rapidly when the host haemolymph was too concentrated. In some cases, endoparasitism causes specific changes in host blood composition (see Fisher, 1971), which may serve to ameliorate conditions for the intruder. On the thermal side, the temperature of the parasitoid will be only slightly buffered by the surrounding host tissues in most cases, and there may be difficulties once the host begins to succumb if it becomes inactive while in an exposed site. Hence the "microclimatic" advantages for an insect parasitoid may be only limited.

Ectoparasites. Insects parasitic on the outer surfaces of mammalian hosts will benefit quite substantially from the relative constancy of temperature. In many cases where the host has a thick layer of hair or other surface modifications the effects can be profound, with a layer of trapped still and humid air providing an ideal microhabitat. Instructive examples of this phenomenon have been described by Davies (1948) for *Lucilia* larvae developing under sheep fleece, and by Wigglesworth (1941) for the human louse; in the latter case the conditions between skin and clothing remained at 28–31°C and 23–70% RH. A further neat demonstration is the work of Morgan (1964) on the horn-fly *Haematobia*, where the microclimatic differences over dark and light patches of the host's skin were sufficient to account for the parasite's distribution on cattle. However, the thermal advantages may be counterbalanced in blood-feeding ectoparasites by problems in achieving osmotic balance, as they periodically take in excess water and/or nitrogen, so that specialised excretory mechanisms may be required (Maddrell, 1980b).

Conspecific aggregations. One of the less recognised advantages of animal aggregations, at least for small arthropods, may be the creation and maintenance of local microclimates around a group, as first demonstrated by Allee (1926) for woodlice. Small animals closely grouped effectively increase their thermal mass and so may "damp out" changes in body temperature as the environment heats up or cools. Furthermore losses of water by evaporation or excretion can contribute to a local humid area around a cluster of animals if moisture is retained within the interstices of the group. In some cases, and particularly for larval Lepidoptera, the effects are accentuated by spinning communal webs or nets which help to create the microclimate. Direct evidence for physiological effects attributable to group-living on this scale is rare; but development rates can certainly be affected by raised temperatures in *Malacosoma* (Sullivan and Wellington, 1953) and *Inachis* (Mosebach-Pukowski, 1937), and water balance in groups of *Pieris* and of *Inachis* is also known to be more easily achieved than in the same animals when solitary (Willmer, 1980a). Even in insects which do not specifically aggregate, significant effects can occur; Pimentel (1958) has shown an increment in temperature in cultures of the flour beetle *Tribolium* which is proportional to the number of insects present.

Further evidence for microclimatic effects of group living comes from the many studies of social insects, and much of this work has been reviewed by Wilson (1971). The literature on this subject is relatively large, and includes good evidence for the control of both temperature and humidity, the degree of control increasing from the loose aggregations of army ant bivouacs up to the highly organised societies of many bees, wasps, ants and termites. Specific techniques may be used to effect this control, such as fanning, collecting and adding water, clustering or "shivering", and these are particularly directed to the brood area where the larvae are especially sensitive to heat loss and desiccation. In some cases the behaviour may be subject to pheromonal control (Ishay, 1972, 1973), although for honey-bees the queen pheromones are unnecessary (Heinrich, 1981). The architecture of the nests of social insects is also an important controlling factor for the internal microclimate, and the structure may be designed and orientated for maximum "air-conditioning".

Associations with excrement. Many insects, particularly in the larval stages, are associated with animal excrement, and the microclimatic stability of dungpats can be a critical factor in their success (Greenham, 1972; Tyndale-Briscoe *et al.*, 1981). Fresh dung provides high temperature and moisture, while older dung patches may develop a hardened crust and remain warm and moist internally. Conditions in cow dung (Greenham, 1972) have been described in relation to the bushflies which utilise it.

2.2.3 *Manufactured microclimates*

This category covers a variety of structures made either by the insects themselves (which would overlap with burrows or nests already described), or by other animals including man. In the first category come such structures as larval cases, the puparia of higher flies, and pupal cases of some Lepidoptera where a very local microhabitat is maintained and may be carried around by an individual insect. Chauvin *et al.* (1979) have convincingly demonstrated the efficacy of the larval case of *Tinea* in restricting water loss and thus contributing to the osmotic physiology of the species. The enclosed capsules in which *Ephestia* pupates within stored flour have a similar effect (Willmer, unpublished) and can reduce the rates of water loss by 70%. The papery linings used by some insects in their burrows, often in the areas where eggs are laid, also assist the water balance of the occupants; this phenomenon is known for many Hymenoptera which are fossorial (Michener, 1974), and the larvae of a solitary bee (*Andrena*) within their lined cells showed only one third of the water loss of larvae laid within soil at the same temperature and humidity (Willmer, unpublished).

On a larger scale, many microhabitats are created by man's activities, and the bioclimatology of warehouses, bulk stores and human habitations is an important aspect of applied entomology which has its own extensive literature, and so will not be considered further here.

2.3 SELECTION OF MICROCLIMATE

All animals show hygrothermal preferenda, and will exhibit choices with respect to temperature and humidity which generally accord with the ranges encountered in their normal environments. This subject has been thoroughly covered by Cloudsley-Thompson (1970), Bursell (1974a, b) and May (1979), so is treated only briefly here. It is intuitively obvious that arctic insects will choose lower temperatures, and desert forms may select humidities further from saturation, than would be expected for their otherwise similar temperate relatives. In this sense physiological processes are adapted to their physical environment, and may even be able to acclimate within the lifetime of an individual as conditions change (see Chapman, 1955; Coenen-Stass, 1976). There may be other interactions of microclimatic choice with physiology though, in that temperature and humidity preferenda often change with the state of hydration of an insect (Gunn, 1934; Roth and Willis, 1951; Rayah, 1970; Cloudsley-Thompson, 1979; Prince and Parsons, 1977). There is therefore considerable variation both within and between species in the nature and the precision of microclimatic choices.

There will also be variation in choice within a species in relation to developmental stages, as requirements and hence distributions change. Gravid females must select oviposition sites appropriate to the needs of the eggs and young larvae; larval stages select their own environment and often that of the pupa; and adult insects in turn select habitats appropriate to their own special needs, whether on the scale of migration en masse, local dispersions, or finely-tuned variations in siting of activities through a day. The selection of a habitat is thus no simple matter, and will require complex interacting sensors both of internal state and external conditions; though the nature of receptors and their role in the control of physiological balance are beyond the scope of the present work.

3 Physiological effects of the microenvironment

The two most important variables in any particular microenvironment are the temperature and the moisture content, in their turn affected by radiation (whether directly from the sun or as modified by reflection or transmission from other surfaces) and by air movements. Consequently the major physiological effects of climate are likely to be related to thermal balance and to water balance, and microclimatic interactions with temperature regulation and osmotic physiology are therefore reviewed in the sections which follow. The work discussed relates principally to rather short-term effects, and it is through these mechanisms that the longer term effects of climate on reproduction, development and life cycles, reviewed by Bursell (1974a, b), must ultimately be determined.

3.1 THERMAL BALANCE

Insects have been conventionally regarded as classic examples of ectotherms, their body temperature reflecting environmental conditions and being "regulated" only by behavioural mechanisms. Where this is true, their thermal balance should bear a direct and obvious relation to microclimate and to their selection of appropriate conditions. However, as recent reviews by Heath *et al.* (1971), Heinrich (1974) and May (1979) have pointed out, a good many cases of endothermy in insects from most of the major orders have now been described, and in these animals the interactions between the microenvironment and physiological mechanisms may be both more complex and more subtle. Furthermore, many insects regulate their body temperatures actively at some times, perhaps just before or during flight, and yet are passive heat exchangers at others, and they are effectively "heterotherms" (c.f. Heinrich, 1974), so that the whole field of insects and temperature is less straightforward than was once thought.

The measurement of thermal balance in small animals presents certain technical difficulties, since body temperature varies rapidly and there may be significant gradients within the insect. Temperature probes must therefore be small and have a good response time, so that they give an indication of the actual tissue temperature rather than of their own thermal responses to the environment. Many earlier studies relied on fine thermocouples, either implanted or affixed to the thoracic tergites, and this is probably still the best solution for smaller insects if the thermocouples are carefully constructed and calibrated (Unwin, 1980), although some problems will always remain and necessitate caution in interpreting results (see Krogh, 1948; Parry, 1951; Baker and Lloyd, 1970). Implanted thermistors are an alternative, and having recently become both smaller and more reliable may be preferred for the larger insects. The use of infrared thermography to estimate surface temperatures (from which a core temperature can be derived) has also become feasible relatively recently; Cena and Clark (1972) used this technique with hive bees, and it has also been applied to basking butterflies (Clark *et al.*, 1973). For some insects wing-beat frequency gives a good indication of temperature (see Sotavalta, 1947; Digby, 1955), and devices to measure such frequencies in the field are available (Unwin and Ellington, 1979) which could be useful for swarming or station-keeping insects. A further possibility which may be considered for environmental work is the use of thermal "paints" (see Meeuse, 1973).

The maintenance of body temperature (T_b) is desirable because physiological and biochemical mechanisms may be subverted at either extreme of the temperature range. Not only does the rate of each enzymic reaction depend upon temperature, but the rates of reaction and of other larger-scale processes (muscle activity, heart-beat etc.) have different Q 10's so that changes of temperature can disturb the balance between otherwise complementary phenomena (c.f. Richards, 1958). Excessively high temperatures may cause a build up of the end products of metabolism, exhaustion of food reserves, inactivation of enzymes and coagulation of structural proteins, disruption of biological membranes and of the physiological exchanges of ions, water and metabolites which depend upon their structural integrity; and, in the special case of terrestrial arthropods, changes in the permeability of epidermal or cuticular layers and possible resultant death from desiccation. The critical upper temperature for such changes of course varies widely; though commonly in the range 40–45°C (Bursell, 1974b), some insects can survive up to 50–55°C (Cloudsley-Thompson, 1962c, 1970), while arctic forms may be killed by the heat of human skin (see Edwards and Nutting, 1950). The same variability applies to lower lethal temperatures; here the critical physiological changes may include a paucity of end-products, build-ups of harmful substrates normally detoxified, structural changes to membranes, and actual freezing of interstitial fluids or cell contents resulting in altered concentrations of solutes and osmotic stresses. Since even moderately low environmental temperatures may preclude locomotory activity for an ectotherm, insects may be stranded in unfavourably cold areas and so succumb gradually to such effects.

To avoid all these dangers, body temperature may be regulated by only two main types of mechanism; either by changes of metabolic heat production and internal redistributions of the heat generated (endothermy), or by alterations of heat exchange with the environment. The interactions of these mechanisms with the microclimate of an insect are considered below.

3.1.1 *Ectothermy and endothermy – Controlling factors*

A small terrestrial animal exchanges heat with the environment by the conventional processes of conduction, convection, radiation and evaporation (Porter and Gates, 1969; Bakken and Gates, 1975). The available evidence suggests that in most environmental conditions insects gain heat principally by radiation (Parry, 1951; Shepherd, 1958) and lose it principally by convection (Digby, 1955; Edney, 1971b; Bursell, 1974a; May, 1979). Conduction is minimal since points of contact with the substrate are very small, and evaporation is usually precluded because insects cannot afford the water loss entailed unless overheating becomes critical (see p. 25). Control of body temperature and thermal physiology by ectothermal means therefore involves behaviour appropriately attuned to the four climatic variables, and particularly taking account of wind and radiation. Behavioural mechanisms allow insects to maintain T_b by finding and using more equable microenvironments in which they may heat up or cool down as required, permitting the correct functioning of their physiological machinery.

As regards endothermy, the essential strategy is the generation of metabolic heat as a means of regulating T_b. All animals produce heat from metabolism, but this is unimportant in quantitative terms for most nonflying insects (Church, 1960a; Heinrich, 1974). During flight, a degree of endothermy may be "obligatory" (Heinrich, 1975), but only in certain cases is excessive heat produced in the absence of flight, often by uncoupling the flight muscles to produce apparent shivering in the thorax. As a distinctive physiological technique this means of thermogenesis is well-known in the birds and mammals, but it is relatively rare elsewhere. This can be largely attributed to scaling factors, since it is much more economical to generate and conserve heat within a large body than in a small structure where any temperature increment is rapidly dissipated at the surfaces; these considerations are pursued in more detail by May (1976a). Nevertheless some large insects show considerable powers of endothermic regulation, and the mechanisms of shivering to achieve warm-up and blood-shunting to permit cooling may be directly triggered by microclimate as perceived by the insect's receptors. Such techniques have been demonstrated in bees (Heinrich, 1972, 1974), beetles (Bartholomew and Casey, 1977b; Bartholomew and Heinrich, 1978; Heinrich and Bartholomew, 1979), moths (Heath and Adams, 1967; Hanegan and Heath, 1970; Heinrich, 1971a, b; Heinrich and Bartholomew, 1971), butterflies (Kammer, 1970), dragonflies (Pond, 1973; May, 1976b) and Orthoptera (Heath and Josephson, 1970; Uvarov, 1977), and even in small syrphid flies (Heinrich and Pantle, 1975). Reviews of this field are presented by Heinrich (1974) and May (1979), and the subject is therefore treated here only insofar as it relates to the issue of microclimatic interactions.

A further example of "internal" controls relating to thermal problems is the development of physiological techniques to ameliorate cold stress and possible freezing. Many temperate and arctic insects, as eggs, larvae or adults, make use of antifreeze compounds in the haemolymph and/or cells to prevent tissue damage (see reviews by Salt, 1969; Asahina, 1969; Downes, 1965), and in some cases this effect is facultative and can be induced by the direct influence of climatic factors (Asahina, 1969).

In all these cases, even where endothermic or physiological mechanisms do exist, the insects invariably rely upon many behavioural regulatory techniques as well, and their immature stages are usually completely ectothermic; hence the dependence of all insects upon the microclimatic framework of their environment is profound.

The various sources of heat gain and loss for an insect are influenced in their turn by several features of the animal itself: its size, shape and surface area; the colour and microsculpture of its surfaces; and its orientation or posture. Since these factors affect the rate of heat exchange they also determine the tolerances the species may have in timing its activities and in seeking suitable microhabitats in which to pursue them. Hence the nature of each insect's microclimatic choices is the end result of the interactions of all these features of environment, physiology and form. Each of the controlling factors can therefore be considered in terms of its contribution to this overall pattern.

(*a*) *Size and shape* The interrelation of such factors as size, shape and surface area of insects with their thermal balance has been extensively studied over the last 30 years. Early studies by Parry (1951) on model insects pinpointed some of the more important relationships, and these observations were extended by Digby (1955) and by Church (1960a, b) with actual specimens covering a considerable size range. The effects of size (whether weight, or some appropriate linear dimension such as thoracic breadth) were thus shown to be clear-cut and predictable. Larger insects attain a higher temperature excess but take longer to reach it under constant conditions (Digby, 1955; May, 1976a; Willmer and Unwin, 1981). At the upper end of the size range, these effects may be modified by the action of endothermic mechanisms; and there is always a shape effect superimposed, the slope of the size/temperature relation being different for elongate "locust-type" insects and for rounded "fly and bee types" (Digby, 1955). These effects are summarised in Fig. 6.

The size of insects therefore affects their needs for appropriate microclimates; larger forms may tolerate less stable environments and intermittently higher radiation, because they will change their T_b more slowly and their size "smoothes out" some of the variation. They may also become active under cooler conditions with limited solar radiation, as their temperature excesses can be greater and they may reach an adequate T_b more readily than a small insect. But they must avoid constant high insolation, which could cause overheating. These trends can be discerned in a wide range of entomological studies, and some examples are described in section 4.

An additional possibility which some insects may exploit is that of changing effective thermal size by aggregation (see section 2). This may allow an increase in T_b if net surface area is decreased, because metabolic heat can be shared and conserved while convective cooling is reduced; this occurs with caterpillars (Mosebach-Pukowski, 1937; Wellington, 1949), sawfly larvae (Seymour, 1974), ants (Jackson, 1957) and locusts (Uvarov, 1977). Clustering may also serve to lessen the effects of environmental change, since rates of temperature change will be reduced by the greater thermal capacity of the group.

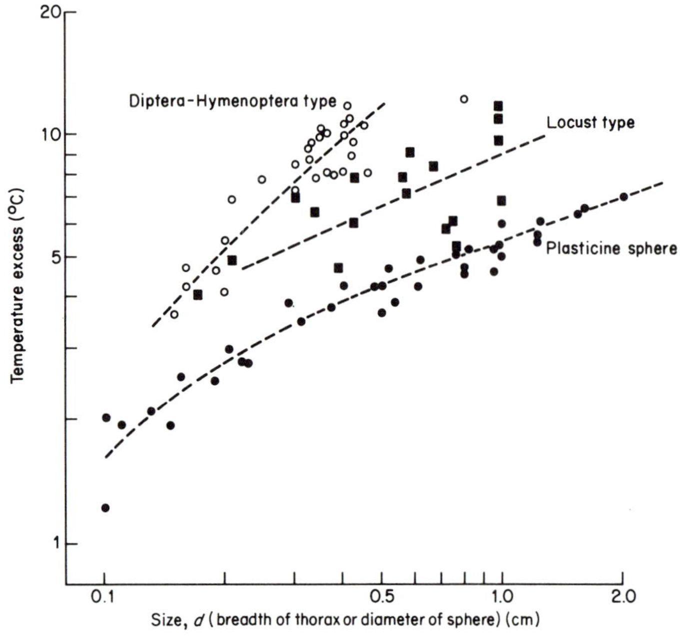

Fig. 6 The relation of size and shape to temperature excess for a range of insects, compared with that of a model plasticine sphere. (From Digby, 1955)

(*b*) *Colour* The effects of colour (surface absorptivity, or reflectivity) on thermal physiology have been somewhat controversial. Digby (1955) found clear effects when insects were "painted", but thought the role of colour to be rather limited since his estimations showed that even a pale and apparently highly reflective insect still absorbed about three-quarters of received radiation. Other authors have felt colour to be of negligible importance (Pepper and Hastings, 1952; Bursell, 1974b). More recently, Edney (1971b) and Hamilton (1975) have argued persuasively for the importance of colour as a thermal strategy in desert insects, where either black or white seem to be preferred according to the time of day when a particular species is active. Thus large black desert beetles were 2.5°C hotter than ones with white elytra similarly exposed to the sun (Edney, 1971b), while for two smaller species the difference was 1°C.

Further direct evidence for the importance of colour comes from those insects which have differently pigmented morphs, where dark forms may have a higher T_b under the same conditions; this has been shown in butterflies (Watt, 1968, 1969; Douglas and Grula, 1978) and in the honey-bee (Cena and Clark, 1972). The same effect can occur if differently coloured parts of the same insect are exposed, as is the case for a cicada which offers its dark dorsum to the sun at low T_b and its white ventral surface if it is too hot (Heath *et al.*, 1972). In a few well-documented cases, insects can actually change their colour. The dragonflies *Austrolestes* (Veron, 1973) and *Diphlebia* (O'Farrell, 1963) may both be mainly pale blue or dark, and T_b rises faster if the latter colour is adopted (O'Farrell, 1963). Certain grasshoppers (Key and Day, 1954) and beetles (Hadley, 1979) show a similar effect. In each of these cases, microclimate seems to have a direct triggering effect, and for some of the insects which change colour humidity would appear to be the most important causative factor (Rowell, 1971; Hadley, 1979). Seasonal changes in colouration in insects such as aphids may be similarly linked to the weather (Dixon, 1972).

The development of a reflectometer which can be simply constructed and used in the field (Willmer and Unwin, 1981) has contributed to the dispute over the practical importance of insect colours, showing that insects of higher reflectivity do heat up more slowly than dark forms of the same size (see Fig. 7). Recent field studies using this instrument confirm the importance of such differences by demonstrating a relationship between the mean surface reflectance of insects active at particular sites and the received radiation at that site (Fig. 8) as it varies through time (Willmer, 1982; and in prep.). These studies also confirm that most temperate insects do have rather low reflectances, ranging up to 25%. However, white forms in deserts may be much more reflective; Rücker (1933) gave values of 74% for white beetles, and Edney (1971b) estimated a value of 79% for a white tenebrionid.

In general, perceived "colour" and the net "reflectivity" measured by the device mentioned above give a good indication of total reflectivity extending into the UV and near IR ranges (Hamilton, 1971; Willmer and Unwin, 1981). But in a few special cases, insects may have unexpectedly high transmittance in the IR sector of the spectrum (Henwood, 1975a), which may be a specific thermally-adaptive strategy to permit warm-up at low sun angles.

(*c*) *Insulation* Insects may be insulated in various ways; by bristles, hairs or scales, and in some dragonflies by internal air sacs. Indeed an external insulating layer could be regarded as a means of creating a very local microclimate. Church (1960b) has shown convincing evidence for a thermal effect of hair in bees and scales in moths, which may be especially important in moving air during flight. The subelytral cavity of beetles may also have a useful role as insulation (Cloudsley-Thompson, 1964; Hadley, 1970), staying 2–7°C hotter than the underlying abdomen. The effects of insulation in reducing rates of temperature change may also be seen from Fig. 7.

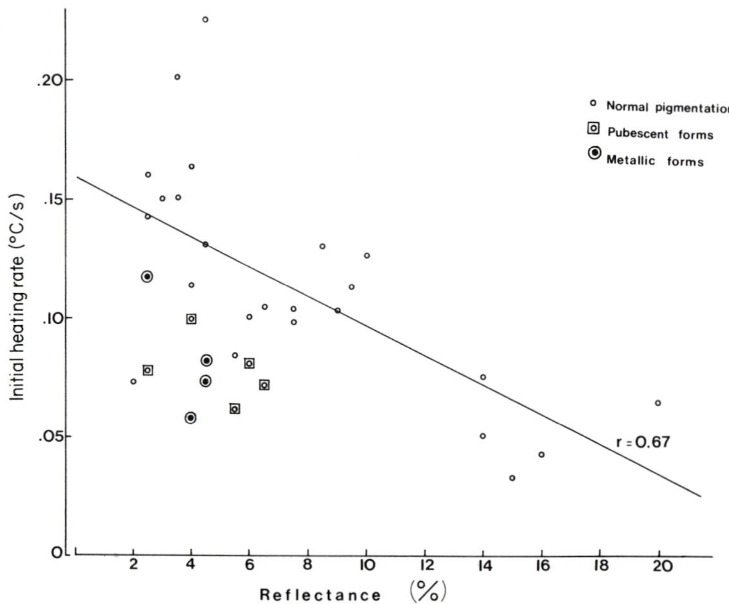

Fig. 7 The relationship between colour (reflectance) and heating rate for temperate insects, and the abnormalities recorded for very pubescent insects and those with metallic colouration. (From Willmer and Unwin, 1981)

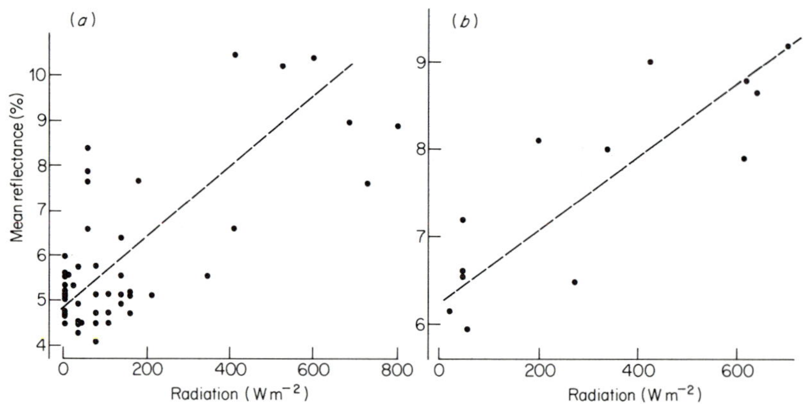

Fig. 8 The interactions of climate and insect colour. Recorded radiation is correlated with the mean reflectance of all insects present at a particular site over a series of observation periods. (*a*) shows the results for a "resting site" on lily pads (Willmer, 1981) and (*b*) shows the same effect for a feeding site at *Tilia* flowers. (Willmer, in prep.)

(*d*) *Behaviour* Posture. Posture for an insect may have a very direct effect upon thermal physiology, especially in those forms which live on very hot surfaces and exhibit "stilting". This phenomenon is recorded for many long-legged insects, including locusts (Waloff, 1963), grasshoppers (Chapman *et al*, 1926; Hafez and Ibrahim, 1964), tenebrionid beetles (Hadley, 1970; Hamilton, 1971; Henwood, 1975b), and tiger beetles (Dreisig, 1980). Simply by raising the body mass a few mm above the sand or rock surfaces, the insect may move to a microclimate some 10 or 20°C cooler (see section 2). By contrast, in conditions of low insolation in the evening locusts may crouch against the warm ground to gain heat by conduction (Waloff, 1963); and certain ants do likewise at times as a means of keeping cool (Gamboa, 1976). The low-slung posture of pond-skaters and other water-surface insects may also help to keep the body in the equable and predictably moist boundary layer (section 2) and thus reduce physiological stress.

Other clear examples of postural control are familiar from studies of the Lepidoptera; adults may "bask" either with wings outspread or side on to the sun to maximise radiative gain, or can adopt the alternative stance of wings closed together over the back and directed into the sun, reducing heat uptake (Vielmetter, 1958; Clench, 1966; Watt, 1968, 1969; Kevan and Shorthouse, 1970; Wasserthal, 1975). The work of Douwes (1976) with *Heodes* shows these effects particularly clearly, as temperature differences could be correlated with wing angle. Comparable thermoregulatory strategies have also been demonstrated in cicadas (Heath, 1967), where the wings are held away from the abdomen at low ambient temperatures, and they may prove to be a common feature of the larger-winged insects.

Orientation. This is an extension of postural control, whereby insects can alter their radiative heat gain by exposing maximum or minimum surface area to the sun. It is especially impressive in insects with elongate bodies (Digby, 1955), and in the Odonata (Hardy, 1966; May, 1976b) and Orthoptera (Fraenkel, 1930; Volkonsky, 1939; Uvarov, 1977) in particular. For example in locusts and stick insects orientation across the windstream during insolation can reduce temperature excess by nearly 50% (Digby, 1955), and at the low wind speeds characteristic of most insect microhabitats even small variations of air movement can therefore be critical (Parry, 1951). Orientation of desert locusts in relation to wind and radiation is summarised in Fig. 9.

Orientation effects also occur in the spruce budworm larva (Shepherd, 1958), and in desert caterpillars (Casey, 1976a), where a ten-fold difference of insolated body area occurred as the larvae moved in three dimensions around its food-plant to stabilise T_b. Even fairly squat desert beetles show important orientation effects (Edney, 1971b), with up to 4.5°C temperature difference between head-on and lateral irradiations.

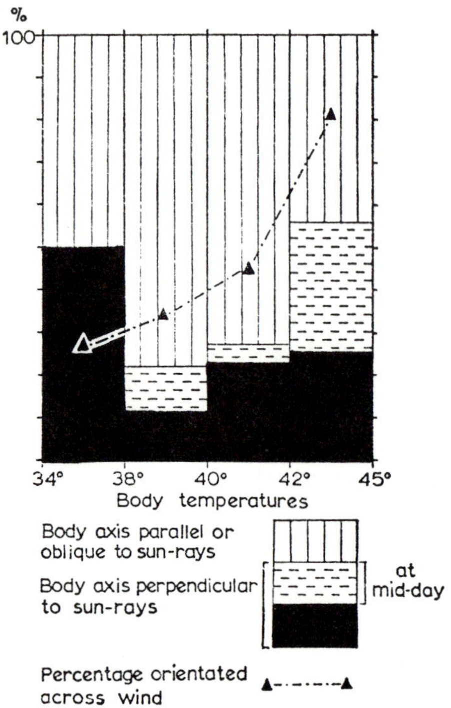

Fig. 9 Orientation with respect to sun and wind at different T_b's in the desert locust, *Schistocerca*. (From Waloff, 1963)

Burrowing. The fossorial habit is well-established as a means of behavioural thermoregulation, especially in deserts where burrows are generally cooler, and are certainly more regular in temperature, than the surface (see section 2). Beetles frequently use burrows to evade high temperatures (Hadley, 1970; Cloudsley-Thompson, 1970; Hamilton, 1971; Henwood, 1975b), but may also burrow into warm sand to help increase T_b when the air is cool (Hamilton, 1971). In temperate conditions burrows provide a thermally desirable retreat for tiger beetles (Dreisig, 1980) and for solitary wasps (Willmer, in prep.). Ant-lion larvae may even move radially within their "burrows" to stay in shaded sectors and prevent excessive increments in T_b (Green, 1955).

Sun-shade alternation and basking. This is probably the most ubiquitous method for regulating T_b, and innumerable examples of the use of sun and shade microclimate are available (see May, 1979). Convincing evidence for the alternation of sun-basking and resting in shade to control T_b is given for cicadas by Heath (1967) and Heath and Wilkin (1970); and for tenebrionid beetles by Edney (1971b), who demonstrated control to within

1°C as the animals moved within their naturally heterogeneous environment. However, "shuttling" between sun and shade may be less common in insects than in other terrestrial ectotherms with a larger thermal mass.

Evaporative behaviour. A number of insects when extremely heat-stressed adopt behaviour patterns which allow for evaporative cooling. Bees may extend their tongues and evaporate collected nectar or saliva ("tongue-lashing") (Lensky, 1964; Esch, 1976; Corbet and Willmer, 1980; Heinrich, 1980a, b) which can cool the head and thorax (Fig. 10) though it may also serve to concentrate the nectar before storage. Some sawfly larvae (Carne, 1962; Seymour, 1974) raise their abdomens and may extrude fluid over their posterior surfaces at temperatures above 37°C, to keep T_b below its upper critical limit. Caterpillars may show a similar trick (Adams and Heath, 1964), expelling droplets over the thorax. The "honeydew panting" of certain aphids has also been construed as a useful evaporative technique (Mittler, 1958; Paul, 1975).

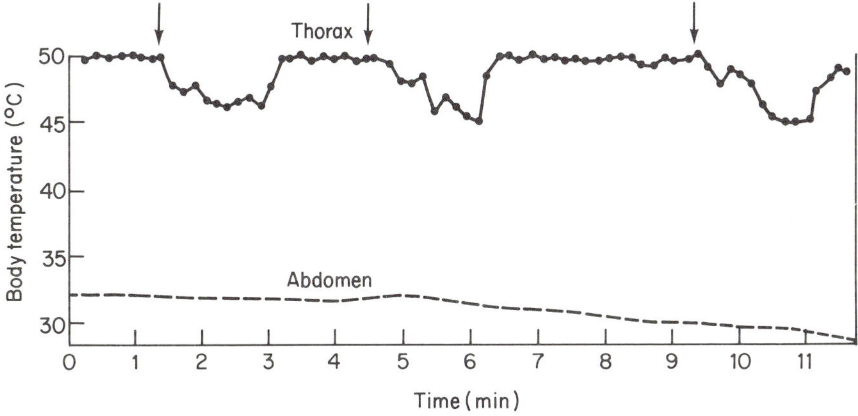

Fig. 10 Evaporative cooling produced by tongue-lashing in bees. Three successive instances of droplet extrusion by honey-bees are shown, cooling the thorax by up to 5°C. (From Heinrich, 1980b)

An alternative approach is an increased opening of the spiracles, allowing moisture loss; this occurs in heat-stressed tsetse flies (Edney and Barrass, 1962), and also in dragonflies (Miller, 1962) and locusts (Loveridge, 1968b; Weis-Fogh, 1967). In all these cases, water loss must be less critical than the alternative of heat overload. The behaviour patterns are rarely seen below about 35°C, so that microclimate has a direct initiating effect in at least some of these examples.

Flight. Flight in insects requires very high metabolic activity in thoracic muscles, and in many larger forms the heat thus generated provides a means of raising T_b (May, 1979). However in smaller insects the extra convective losses during flying may exceed any heat gains, and the insect may actually cool down slightly (Digby, 1955; F. S. Gilbert, personal communication). Hence brief periods of flight, even if not directed towards other ends, may have a thermoregulatory role; and clearly they may also serve to take the insect rapidly into more favourable microclimatic zones.

By a combination of these various behavioural techniques, many insects can achieve a remarkable degree of thermal stability in their natural environments. Some examples of the control of T_b effected by behaviour alone are given for larvae and adults in Fig. 11, the latter being generally more efficient regulators due to their greater mobility and perhaps particularly to the presence of wings.

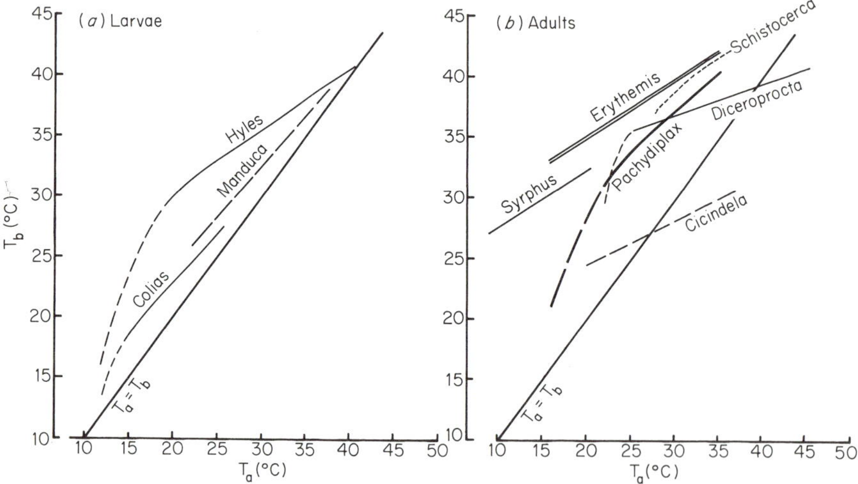

Fig. 11 The regulation of T_b by behavioural means in a range of insects. T_b is compared with T ambient in all cases except *Cicindela* (which is related to soil surface temperature). Data are derived from: May, 1976b (*Pachydiplax* and *Erythemis*); Heath and Wilkin, 1970 (*Diceroprocta*); Dreisig, 1980 (*Cicindela*); Heinrich and Pantle, 1975 (*Syrphus*); Waloff, 1963 (*Schistocerca*); Casey, 1976a (*Hyles* and *Manduca*); Shermann and Watt, 1973 (*Colias*)

3.2 WATER BALANCE.

Just as with temperature control, insects may regulate their water balance both by internal metabolic or physiological effects and by behaviour appropriately coordinated with their environment. In the former category are such mechanisms as the production of water from metabolism, particularly of stored fats (though the significance of this is somewhat controversial; see Mellanby, 1942; Edney, 1957; Barton Browne, 1964); and the redistribution of solutes or water between different compartments of the body to minimise osmotic stress on critical tissues. In the latter group are behaviour patterns closely allied to those for controlling temperature, all resulting in the occupation of more tolerable microniches. Several thorough reviews of the water relations and osmoregulation of arthropods are available (Barton Browne, 1964; Berridge, 1970; Stobbart and Shaw, 1974; Edney, 1977), all pointing to the extreme importance of conserving water to most insects at most times. The water activity (Aw) of insect haemolymph (osmotic concentration around 300–600 mOsm) is equivalent to 99.5–99.8% RH, (Wharton and Richards, 1978) so that the net gradient for water movements is almost invariably outwards, except in saturated atmospheres. Hence, whereas with temperature insects may change their behaviour and physiology to permit either heating or cooling, with water balance the problem is generally that of retaining water and replacing the inevitable losses, and in this sense most environmental conditions are to varying degrees "unfavourable" (c.f. Kühnelt, 1936).

The loss of water is primarily related to air moisture content and the severity of the gradient between air and insect fluids. This can be only marginally altered by internal changes in osmotic concentration, so that the microclimate in which an insect lives is the critical parameter. Relative humidity (or saturation deficit) will be the central variable, but this is of course modified by the air temperature and by wind; and since the surface temperature of the insect will affect local gradients of heat and moisture and the rate of evaporation, radiation is also directly related to water loss. Thus in general insects must seek rather cool and very moist conditions to limit losses of water, though these requirements will inevitably be moderated by the conflicting needs of thermal balance and by interrelations with the insects' other necessary pursuits. There is also a danger of increased pathogenic attack at high RH, so that near-saturated habitats are in practice often avoided.

Studies of the environmental physiology of water balance in insects have been much assisted by recent improvements in technology and methodology. Analysis of water losses requires some estimate of either body weight or of internal fluid concentrations, and neither of these variables could be readily measured in the field until quite recently. Consequently water loss had to be calculated indirectly by extrapolating from known rates of change of weight in laboratory simulations, and osmoregulation was only accessible to analysis if samples were collected in the field and stored for later measurement, at best a dubious technique (Unwin and Willmer, 1978). Now however some of the more elegant commercial microbalances can be adapted for battery-powered field use, and a simple but accurate mechanical balance has also been designed specifically for such uses (see Unwin, 1980). These devices allow weight changes to be measured for insects taken directly from their freely-chosen microhabitats and returned thence. A further development has been the advent of a field osmometer (Unwin and Willmer, 1978), whereby small fluid samples can be removed from an insect and measured in the field within seconds by a freezing-point technique. This new machine has been used to show direct microclimatic effects upon water balance (Willmer, 1980a). Other techniques to give further information on changes in body fluid composition have also been devised; in particular simple refractometers and modified pH indicator systems assist the analysis of solute changes (Willmer, 1980a, and unpublished), and commercial miniature TLC systems can be used to analyse sugar and amino acid compositions (some of these techniques are described in Corbet *et al.*, 1979 and in Willmer, 1980b in relation to nectar analyses). As yet no simple field ion-photometer has been devised, though this might be a possibility. In all cases, these techniques are designed to avoid the problems of storing samples of insect haemolymph, where evaporation and contamination from the storage vessels seem almost unavoidable (Unwin and Willmer, 1978).

A further technical development which may prove extremely valuable is the use of tritiated water to give a direct assay of water losses and exchanges. This subject has been reviewed by Lifson and McClintock (1966) and by Nagy (1975), and some early field studies on crickets (van Hook and Deal, 1972), beetles (Bohm and Hadley, 1977) and flies (Arlian and Eckstrand, 1975) indicate the potential of this approach.

Apart from studies of the changes of water balance in the insect, it is also essential to know air humidity within the microniche. Here again new techniques are available; in particular, the use of tiny droplets of equilibrated solutions of potassium acetate or other appropriate solutes (see Corbet *et al.*, 1979; Unwin, 1980), which can be measured refractometrically or may be continuously monitored electrically. Capacitance hygrometers are also being produced in smaller and faster-responding forms now, though they are still expensive and delicate for many purposes; and infrared psychrometers may prove useful in the future (Sheriff, 1973).

On average through time, the water content of an insect must remain constant. For most insects there is a continuous net passive loss of water, through the excretory orifices, spiracles and general cuticular surface. This must be balanced by water intake from food and drink, by "oxidation" water from metabolism, and in a few cases by active uptake of water, whether from the air or as a liquid, usually through rectal or salivary gland sites (see Noble-Nesbitt, 1976; Edney, 1975, 1977). If an insect fails to balance its losses and so dehydrates, the inevitable consequence is osmotic concentration of its fluids, both extra- and intracellularly, and these changes can only be partially relieved by osmoregulatory mechanisms to limit cellular effects. The results of such osmotic stress are widespread: enzyme function may be restricted, cell membranes damaged, and the structural integrity of macromolecules affected by ion changes and by shrinkage or swelling. Nerves and muscles, which rely on specific ionic gradients, may be particularly vulnerable. In the special case of insects living in very damp soils (Galbreath, 1975) or in local pockets of high RH such as caves (Buxton, 1932b; Howarth, 1980), there may be an excessive water intake and the usual problems are reversed with body fluids becoming dangerously diluted. Thus it is desirable for all active insects to keep their water content and the osmolality of their fluids at a steady controlled value, and most studies of osmoregulation indicate a remarkable ability to do this. However, if extreme conditions persist, many insects can and do survive quite large changes in haemolymph concentration, though usually at the expense of activity (e.g. Hinton, 1960; Buck, 1965; Jones, 1975).

3.2.1 *Water balance – Controlling factors*

As with body temperature T_b, a number of features of the insect itself and of its behaviour will affect its water content W_b in relation to its environment. In some cases these controlling factors are the same, since thermal and osmotic problems are so intimately linked, but the particular effects of each variable may be rather different.

(*a*) *Size and shape* The difficulty of maintaining W_b for an insect is essentially related to the properties of its exposed surfaces, and hence the parameters of surface area, shape and size are critical to the loss of water and in turn to the degree of microclimatic control necessary. Of two insects differing only in scale, the larger will be at an advantage in terms of percentage water loss and its body fluids will therefore become concentrated less rapidly. Similarly a compact shape without elaborate or flattened appendages will be an advantage. There is of course always the opposing need for reduced size to allow exploitation of small humid zones.

Direct evidence for the effects of size and surface area on water relations is rather difficult to obtain, since other parameters such as permeability and surface ultrastructure may also vary, and it may strictly only be obtainable within a species (e.g. Loveridge, 1968a). Studies by Bohm and Hadley (1977) on beetles gave somewhat inconclusive results on this point, as water loss also varied with the food and water reserves with size effects masked by greater consumption and hence defaecation. Smaller third instar caterpillars of *Pieris* lost water more rapidly than the fifth instars (Willmer, 1980a), though here there was some indication that permeability was decreasing as the larvae aged. More rigorous testing of this issue is still required.

Again as with temperature control, there is the possibility of increasing effective size by aggregation. The efficacy of this in terms of water balance may be seen in the work of Verhoeff and Witteveen (1980) for collembolans and of Willmer (1980a) for caterpillars of *Pieris* and *Inachis* (see Fig. 12).

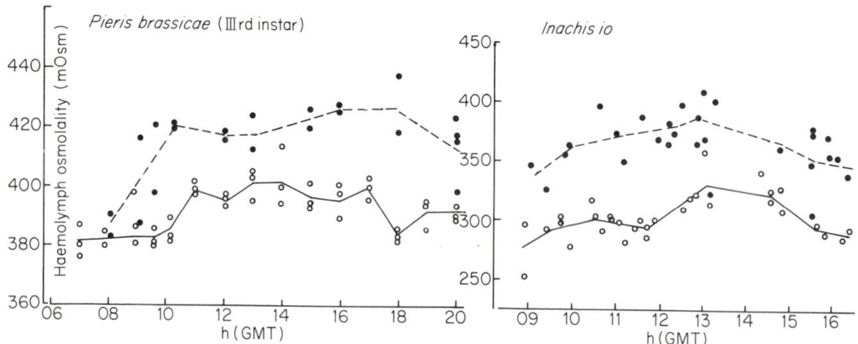

Fig. 12 The effects of the environment on haemolymph concentration in solitary (●) and gregarious (○) butterfly larvae. Haemolymph osmolality is shown as it varied through a day of decreasing and subsequently rising RH, the increments being larger for solitary forms for both *Pieris* (segregated experimentally) and for *Inachis* (naturally both solitary and grouped). (Data from Willmer, 1980a)

(*b*) *Permeability* Perhaps the single most critical factor controlling the water balance of terrestrial insects is the permeability of the exposed cuticular surfaces, and the literature on this subject is now enormous; recent reviews by Berridge (1970), Ebeling (1974), Beament (1976) and Gilby (1980) are particularly informative. Most insects have highly impermeable cuticles, but the range of values measured for this material is very large (c.f. Bursell, 1974a) and there seems little doubt that permeability is well correlated with habitat (Bursell, 1974a; Edney, 1976b, 1980), xeric insects having the highest recorded values and suffering only around 1 μg water loss/cm^2/h/mmHg, (Bursell, 1958, 1974a).

Beyond the essential impermeability of terrestrial insect cuticles, though, there is scope for further refinement. Firstly, permeability may vary widely in different areas of any one insect (Beament, 1976) and may also change through the various stages of the life cycle (Beament, 1959). A number of studies also provide evidence that permeability can be modified by other factors, both internal and external. The profound effects of temperature upon cuticular properties are well-known, though their nature is still the subject of some controversy (see Beament, 1964, 1976; Toolson, 1978; Machin, 1980; Gilby, 1980). There can be little doubt that both ambient temperature and the surface temperature of the cuticle itself as modified by radiation will have significant effects upon water loss because of such mechanisms. It is also likely that the relative humidity of the air can affect cuticle permeability directly (Loveridge, 1968a; Humphreys, 1975) though the mechanism of this is not clear.

(*c*) *Physiological controls* Interacting with the effects discussed above there may be physiological changes due to internal mechanisms. Winston and Beament (1969) have shown that water activity in the cuticle is lower than that in the blood, suggesting "active" mechanisms to regulate water gradients and perhaps a homeostatic feedback system for water control. A related issue is the demonstration of hormonal control of integumentary water loss in *Periplaneta* (Treherne and Willmer, 1975), and more recently in *Leucophaea* (Franco, see Edney, 1980), suggesting important possibilities for feedback controls of both temperature and water balance. Bees are also believed to control water balance hormonally (Altmann, 1953, 1956), and many cases of hormonally-mediated excretory control are now known, both at Malpighian tubule (see Maddrell, 1971, 1980a) and rectal (Cazal and Girardie, 1968; Mordue, 1969; Noble-Nesbitt, 1978; Phillips, 1980) sites. These mechanisms allow integrated neuroendocrine control of water balance whereby extra transpiration losses can be compensated by producing drier excreta; the whole field of control of water relations is reviewed by Maddrell (1980b).

(*d*) *Behaviour* Most of the behaviour patterns which can help to regulate T_b will also affect W_b, and those which are used to raise the insect's temperature (sunbasking, and its interrelations with posture and orientation) will usually have adverse effects on water balance. However, most heat-avoidance strategies will take the insect into cooler more humid zones where water loss will be reduced, whether this involves stilting in deserts, burrowing, or shade-seeking. Some specific examples of these effects are discussed in section 4, from which it may be seen that in many cases thermal considerations appear to outweigh the needs of water balance in determining microhabitat choice.

There are also some specific behaviour patterns which affect insect water balance. Mention has been make of techniques for increasing evaporation (section 3.1) which may be used as an emergency strategy to lower T_b. Related to this are techniques such as spiracle opening and closing, or changes of the ventilation rate, which may be controlled in relation to ambient temperature, or more directly to humidity and water balance (Bursell, 1957a; Miller, 1964; Loveridge, 1968b; Krafsur, 1971a, b); in very dry air many insects will close their spiracles and reduce their oxygen consumption (Bursell, 1957a; Krafsur, 1971a). A further example is the extrusion of salivary or excretory fluids, which again may help to lower T_b (see above). Alternatively some insects have the ability to take up moisture from non-saturated air (Edney, 1975) either by exposing or everting special saliva-coated sacs (O'Donnell, 1978) or by mechanisms involving rectal pumping (Noble-Nesbitt, 1978; Machin, 1979), and such behaviours may therefore be diagnostic of water-stressed insects. In at least some cases these techniques are a direct consequence of microclimate, and only switch on at some specific RH, the "equilibrium humidity" of the particular species (Knülle and Wharton, 1964; Wharton and Richards, 1978; Machin, 1980).

Thus, as with temperature control, the water balance of insects is determined by a range of internal mechanisms and external changes, mediated at the integumentary surfaces and interacting closely with each other. Microclimate plays a causal role in establishing the water activity gradient along which water is lost from the insects' fluids, and in triggering physiological and behavioural mechanisms; and features of the insect in turn affect the nature of, and necessity for, its choices of microclimate through time and space.

4 Activity, hygrothermal control and microclimate: case histories

The various factors discussed above in relation to hygrothermal balance interact to determine the range of microclimates suitable for a particular insect. Hence they will in turn affect the patterning and overall level of its activities. Those activities which are particularly energy consuming (flight, courtship and foraging, for example) may only be possible when ambient conditions are warmer, whilst resting and certain more domestic pursuits may be more appropriate when the ambient conditions are either too hot or too dry or too cool, so long as more favourable microhabitats can then be found to retreat to. There is ample evidence from early studies that "insect activity" can be directly correlated with temperature or with humidity, and some elaborate predictive models are now available for vertebrates (Porter *et al.*, 1973; Bakken and Gates, 1975). It is the job of the environmental physiologist to explain the patterning of activities, whether seasonal or diurnal, in terms of predictable physiological and climatic relations, and the more successful attempts at such explanations are reviewed in the section which follows. There will of course be many determinants of a particular insect's activities, including intrinsic timing factors and external features such as food availability and the behaviour of mates or predators; in the best of the available works all of these considerations can be interlinked.

4.1 COLEOPTERA

Some of the most elegant and comprehensive studies of environmental physiology yet available have concerned the lives of beetles, particularly the desert tenebrionids and their strategies for maintaining a favourable hygrothermal state. The early studies of Hadley (1970) laid much of the groundwork, in collecting long-term micrometeorological data and relating it to the activity and temperature of a burrowing beetle *Eleodes armata* in Arizona deserts. This work provided unequivocal evidence for the efficacy of burrowing to escape hot dry conditions at the surface to allow a choice of microclimates according to vertical position in the burrow. It also permitted the construction of a heat-exchange budget, as did the work by Henwood (1975b) on *Onymacris*, both cases underlining the necessity for avoidance of surface conditions through most of the daylight hours. More recently, Bohm and Hadley (1977) have begun the analysis of water budgets in *Eleodes* using tritiated water. Edney (1971b) has pursued similar ideas in work with Namib desert species of *Onymacris*, and has also produced good evidence for the effects of colour, orientation and behaviour on body temperatures. Both Edney (1971b) and Hamilton (1975) note the prevalence and importance of black and white colours, supplementing field observations with laboratory work on appropriately matched species. Henwood (1975a) has further suggested a specific high IR transmittance in *Onymacris* to keep T_b high at dusk or dawn; while Hadley (1979) reports colour changes linked to hygrothermal stress in *Cryptoglossa*. Edney's early work was extended by Holm and Edney (1973) into a full consideration of activity patterns in relation to climate (Fig. 13), showing that the unimodal daily activity of *Onymacris rugatipennis* became a bimodal regime in the summer to avoid hot extremes and maintain a reasonably constant T_b with limited loss of water. Edney (1971a) has also published specific water balance data for these species, correlating rates of water loss with the degree of exposure to desiccating conditions normally experienced by different summer and winter species. Other workers have also considered the control of water loss (Ahearn and Hadley, 1969; Ahearn, 1970), and some specific studies of osmoregulatory mechanisms are available (Riddle *et al.*, 1976; Broza *et al.*, 1976; Nicolson, 1980), the last of these showing the extreme capacity for enduring water loss without damage to cells by appropriate solute redistributions in *Onymacris*. Perhaps this osmotic tolerance explains the findings of Hamilton (1971) which suggest that thermal balance overrides considerations of water loss in determining the activity of *Cardiosis* beetles, with surface and ambient temperatures as the critical microclimatic parameters; these beetles behaved as "maxitherms", seeking to maintain T_b at 37.5–40°C for as much of the day as possible. In many species, the loss of water which inevitably results from such strategies may be balanced by specialised drinking habits, using dew and fog (Louw, 1972; Seely, 1979) which as they condense may be trapped and channelled on the body by basking (Hamilton and Seely, 1975) or collected from specially constructed catchment trenches (Seely and Hamilton, 1976). The subelytral cavity may act as a storage space permitting body expansion after imbibing such large and intermittent water sources (Slobodchikoff and Wismann, 1981). In some cases tenebrionids may even benefit from evaporative cooling, as both *Eleodes* and *Cryptoglossa* produce a phenolic defensive secretion when heat-stressed (Ahearn and Hadley, 1969). Thus almost all the behavioural or intrinsic features which can affect hygrothermal balance have been demonstrated in the tenebrionid beetles, together with some unique adaptations of their own appropriate to their extreme habitats.

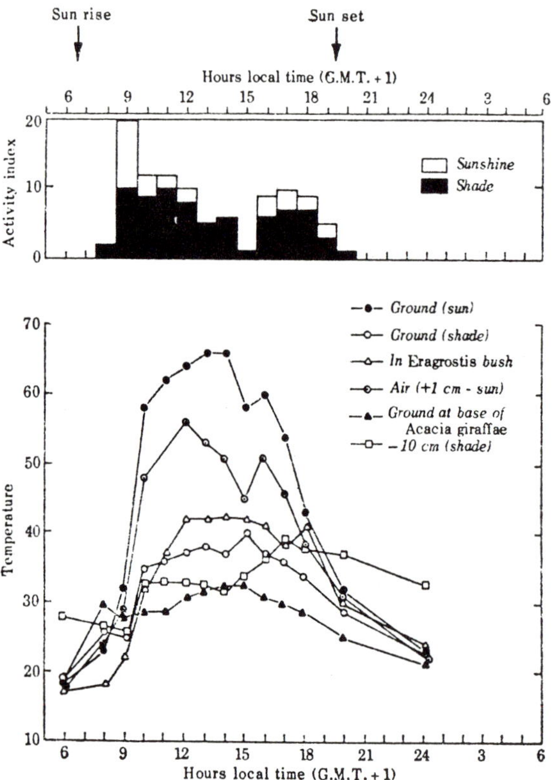

Fig. 13 The activity of the tenebrionid *Onymacris rugatipennis* in relation to the microclimate of its habitat during a 24 h summer cycle. (From Holm and Edney, 1973)

In less severe climates, beetles have still provided useful test animals for environmental physiologists. An early study by Baker and Lloyd (1970) showed the critical roles of radiative gains and convective losses in the heat budget of a boll weevil in varying climatic regimes simulated in the laboratory. More recently, work on dung beetles (Heinrich and Bartholomew,

1979) and other large species (Bartholomew and Casey, 1977a, b) has produced the first confirmation of endothermy in the Coleoptera, and the physiological ability to regulate T_b has been neatly correlated with ecological strategies and competitive ability. In some cases the regulation of T_b has been specifically related to size (Bartholomew and Casey, 1977a; Heinrich and Bartholomew, 1979). A further example is the comprehensive study of tiger beetles (*Cicindela*) by Dreisig (1980); this provides direct analyses of the interactions of diurnal activity patterns, thermal regulation, water loss and microclimate. These beetles used burrowing, stilting and basking techniques to maintain T_b at around 35°C, and perhaps also used brief flights to lower T_b when hot. Dreisig's estimates of daily water loss again suggested that water balance was only a secondary consideration, losses being readily replenished from food, so that temperature may have been more important than humidity as a determinant of activity (Fig. 14).

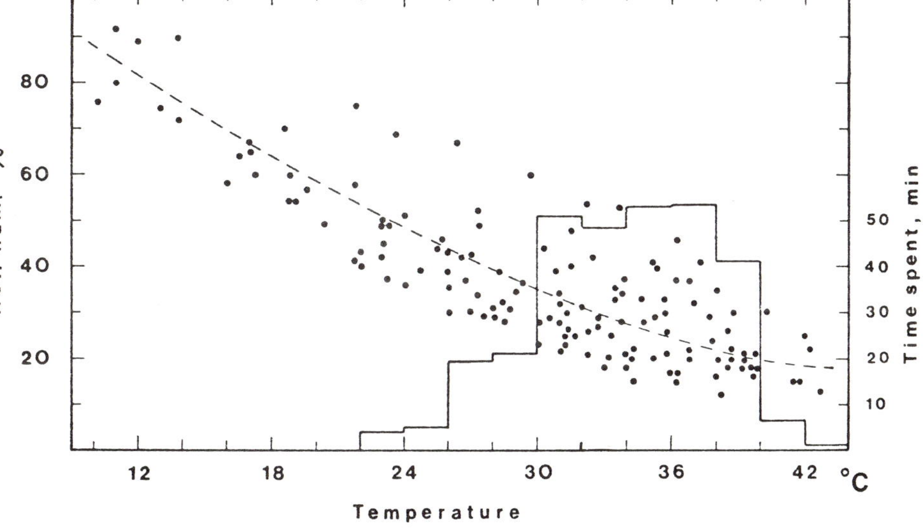

Fig. 14 Temperatures and humidities 5 mm above the soil surface in *Cicindela* habitats during the activity period of the beetles. The bars show calculated mean times spent at each combination through an average day, with favoured temperatures largely overriding the hazards of low humidity in determining choices. (From Dreisig, 1980)

4.2 DIPTERA

There has been relatively little work on the environmental physiology of flies, perhaps due to their generally small size and versatility in flight; although their economic importance makes them appropriate targets for such studies. A notable exception which contributed much to the early study of applied environmental physiology was the caucus of work on the tsetse fly by Bursell, considering both the pupae (1958) and adults (1957b, 1959). For this insect, distribution, dispersal and activity could be correlated with water balance requirements and microclimate. Similar correlations have been suggested for *Phlebotomus* (Theodor, 1936), and more rigorously for *Aedes* (Platt *et al.*, 1958) which is distributed principally in relation to RH. The water balance of the fruit fly *Drosophila* has been examined using tritiated water (Arlian and Eckstrand, 1975), and again the importance of RH in determining ecological patterns is indicated. In rather different climatic regimes, Corbet (1966) has shown the critical role of microclimate in the patterning of behaviour and development of arctic mosquitoes, which must achieve a threshold T_b even at very low ambient temperatures by appropriate thermoregulatory behaviour; and their additional strategy of flower basking as adults will contribute to this (Hocking and Sharplin, 1965).

Rather more specific tests of microclimatic effects have been shown in studies of flower-feeding flies (Maier and Waldbauer, 1979; Willmer, in prep.). In the latter study, the diurnal patterning of different species could be correlated with their size and colour through the effects on T_b (Fig. 8*b*). Similarly, for resting flies on lily pads the larger and darker species appeared earlier and later in the day while only small bright dolichopodids persisted through the hours of peak radiation (Willmer, 1981 and see Fig. 8*a*). In both these cases, activity regimes for larger species changed from bimodal to unimodal patterns on days which remained overcast, confirming the importance of radiation rather than time of day.

In a few cases, flies may augment behavioural regulation of their temperature with endothermic mechanisms. *Calliphora* (Digby, 1955) and *Syrphus* (Heinrich and Pantle, 1975) use this strategy, though in both cases it may only be a reserve mechanism for use in shady conditions; while *Gasterophilus* routinely uses endothermy to maintain T_b (Humphreys and Reynolds, 1980).

4.3 LEPIDOPTERA

Several early studies of adult Lepidoptera showed the importance of behavioural patterns for their thermal physiology (Clench, 1966; Watt, 1968; Kevan and Shorthouse, 1970), and there can be little doubt that the use of basking and orientation techniques are crucial factors for large-winged forms. Watt (1968) also showed the effectiveness of colour in *Colias* butterflies in relation to thermoregulation. More recent and extensive work by Douwes (1976) on *Heodes* and by Casey (1976b) on *Hyles* has provided evidence for the interactions of radiation and body temperature on activity patterns.

Heodes has an optimal T_b of around 35°C, basking at lower temperatures and only feeding and flying when radiation intensity is high (Douwes, 1976) (see Fig. 15). *Colias* flying activity is similarly correlated with climate (Leigh and Smith, 1959). In many moths and a few butterflies, a capacity for endothermic warm-up is superimposed on these patterns (Heath and Adams, 1967; Hanegan and Heath, 1970; Kammer, 1970; Heinrich, 1970, 1971a, b; Heinrich and Bartholomew, 1971), so that flight can be achieved at lower levels of solar input.

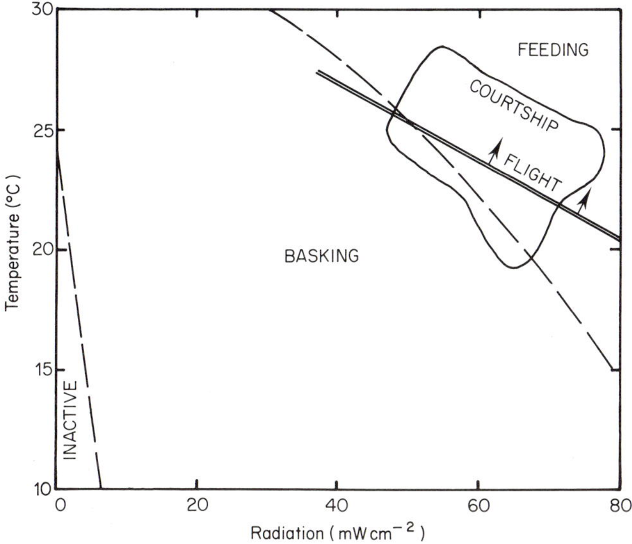

Fig. 15 The hierarchy of activities in relation to the thermal environment for the butterfly *Heodes virgaureae*. (From Douwes, 1976)

There have also been studies of larval Lepidoptera, which are of interest in highlighting the variation even within the lifecycle of a single species in the effects and choices of microclimate. The problems for caterpillars clearly differ from those of the adult, since moist food and shelter are generally available and flight impossible. Larvae often live in groups, thus raising their net temperature and limiting physiological stress (Mosebach-Pukowski, 1937; Wellington, 1949; Sullivan and Wellington, 1953). For solitary forms, the study by Casey (1976a) of *Hyles* and *Manduca* caterpillars is interesting for the parallels it provides with his work on *Hyles* adults; the larvae regulated T_b by exploiting the microclimatic variety of the foodplant, and their locomotory patterns and feeding rates thereon were in turn a function of T_b. Comparable work by Sherman and Watt (1973) on *Colias* larvae, using implanted thermistors, showed a similar regulation of T_b but at a level 10–15°C lower than the optimum for adults (Watt, 1968, 1969). On the water balance side, studies of caterpillars are rare, and problems are usually not extreme; but Willmer (1980a) provides evidence for microclimatic effects on fluctuations in body fluid concentration, with highly protected species (in flower buds or leaf rolls) being much less affected by ambient changes than those species living in more exposed sites (Fig. 16). Microhabitat in fact interacts critically with physiology in these cases, in that the larvae normally protected within zones of high RH proved much more susceptible to low humidity when removed from their microhabitats, and showed lesser capacities for osmoregulation and homeostasis (Fig. 16). This study also demonstrated the uses of large size and of aggregative behaviour in limiting the changes of body fluid concentration.

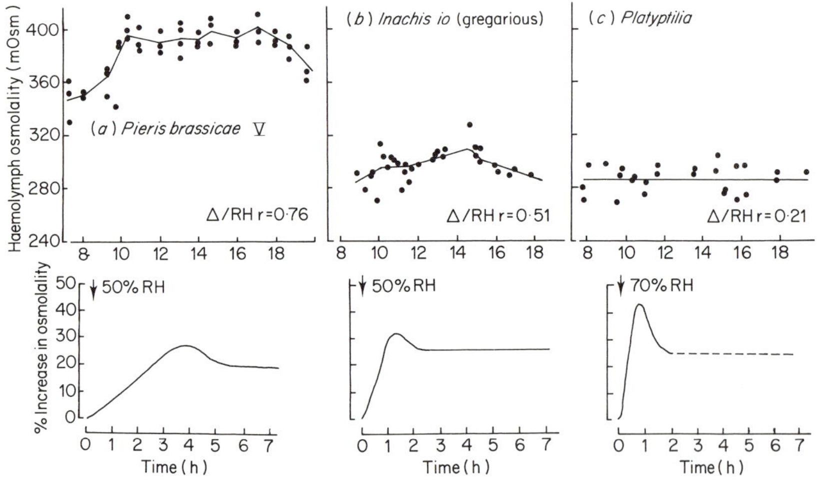

Fig. 16 Haemolymph concentration changes in various caterpillars. (*a*)–(*c*) show changes through a day in the natural environment as the humidity changed: the species live on cabbage leaves, on nettles in conspecific aggregations, and in the flower buds of *Justicia* (Fig. 5) respectively, showing a series of progressively more protected and humid microhabitats. Lower diagrams show the percentage changes in body fluid concentration when larvae were removed from their normal habitats and artificially exposed to low humidity. (Data modified from Willmer, 1980a)

4.4 HYMENOPTERA

Without doubt the single most impressive account of the environmental physiology of an insect is that of Heinrich (1979), collecting together and amplifying the results of his many earlier papers on bumble-bees, and thus encompassing foraging strategies, flight control, thermoregulation, energetics and ultimately reproductive success. In these insects the control of T_b, both for individuals and for the nest and its brood, plays a central role in their ecological success, and involves both behavioural and endothermic techniques. In accordance with the principles discussed earlier in this review, the large size, good insulation, and intrinsic warm-up capacity of bumble-bees allow them to be active early in the season and earlier on a particular day, pre-empting other foragers at many flowers. T_b is maintained continuously when using high-reward flowers, but is allowed to drop when walking between clumped small florets with limited nectar, underlining the high costs of thermoregulation and the necessity for the insect to work out the economics of foraging "correctly". There are still some inconclusive aspects of the bumble-bee story; some controversy remains over the role of colour in terms of thermal regulation for the genus (see Stiles, 1979; Plowright and Owen, 1980), but it is probable that for such large insects colour can be of only limited importance. On the water balance side too there are gaps in our knowledge, with little information available except in terms of the nest water economy (Wojtowski, 1963; and see Lindauer, 1955), and studies of foraging in relation to chosen nectar concentrations, drinking of water, and likely evaporation during flight or after deposition of stores in the nest would be of some interest. Some information of this nature has been obtained for solitary wasps (Willmer, in prep.), suggesting that water may be a more important determinant of foraging than most studies (usually concentrating on energy supplies from collected food) have implied.

Several studies of bees and wasps may be cited as good indicators of the role of insect size in prescribing diurnal activity patterns. A study by Heinrich (1976) showed bumble-bees foraging in cooler conditions than was possible for smaller solitary bees, thus partitioning the available resource; and similar effects can be seen in the recorded visits to Convolvulaceae described by Schlising (1970). Size was also a factor in the foraging strategies of the three bee genera discussed by Schaffer *et al.* (1979), due to its crucial effects on thermal gains and losses, and of species of *Trigona* foraging on *Justicia* flowers (Willmer and Corbet, 1981), (see Fig. 17).

Finally there have been some useful studies on the activity patterns of ants, where the problems differ due to the general absence of flight, restricting the insects to the often more severe surface microclimates. Work on the individual and collective hygrothermal economies of ants is as yet in its infancy, but the limiting effects of high temperature have been made clear for leaf-cutter ants (Gamboa, 1976), honey ants (Kay and Whitford, 1978), and for a number of temperate species (Bernstein 1979) where distribution and foraging regimes may be curtailed by the insects' thermal physiology. In the tropical ants, activity may be almost continuous, but there is a predictable avoidance of highly insolated areas in at least some species (Willmer and Corbet, 1981) (Fig. 17).

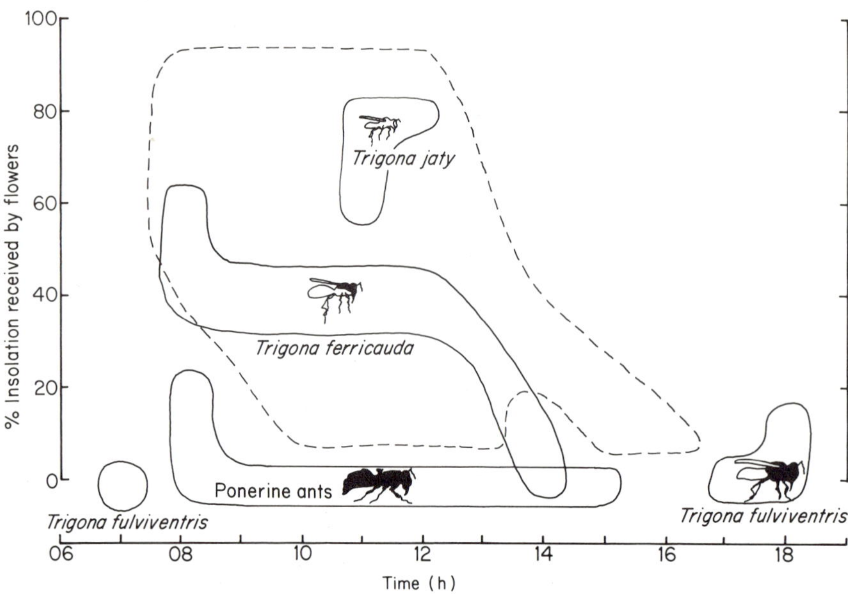

Fig. 17 Distribution of Hymenoptera of different sizes and colours feeding on nectar and pollen of *Justicia*. Times and zones of peak activity for each species are shown by the contours, and may be related to both time and microclimate (here defined by insolation received by different flowers). The dashed line shows the zones of high nectar reward in both time and space. (From Willmer and Corbet, 1981)

4.5 ODONATA

Though often regarded as rather "primitive" insects, the Odonata show a marked ability to regulate their own hygrothermal physiology. The most obvious mechanism for this is the choice of flying area or of perch. For example, dragonflies may fly up above trees and bushes early in the morning to receive maximum insolation, descending to lower levels later in the day (May, 1978). Similarly they may perch on or near sandy surfaces or low sunlit leaves at temperatures below about 28°C, but seek higher cooler levels when the air and ground are hotter (May, 1976b). The alternative strategies of "perching" or "flying" when seeking a mate have been fully analysed in terms of heat balance by Heinrich and Casey (1978); and Lutz and Pittman (1970) have shown the effects of characteristic minimum and maximum temperatures for flight in different species in determining the pattern of a dragonfly community. In view of this close dependence of activity upon ambient temperature, it is perhaps not surprising that some dragonflies can change colour to effect some control of T_b, perhaps by a hormonal mechanism (Veron, 1973). Furthermore under extreme heat stress some species can increase their ventilation rate to allow evaporative cooling (Miller, 1962).

4.6 ORTHOPTEROIDS

Much of the recent work on the ecophysiology of grasshoppers and locusts has been reviewed by Uvarov (1977), and the mechanisms are generally rather similar to those of the Odonata, which share a similar body form. The insects may perch at the top of plants around sunrise, descending later in the day (Chapman, 1959; Waloff, 1963), and may climb again if the ground becomes too hot. They may also aggregate in heaps, to limit thermal stress (Chapman, 1955). Colour changes occur, probably with significant thermal effects, and dark colouration is accompanied by the "basking" posture (Key and Day, 1954). In many of these cases, T_b is maintained remarkably constant (see also Stower and Griffiths, 1966), and when heat stress does occur locusts may resort to the further strategy of stilting (Uvarov, 1977) or may increase their ventilation (Loveridge, 1975).

Studies of this type on the acridid insects have been supplemented by physiological studies of flight and the accompanying effects on water loss and body temperature, as these matters are of some applied importance in the locust in relation to feeding habits (Rainey *et al.*, 1957; Weis-Fogh, 1967). *Schistocerca* can fly for hours at 25–35°C, and can choose its height and air temperature appropriately to balance water gained from metabolism with water losses incurred; at 25°C a humidity of only 35% is adequate for sustained flight, the limit being set only by fuel reserves (Weis-Fogh, 1967). Thus reasonably large desert insects do achieve hygrothermal balance even under fairly extreme conditions; and reserves can always be topped up from green foodstuffs and from early morning dews.

An alternative water source permitting continuous active life in deserts is shown by another representative of the orthopteroid orders, the desert cockroach *Arenivaga*. This insect can absorb water vapour from air above 82% RH, (Edney, 1966; O'Donnell 1978) and Edney *et al.* (1974) have shown that these conditions can be found only 45 cm below the surface in the habitat where it makes its sand burrows, even in summer. The vertical distribution of this roach showed a good correlation with its preferred temperature and RH range; and even if it does get caught on the surface and subjected to stress it shows very high tolerance of osmotic change and good regulatory responses (Edney, 1968).

A final example drawn from the Orthoptera is the work on cicadas by Heath and Wilkin (1970), demonstrating a series of temperature thresholds for successively more energy-requiring activities (Fig. 18). This insect (*Diceroprocta*) also showed elaborate movements around its natural vegetation to regulate T_b within narrow limits by microclimatic choice.

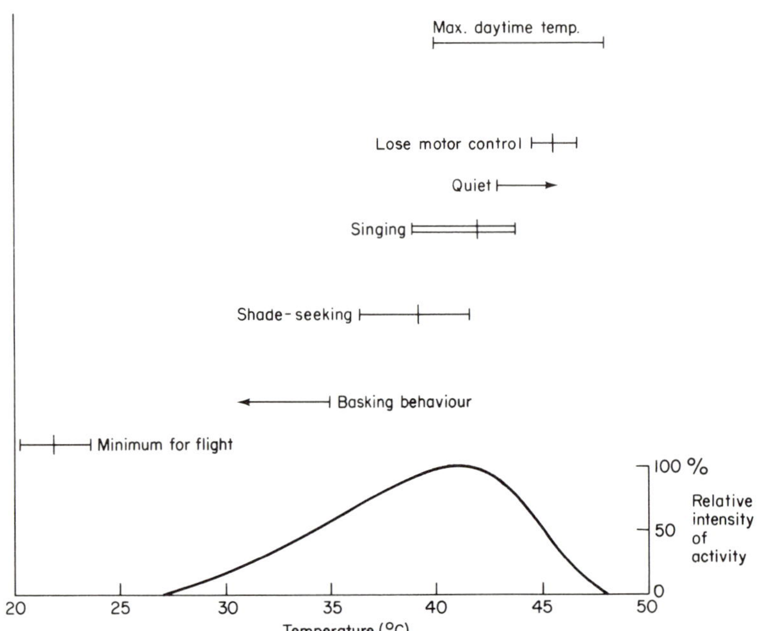

Fig. 18 The temperature-related activities of the cicada, *Diceroprocta apache*, showing the mean T_b and range for each activity. The curve relates the subjective intensity of net activity to ambient temperature. (From Heath and Wilkin, 1970)

5 Concluding remarks

Climate plays a critical role in the life of terrestrial insects. It affects their geographical and ecological locations, the site and timing of their activities, the success of oviposition and hatching, and the duration of developmental stages; thus ultimately it is often a key factor in the selective processes acting on insects, to a far greater extent than for the much larger terrestrial vertebrates. The effects of the physical environment on insects must be mediated through the biochemistry and physiology of the individual, whether as egg, larva or adult, and this is generally expressed via changes in the microenvironment of the fluids in the tissues and cells of the insect (particularly their temperature and concentration) upon which basic life processes depend. Ultimately these processes are conservative in their requirements, although in different animals they may be adapted to function over varying optimal ranges. It is therefore the apparently enormous variety of the environments inhabited by insects which seems to require explanation, and this may be where the concept of microclimate provides the key; for most of the localised sites where insects choose to spend lengthy periods do conform to certain limits of what is tolerable for achieving hygrothermal balance, and of what inflicted stresses the cuticular and excretory systems can safely handle.

Many of the factors discussed in this review which may interact with climatic parameters impose conflicting requirements on insects. Characteristics which help to regulate T_b within acceptable bounds may have adverse effects on water balance, and the larger size which favours slower and more readily-managed internal changes when under stress simultaneously precludes the use of many of the smallest and most favourable microniches. Hence there is no single optimal solution to the problem of integrating microclimate and physiological functioning; the chosen regimes of humidity, temperature, radiation and wind will be determined both by the intrinsic properties of the species and by the biotic and physical characteristics of available niches and their use by competing species. Because of these complex interactions, it has always proved difficult to draw conclusions about structural and physiological relations with climate on a global scale, the diversity of insects in all climates being such as to defeat attempts at generalisation. It might be predicted for instance that insects would in general be darker-coloured in colder areas, and paler in deserts or the tropics, or that mean sizes of insects would be correlated with climate, (even though diurnal activity patterns will differ and obscure any very gross differences). But few studies are available which address these problems, and those so far published are rather inconclusive (e.g. Schoener and Janzen, 1968); (although familiarity with taxonomic works on insects from many groups will reveal the recurring theme of specimens from the warmer parts of the range being generally brighter and paler than those from the same taxon collected in cool areas). Even for a physiological parameter such as permeability there are no ready correlations with climate, for while many desert species are indeed highly impermeable there are at least equal numbers of arthropods living within the humid microhabitats of even the severest deserts which are by any standards rather "leaky"; woodlice are perhaps the classic example of this apparent anomaly.

Thus any broader view of the problem of insects and climate forces an appreciation of the interactive features of all physiological adaptation with microclimatic patterns and ecological strategies; and so underlines the vital role of carefully planned and integrated studies of individual cases. It is greatly to be hoped that new techniques and approaches will contribute to a continuing growth in entomological ecophysiology, upon which future syntheses may be based.

Acknowledgements

I would like to thank D. M. Unwin, O. E. Prŷs-Jones, F. S. Gilbert, Dr J. E. Treherne and Dr S. A. Corbet for the many hours of discussion which inspired this review. My particular gratitude is due to New Hall, Cambridge, for continuing financial support; and to the Fellows thereof who made this work possible.

NOTE ADDED IN PROOF

Since this review was written, the book *Insect Thermoregulation* (1981), (B. Heinrich, ed.) Wiley Interscience, New York has appeared; and the chapter therein by T. Casey covers similar ground to that of section 3.1 of the present work. This volume also contains valuable surveys of microclimatic conditions in social insect nests (T. Seeley and B. Heinrich), and of the physiological mechanisms of thermoregulation (A. Kammer).

References

Adams, P. A. and Heath, J. E. (1964). An evaporative cooling mechanism in *Pholus achemon*. *J. Res. Lepid.* **3,** 69–72

Ahearn, G. A. (1970). The control of water loss in desert tenebrionid beetles. *J. exp. Biol.* **53,** 573–595

Ahearn, G. A. and Hadley, N. F. (1969). The effect of temperature and humidity on water loss in 2 desert tenebrionid beetles, *Eleodes armata* and *Cryptoglossa verrucosa*. *Comp. Biochem. Physiol.* **30,** 739–749

Allee, W. C. (1926). Studies in animal aggregations: causes and effects of bunching in land isopods. *J. exp. Zool.* **45,** 255–277

Altmann, G. (1953). Hormonale Regelung des Wasserhaushaltes der Honigbiene. *Z. Bienenforsch.* **2,** 11–16

Altmann, G. (1956). Die Regulation des Wasserhaushaltes der Honigbiene. *Insectes Sociaux* **3,** 33–40

Andrewartha, H. G. and Birch, L. C. (1954). *The Distribution and Abundance of Animals*. University of Chicago Press, Chicago and London

Arlian, L. G. and Eckstrand, I. A. (1975). Water balance in *Drosophila pseudoobscura*, and its ecological implications. *Ann. Ent. Soc. Am.* **68,** 827–832

Asahina, E. (1969). Frost resistance in insects. *Adv. Insect Physiol.* **6,** 1–49

Baker, D. N. and Lloyd, E. P. (1970). An energy balance for the boll weevil, *Anthomis grandis*. *Ann. Ent. Soc. Am.* **63,** 104–107

Baker, H. G. and Hurd, P. D. (1968). Intrafloral ecology. *Ann. Rev. Ent.* **13,** 385–414

Bakken, G. S. and Gates, D. M. (1975). Heat-transfer analysis of animals: some implications for field ecology, physiology and evolution. In: *Perspectives of Biophysical Ecology* (D. M. Gates and R. B. Schmerl, eds) pp. 255–290. Springer Verlag, Berlin

Bartholomew, G. A. and Casey, T. M. (1977a). Body temperatures and oxygen consumption during rest and activity in relation to body size in some tropical beetles. *J. Therm. Biol.* **2,** 173–176

Bartholomew, G. A. and Casey, T. M. (1977b). Endothermy during terrestrial activity in large beetles. *Science* **195,** 882–883

Bartholomew, G. A. and Heinrich, B. (1978). Endothermy in African dung beetles during flight, ball making and ball rolling. *J. exp. Biol.* **73,** 65–83

Barton Browne, L. B. (1964). Water regulation in insects. *Ann. Rev. Ent.* **9,** 63–82

Beament, J. W. L. (1959). The waterproofing mechanism of arthropods. I. The effect of temperature on cuticle permeability in terrestrial insects and ticks. *J. exp. Biol.* **36,** 391–422

Beament, J. W. L. (1964). The active transport and passive movement of water in insects. *Adv. Insect Physiol.* **2,** 67–129

Beament, J. W. L. (1976). The ecology of cuticle. In: *The Insect Integument* (H. R. Hepburn, ed.) pp. 359–374. Elsevier/North-Holland, Amsterdam

Bernstein, R. A. (1979). Schedules of foraging activity in species of ants. *J. Anim. Ecol.* **48,** 921–930

Berridge, M. J. (1970). Osmoregulation in terrestrial arthropods. In: *Chemical Zoology* (M. Florkin and B. T. Scheer, eds) Vol. 5, pp. 287–319. Academic Press, New York and London

Berry, L. and Cloudsley-Thompson, J. L. (1960). Autumn temperatures in the Red Sea hills. *Nature* **188,** 843

Bohm, B. C. and Hadley, N. F. (1977). Tritium-determined water flux in the free roaming desert tenebrionid beetle, *Eleodes armata. Ecology* **58,** 407–414

Broadbent, L. (1950). The microclimate of the potato crop. *Quart. J. Roy. Meteor. Soc.* **76,** 439–454

Broza, M., Borut, A. and Pener, M. (1976). Osmoregulation in the desert tenebrionid beetle, *Trachyderma philistina* Ruche during dehydration and subsequent rehydration. *Israel J. Med. Sci.* **12,** 868–871

Buck, J. (1965). Hydration and respiration in chironomid larvae. *J. Ins. Physiol.* **11,** 1503–1516

Büdel, A. (1956). Das Mikroklima in einer Blüte. *Z. Bienenforsch.* **3,** 185–190

Bull, E. and Mitchell, R. W. (1972). Temperature and relative humidity responses of two Texas cave-adapted millipedes *Cambala speobia* (Cambalida: Cambalidae) and *Speodesmus bicornourus* (Polydesmida: Vanhoeffeniidae). *Int. J. Speleol.* **4,** 365–393

Burrage, S. W. (1976). Aerial microclimate around plant surfaces. In: *Microbiology of Aerial Plant Surfaces* (C. H. Dickinson and T. F. Preece, eds) pp. 173–184. Academic Press, London and New York.

Bursell, E. (1957a). Spiracular control of water loss in the tsetse fly. *Proc. Roy. Entomol. Soc. London A* **32,** 21–29

Bursell, E. (1957b). The effect of humidity on the activity of tsetse flies. *J. exp. Biol.* **34,** 42–51

Bursell, E. (1958). The water balance of tsetse pupae. *Phil. Trans. Roy. Soc. London B* **241,** 179–210

Bursell, E. (1959). The water balance of tsetse flies. *Trans. Roy. Entomol. Soc. London* **111,** 205–235

Bursell, E. (1974a). Environmental aspects–humidity. In: *The Physiology of Insecta* (M. Rockstein, ed.) Vol. 2, pp. 43–84. Academic Press, New York and London

Bursell, E. (1974b). Environmental aspects–temperature. In: *The Physiology of Insecta* (M. Rockstein ed.) Vol. 2, pp. 1–41. Academic Press, New York and London

Buxton, P. A. (1932a). Terrestrial insects and the humidity of the environment. *Biol. Rev.* **7,** 275–320

Buxton, P. A. (1932b). Climate in caves and similar places in Palestine. *J. Anim. Ecol.* **1,** 152–159

Carne, P. B. (1962). The characteristics and behaviour of the sawfly *Perga affinis affinis* (Hymenoptera). *Aust. J. Zool.* **10,** 1–34

Casey, T. M. (1976a). Activity patterns, body temperature and thermal ecology in two desert caterpillars. *Ecology* **57,** 485–497

Casey, T. M. (1976b). Flight energetics in sphinx moths: heat production and heat loss in *Hyles lineata* during free flight. *J. exp. Biol.* **64,** 545–560

Cazal, M. and Girardie, A. (1968). Contrôle humoral de l'équilibre hydrique chez *Locusta migratoria migratorioides. J. Ins. Physiol.* **14,** 655–668

Cena, K. and Clark, J. A. (1972). Effect of solar radiation on temperatures of working honey bees. *Nature* (*New Biol.*) **236,** 222–223

Chapman, R. N., Michel, C. E., Parker, J. R., Miller, G. E. and Kelly, E. G. (1926). Studies in the ecology of sand dune insects. *Ecology* **7,** 416–426

Chapman, R. F. (1955). Some temperature responses of nymphs of *Locusta migratoria migratorioides* (R&F) with special reference to aggregation. *J. exp. Biol.* **32,** 126–139

Chapman, R. F. (1959). Field observations on the behaviour of hoppers of the red locust (*Nomadacris semifasciata* Serville). *Anti-Locust Bull.* **33,** 1–51

Chauvin, G., Vannier, G. and Gueguen, A. (1979). Larval case and water balance in *Tinea pellionella. J. Ins. Physiol.* **25,** 615–619

Cheung, W. W. K. and Marshall, A. T. (1973). Water and ion regulation in cicadas in relation to xylem feeding. *J. Ins. Physiol.* **19,** 1801–1816

Church, N. S. (1960a). Heat loss and the body temperature of flying insects. I. Heat loss by evaporation of water from the body. *J. exp. Biol.* **37,** 171–185

Church, N. S. (1960b). Heat loss and the body temperature of flying insects. II. Heat conduction within the body and its loss by radiation and convection. *J. exp. Biol.* **37,** 186–212

Clark, J. A., Cena, K. and Mills, N. J. (1973). Radiative temperatures of butterfly wings. *Z. angew. Entomol.* **73,** 327–332

Clench, H. K. (1966). Behavioural thermoregulation in butterflies. *Ecology* **47,** 1021–1034

Cloudsley-Thompson, J. L. (1956). Studies in diurnal rhythms. VI. Bioclimatic observations in Tunisia and their significance in relation to the physiology of the fauna, especially woodlice, centipedes, scorpions and beetles. *Ann. Mag. Nat. Hist.* **9,** 305–329

Cloudsley-Thompson, J. L. (1962a). Microclimates and the distribution of terrestrial arthropods. *Ann. Rev. Ent.* **7,** 199–222

Cloudsley-Thompson, J. L. (1962b). Bioclimatic observations in the Red Sea hills and coastal plain, a major habitat of the desert locust. *Proc. Roy. Entomol. Soc. London A*, **37,** 27–34

Cloudsley-Thompson, J. L. (1962c). Lethal temperatures of some desert arthropods and the mechanism of heat death. *Ent. exp. Appl.* **5,** 270–280

Cloudsley-Thompson, J. L. (1964). On the function of the sub-elytral cavity in desert Tenebrionidae (Col.). *Ent. Monthly Mag.* **100,** 148–151

Cloudsley-Thompson, J. L. (1967). *Micro-ecology.* Edward Arnold, London

Cloudsley-Thompson, J. L. (1970). Terrestrial invertebrates. In: *Comparative Physiology of Thermoregulation*, Vol. I. Invertebrates and non-mammalian vertebrates (G. C. Whittow, ed.) pp. 15–77. Academic Press, New York

Cloudsley-Thompson, J. L. (1975). Adaptations of Arthropoda to arid environments. *Ann. Rev. Ent.* **20,** 261–283

Coenen-Stass, D. (1976). Vorzugstemperatur und Vorzugsluftfeuchtigkeit der beiden Schabenarten *Periplaneta americana* und *Blaberus trapezoideus. Ent. exp. Appl.* **20,** 143–153

Corbet, P. S. (1966). Diel patterns of mosquito activity in a high arctic locality: Hazen Camp, Ellesmere Island, N.W.T. *Can. Entomol.* **98,** 1238–1252

Corbet, P. S. (1972). The microclimate of arctic plants and animals, on land and in fresh water. *Acta Arctica* **18,** 1–43

Corbet, S. A. (1968). The influence of *Ephestia kuehniella* on the development of its parasite *Nemeritis canescens. J. exp. Biol.* **48,** 291–304

Corbet, S. A. and Willmer, P. G. (1980). Pollination of the yellow passion-fruit: nectar, pollen and carpenter bees. *J. Agric. Sci.* **95,** 655–666

Corbet, S. A. and Willmer, P. G. (1981). The nectar of *Justicia* and *Columnea*: composition and concentration in a humid tropical climate. *Oecologia* **51,** 412–418

Corbet, S. A., Unwin, D. M. and Prŷs-Jones, O. E. (1979). Humidity, nectar and insect visits to flowers, with special reference to *Crataegus*, *Tilia* and *Echium. Ecol. Ent.* **4,** 9–22

Corbet, S. A., Willmer, P. G., Beament, J. W. L., Unwin, D. M. and Prŷs-Jones, O.E. (1979). Post-secretory determinants of sugar concentration in nectar. *Plant Cell & Environment* **2,** 293–308

Dafni, A., Ivri, Y. and Brantjes, N. B. M. (1981). Pollination of *Serapis vomeracea* Brig. (Orchidaceae) by imitation of holes for sleeping solitary male bees (Hymenoptera). *Acta Bot. Neerl.* **30,** 69–73

Davies, L. (1948). Observations on the development of *Lucilia sericata* (Mg) eggs in sheep fleeces. *J. exp. Biol.* **25,** 86–102

Digby, P. S. B. (1955). Factors affecting the temperature excess of insects in sunshine. *J. exp. Biol.* **32,** 279–298

Dixon, A. F. G. (1972). Control and significance of the seasonal development of colour forms in the sycamore aphid, *Drepanosiphum platanoides* (Schr.). *J. Anim. Ecol.* **41,** 689–697

Douglas, M. M. and Grula, J. W. (1978). Thermoregulatory adaptations allowing ecological range expression by the pierid butterfly, *Nathalis iole* Boisduval. *Evolution* **32,** 776–783

Douwes, P. (1976). Activity in *Heodes virgaureae* in relation to air temperature, solar radiation and time of day. *Oecologia* **22,** 287–298

Downes, J. A. (1965). Adaptations of insects in the Arctic. *Ann. Rev. Ent.* **10,** 257–274

Downing, N. (1980). The regulation of sodium, potassium and chloride in an aphid subjected to ionic stress. *J. exp. Biol.* **87,** 343–349

Dreisig, H. (1980). Daily activity, thermoregulation and water loss in the tiger beetle, *Cicindela hybrida. Oecologia* **44,** 376–389

Ebeling, W. (1974). Permeability of insect cuticle. In: *The Physiology of Insecta* (M. Rockstein, ed.) 2nd edn., Vol. 6, pp. 271–343. Academic Press, New York and London

Edney, E. B. (1957). *The Water Relations of Terrestrial Arthropods.* Camb. Monogr. Exp. Biol. No. 5. Cambridge University Press, London and New York

Edney, E. B. (1960). The survival of animals in hot deserts. *Smithsonian Reports for 1959*, 407–425

Edney, E. B. (1966). Absorption of water vapour from unsaturated air by *Arenivaga* sp. (Polyphagidae, Dictyoptera). *Comp. Biochem. Physiol.* **19,** 387–408

Edney, E. B. (1967a). Water balance in desert arthropods. *Science* **156,** 1059–1066

Edney, E. B. (1967b). The impact of the atmospheric environment on the integument of insects. *Proc. Int. Biometerorol. Congr.* **4,** 71–81

Edney, E. B. (1968). The effect of water loss on the haemolymph of *Arenivaga* sp. and *Periplaneta americana. Comp. Biochem. Physiol.* **25,** 149–158

Edney, E. B. (1971a). Some aspects of water balance in tenebrionid beetles and a thysanuran from the Namib desert of South Africa. *Physiol. Zool.* **44,** 61–76

Edney, E. B. (1971b). The body temperature of tenebrionid beetles in the Namib desert of Southern Africa. *J. exp. Biol.* **55,** 253–272

Edney, E. B. (1974). Desert arthropods. In: *Desert Biology* (G. W. Brown, ed.) Vol. II, pp. 311–384. Academic Press, New York

Edney, E. B. (1975). Absorption of water vapour from unsaturated air. In: *Physiological Adaptation to the Environment* (J. Vernberg, ed.) pp. 77–97. Intext Educational Publishers, New York

Edney, E. B. (1977). *Zoophysiology and Ecology*, Vol. 9. Water balance in land arthropods. Springer Verlag, Berlin, Heidelberg, New York

Edney, E. B. (1980). The components of water balance. In: *Insect Biology in the Future* (M. Locke and D. S. Smith, eds) pp. 39–58. Academic Press, New York

Edney, E. B. and Barrass, R. (1962). The body temperature of the tsetse fly, *Glossina morsitans* Westwood (Diptera; Muscidae). *J. Ins. Physiol.* **8,** 469–481.

Edney, E. B., Haynes, S. and Gibo, D. (1974). Distribution and activity of the desert cockroach *Arenivaga investigata* (Polyphagidae) in relation to microclimate. *Ecology* **55,** 420–427

Edwards, G. A. and Nutting, W. L. (1950). The influence of temperature upon the respiration and heart activity of *Thermobia* and *Grylloblatta*. *Psyche* **57,** 33–44

Esch, H. (1976). Body temperature and flight performance of honey bees in a servomechanically controlled wind tunnel. *J. Comp. Physiol.* **109,** 265–277

Fisher, R. C. (1971). Aspects of the physiology of endoparasitic Hymenoptera. *Biol. Rev.* **46,** 243–278

Flitters, N. E. (1968). Microelectrothermometry for psychrometric determinations of insect leaf microhabitats. *Ann. Ent. Soc. Am.* **61,** 923–926

Fraenkel, G. (1930). Die Orientierung von *Schistocerca gregaria* zu strahlender Wärme. *Z. Vergl. Physiol.* **13,** 300–313

Galbreath, R. A. (1975). Water balance across the cuticle of a soil insect. *J. exp. Biol.* **62,** 115–120

Gamboa, G. J. (1976). Effects of temperature on the surface activity of the desert leaf-cutter ant, *Acromyrmex versicolor versicolor* (Pergarde). *Am. Midl. Nat.* **95,** 485–491

Geiger, R. (1965). *The Climate near the Ground.* Harvard University Press, Cambridge, Mass. and London

Gilby, A. R. (1980). Transpiration, temperature and lipids in insect cuticle. *Adv. Insect Physiol.* **15,** 1–33

Green, G. W. (1955). Temperature relations of ant-lion larvae (Neuroptera: Myrmeleontidae). *Can. Ent.* **87,** 441–459

Greenham, P. M. (1972). The effects of the variability of cattle dung on the multiplication of the bushfly (*Musca vetustissima* Walk.). *J. Anim. Ecol.* **41,** 153–165

Grimm, H. (1937). Kleintierwelt, Kleinklima and Mikroklima. *Z. für. angew. Met.* **54,** 25–31

Gunn, D. L. (1934). The temperature and humidity relations of the cockroach. II. Temperature preference. *Z. Vergl. Physiol.* **20,** 617–623

Gunn, D. L. (1942). Body temperature in poikilothermal animals. *Biol. Rev.* **17,** 293–314

Haarløv, N. and Peterson, B. B. (1952). Measurements of temperature in bark and wood of Sitka spruce (*Picea sitchensis*) with special reference to temperature in galleries of *Hylesinus* (Dendroctonus) *micans*. *Forstl. Forsøksvoesen Den.* **21,** 1–91

Hadley, N. F. (1970). Micrometeorology and energy exchange in two desert arthropods. *Ecology* **51,** 434–444

Hadley, N. F. (1979). Wax secretion and color phases of the desert tenebrionid beetle *Cryptoglossa virrucosa* (Le Conte). *Science* **203,** 367–369

Hafez, M. and Ibrahim, M. M. (1964). Studies on the behaviour of the desert grasshopper *Sphingonotus carinatus* Sauss., toward humidity and temperature. *Bull. Soc. Entomol. Egypt* **48,** 229–243

Hamilton, W. J. (1971). Competition and thermoregulatory behaviour of the Namib desert Tenebrionid beetle genus *Cardiosis*. *Ecology* **52,** 810–822

Hamilton, W. J. (1975). Coloration and its thermal consequences for diurnal desert insects. In: *Environmental Physiology of Desert Organisms* (N. F. Hadley, ed.) pp. 67–89. Dowden, Hutchinson and Ross, London, Stroudsberg, Penn.

Hamilton, W. J. and Seely, M. K. (1975). Fog basking by the Namib Desert beetle *Onymacris unguicularis*. *Nature* **262,** 284–285

Hanegan, J. L. and Heath, J. E. (1970). Activity patterns and energetics of the moth, *Hyalophora cecropia*. *J. exp. Biol.* **53,** 611-627

Hardy, H. T. (1966). The effect of temperature and sunlight on the posture of *Perithemis tenera*. *Proc. Okla. Acad. Sci.* **46,** 41–45

Haufe, W. O. and Burgess, L. (1956). Development of *Aedes* (Diptera: Culicidae) at Fort Churchill, Manitoba, and prediction of dates of emergence. *Ecology* **37,** 500–519

Heath, J. E. (1967). Temperature responses of the periodical "17-year" cicada, *Magicicada cassini*. *Am. Midl. Nat.* **77,** 64–76

Heath, J. E. and Adams, P. A. (1967). Regulation of heat production of large moths *J. exp. Biol.* **47,** 21–33

Heath, J. E. and Josephson, R. K. (1970). Body temperature and singing in the katydid, *Neoconocephalus robustus*. *Biol. Bull.* **138,** 272–285

Heath, J. E. and Wilkin, P. J. (1970). Temperature responses of the desert cicada, *Diceroprocta apache*. *Physiol. Zool.* **43,** 145–154

Heath, J. E., Wilkin, P. J. and Heath, M. S. (1972). Temperature responses of the cactus dodger *Cacama valvata*. *Physiol. Zool.* **45,** 238–246

Heath, J. E., Hanegan, J. L., Wilkin, P. J. and Heath, M. S. (1971). Adaptation of the thermal responses of insects. *Am. Zool.* **11,** 145–158

Heinrich, B. (1970). Thoracic temperature stabilization by blood circulation in a free-flying moth. *Science* **168,** 580–582

Heinrich, B. (1971a). Temperature regulation of the sphinx moth, *Manduca sexta*. I. Flight energetics and body temperature during free and tethered flight. *J. exp. Biol.* **54,** 141–152

Heinrich, B. (1971b). Temperature regulation of the sphinx moth, *Manduca sexta*. II. Regulation of heat loss by control of blood circulation. *J. exp. Biol.* **54,** 153–166

Heinrich, B. (1972). Energetics of temperature regulation and foraging in a bumblebee, *Bombus terricola* Kirby. *J. Comp. Physiol.* **77,** 49–64

Heinrich, B. (1974). Thermoregulation in insects. *Science* **185,** 747–756

Heinrich, B. (1975). Thermoregulation and flight energetics of desert insects. In: *Environmental Physiology of Desert Organisms* (N. F. Hadley, ed.), pp. 90–105. Dowden, Hutchinson and Ross, London, Stroudsberg, Penn.

Heinrich B. (1976). Resource partitioning among some eusocial insects: bumblebees. *Ecology* **57,** 874–889

Heinrich, B. (1979). *Bumblebee Economics*. Harvard University Press, Cambridge, Mass.

Heinrich, B. (1980a). Mechanisms of body temperature regulation in honeybees, *Apis mellifera*. I. Regulation of head temperature. *J. exp. Biol.* **85,** 61–72

Heinrich, B. (1980b). Mechanisms of body temperature regulation in honeybees, *Apis mellifera*. II. Regulation of thoracic temperature at high air temperatures. *J. exp. Biol.* **85,** 73–87

Heinrich, B. (1981). The mechanisms and energetics of honeybee swarm temperature regulation. *J. exp. Biol.* **91,** 25–55

Heinrich, B. and Bartholomew, G. A. (1971). An analysis of pre-flight warm-up in the sphinx moth, *Manduca sexta*. *J. exp. Biol.* **55,** 223–239

Heinrich, B. and Bartholomew, G. A. (1979). Roles of endothermy and size in inter- and intraspecific competition for elephant dung in an African dung beetle, *Scarabaeus laevistriatus*. *Physiol. Zool.* **52,** 484–496

Heinrich, B. and Casey, T. M. (1978). Heat transfer in dragonflies: 'fliers' and 'perchers'. *J. exp. Biol.* **74,** 17–36

Heinrich, B. and Pantle, C. (1975). Thermoregulation in small flies (*Syrphus* sp.); basking and shivering. *J. exp. Biol.* **62,** 599–610

Henson, W. R. (1958). The effects of radiation on habitat temperatures of some poplar-inhabiting insects. *Can. J. Zool.* **36,** 463–478

Henson, W. R. (1968). Some recent changes in the approach to studies of climatic effects on insect populations. *Symp. Roy. Entomol. Soc. London* **4,** 37–46

Henwood, K. (1975a). Infrared transmittance as an alternative thermal strategy in the desert beetle *Onymacris plana*. *Science* **189,** 993–994

Henwood, K. (1975b). A field-tested thermoregulation model for two diurnal Namib Desert tenebrionid beetles. *Ecology* **56,** 1329–1342

Hinton, H. E. (1960). A fly larva that tolerates dehydration and temperatures of $-270°$ to $+102°C$. *Nature* **188,** 336–337

Hocking, B. and Sharplin, C. D. (1965). Flower basking by arctic insects. *Nature* **206,** 215

Hoffmann, J. H., Moran, V. C. and Webb, J. W. (1975). The influence of host plant and saturation deficit on the temperature tolerance of a psyllid. *Ent. exp. Appl.* **18,** 55–67

Holm, E. and Edney, E. B. (1973). Daily activity of Namib Desert beetles in relation to climate. *Ecology* **54,** 45–56

Hook, R. I. van and Deal, S. L. (1972). Tritium uptake and elimination by tissue-bound and body-water components in crickets (*Acheta domesticus*). *J. Ins. Physiol.* **19,** 681–687

Howarth, F. G. (1980). The zoogeography of specialized cave animals: a bioclimatic model. *Evolution* **34,** 394–406

Humphreys, W. F. (1975). The influence of burrowing and thermoregulatory behaviour on the water relations of *Geolycosa godeffroyi* (Araneae: Lycosidae), an Australian wolf spider. *Oecologia* **21,** 291–311

Humphreys, W. F. and Reynolds, S. E. (1980). Sound production and endothermy in the horse bot-fly, *Gasterophilus intestinalis*. *Physiol. Entomol.* **5,** 235–242

Ishay, J. (1972). Thermoregulatory pheromones in wasps. *Experientia* **28,** 1185–1187

Ishay, J. (1973). Thermoregulation by social wasps: behaviour and pheromones. *Trans. N.Y. Acad. Sci.* **35,** 447–462

Jackson, W. B. (1957). Microclimatic patterns in army ant bivouacs. *Ecology* **38,** 276–285

Jones, R. E. (1975). Dehydration in an Australian rockpool chironomid larva (*Paraborniella tonnoiri*). *J. Entomol.* **49A,** 111–119

Kammer, A. E. (1970). Thoracic temperature, shivering and flight in the monarch butterfly, *Danaus plexippus* L. *Z. vergl. Physiol.* **68,** 334–344

Kay, C. A. and Whitford, W. G. (1978). Critical thermal limits of desert honey ants: possible ecological implications. *Physiol. Zool.* **51,** 206–213

Kennedy, J. S. and Booth, C. O. (1959). Responses of *Aphis fabae* Scop. to water shortage in host plants in the field. *Ent. exp. Appl.* **1,** 274–291

Kevan, P. G. (1975). Sun-tracking solar furnaces in high arctic flowers: significance for pollination and insects. *Science* **189,** 723–726

Kevan, P. G. and Shorthouse, J. D. (1970). Behavioural regulation by high arctic butterflies. *Arctic* **23,** 268–279

Key, K. H. L. and Day, M. F. (1954). A temperature controlled physiological colour response in the grasshopper *Kosciuscola tristis* Sjöst (Orthoptera: Acrididae). *Aust. J. Zool.* **2,** 309–339

Klomp, H. (1962). The influence of climate and weather on the mean density level, the fluctuations and the regulation of animal populations. *Arch. néerl. Zool.* **15,** 68–109

Knülle, W. and Wharton, G. W. (1964). Equilibrium humidities in arthropods and their ecological significance. *Acarologia* **6,** 299–306

Krafsur, E. S. (1971a). Behaviour of thoracic spiracles of *Aedes* mosquitoes in controlled relative humidities. *Ann. Ent. Soc. Am.* **64,** 93–97

Krafsur, E. S. (1971b). Influence of age and water balance on spiracular behaviour in *Aedes* mosquitoes. *Ann. Ent. Soc. Am.* **64,** 97–102

Kristensen, K. J. (1959). Temperature and heat balance of soil. *Oikos* **10,** 103–120

Krogh, A. (1948). Determination of temperature and heat production in insects. *Z. vergl. Physiol.* **31,** 274–280

Kühnelt, W. (1934). Die Bedeutung des Klimas für die Tierwelt. *Biokl. B* **1,** 120–125

Kühnelt, W. (1936). Der Einfluss des Klimas auf den Wasserhaushalt der Tiere. *Biokl. B* **3,** 11–15

Lack, A. J. (1976). Flower-basking by insects in Britain. *Watsonia* **11,** 143–144

Leigh, T. F. and Smith, R. F. (1959). Flight activity of *Colias philodice eurytheme* in response to its physical environment. *Hilgardia* **28,** 569–624

Lensky, Y. (1964). L'économie des liquides chez les abeilles aux temperatures elevées. *Insectes Soc.* **11,** 207–222

Lewis, T. (1962). The effects of temperature and relative humidity on mortality in *Limothrips cerealium* Haliday (Thysanoptera) overwintering in bark. *Ann. Appl. Biol.* **50,** 313–326

Lifson, N. and McClintock, R. (1966). Theory of use of the turnover rates of body water for measuring energy and material balance. *J. Theoret. Biol.* **12,** 46–74

Lindauer, M. (1955). The water economy and temperature regulation of the honeybee colony. *Bee World* **36,** 62–72, 81–92, 105–111

Louw, G. N. (1972). The role of advective fog in the water economy of certain Namib Desert animals. *Symp. Zool. Soc. London* **31,** 297–314

Louw, G. N. and Hamilton, W. J. (1972). Physiological and behavioural ecology of the ultrapsammophilous Namib Desert tenebrionid, *Lepidochora argentogrisea. Madoqua 11* **1,** 87–95

Loveridge, J. P. (1968a). The control of water loss in *Locusta migratoria migratorioides* R. & F. I. Cuticular water loss. *J. exp. Biol.* **49,** 1–13

Loveridge, J. P. (1968b). The control of water loss in *Locusta migratoria migratorioides* R. & F. II. Water loss through the spiracles. *J. exp. Biol.* **49,** 15–29

Loveridge, J. P. (1975). Studies on the water balance of adult locusts. III. The water balance of non-flying locusts. *Zoologica Africana* **10,** 1–28

Lowry, W. P. (1967). *Weather and Life–An Introduction to Biometeorology*. Academic Press, London and New York

Ludwig, D. (1945). The effects of atmospheric humidity on animal life. *Physiol. Zool.* **18,** 103–135

Lutz, P. E. and Pittman, A. R. (1970). Some ecological factors influencing a community of adult Odonata. *Ecology* **51,** 279–284

MacFadyen, A. (1967). Methods of investigation of productivity of invertebrates in terrestrial ecosystems. In: *Secondary Productivity of Terrestrial Ecosystems (Principles and Methods)* (K. Petrusewicz, ed.) pp. 383–412. Polish Academy of Science, Warszawa-Krakow

MacHattie, L. B. and McCormack, R. J. (1961). Forest microclimate: a topographic study in Ontario. *J. Ecol.* **49,** 301–323

Machin, J. (1979). Compartmental osmotic pressures in the rectal complex of *Tenebrio* larvae: evidence for a single tubular pumping site. *J. exp. Biol.* **82,** 123–137

Machin, J. (1980). Atmospheric water absorption in arthropods. *Adv. Insect Physiol.* **14,** 1–48

Maddrell, S. H. P. (1971). The mechanisms of insect excretory systems. *Adv. Insect Physiol.* **8,** 199–331

Maddrell, S. H. P. (1980a). The control of water relations in insects. In: *Insect Biology in the Future* (M. Locke and D. S. Smith, eds) pp. 179–199. Academic Press, New York and London

Maddrell, S. H. P. (1980b). Characteristics of epithelial transport in insect Malpighian tubules. *Current Topics in Membranes and Transport* **14,** 427–463

Madge, P. E. (1956). The ecology of *Oncopera fasciculata* (Walker) (Lepidoptera: Hepialidae) in South Australia. *Aust. J. Zool.* **4,** 327–345

Maier, C. T. and Waldbauer, G. P. (1979). Diurnal activity patterns of flower flies (Diptera: Syrphidae) in an Illinois sand area. *Ann. Ent. Soc. Am.* **72,** 237–245

Mail, G. A. (1932). Winter temperature gradients as a factor in insect survival. *J. Econ. Entomol.* **25,** 1049–1053

Marshall, A. T. and Cheung, W. W. K. (1974). Studies on water and ion transport in homopteran insects: ultrastructure and cytochemistry of the cicadoid and cecropoid Malpighian tubules and filter chamber. *Tissue & Cell* **6,** 153–171

Marshall, A. T. and Cheung, W. W. K. (1975). Ionic balance of Homoptera in relation to feeding site and plant sap composition. *Ent. exp. Appl.* **18,** 117–120

Mattsson, J. O. (1965). Brief fluctuations of temperature in and over a potato field and over fallow ground. *Lund Studies in Geogr.* Ser. A. No. 31, Lund

May, M. L. (1976a). Warming rates as a function of body size in periodic endotherms. *J. Comp. Physiol.* **111,** 55–70

May, M. L. (1976b). Thermoregulation and adaptation to temperature in dragonflies (Odonata: Anisoptera). *Ecol. Monogr.* **46,** 1–32

May, M. L. (1978). Thermal adaptations of dragonflies. *Odonatologica* **7,** 27–47

May, M. L. (1979). Insect thermoregulation. *Ann. Rev. Entomol.* **24,** 313–349

Mazek-Fialla, K. (1941). Die Körpertemperatur poikilothermer Tiere in Abhängigkeit vom Kleinklima. *Z. wiss. Zool.* **154,** 170–246

Meeuse, B. J. D. (1973). Films of liquid crystals as an aid in pollination studies. In: *Pollination and Dispersal* (N. B. M. Brantjes and H. F. Linskens, eds) pp. 19–20. Department of Botany, University of Nijmegen

Mellanby, K. (1942). Metabolic water and desiccation. *Nature* **150,** 21

Michener, C. D. (1974). *The Social Behaviour of the Bees*. Belknap Press, Harvard, Cambridge, Mass.

Michener, C. D., Lange, R. B., Bigarella, J. J. and Salamuni, R. (1958). Factors influencing the distribution of bees' nests in earth banks. *Ecology* **39,** 207–217
Miller, P. L. (1962). Spiracle control in adult dragonflies. *J. exp. Biol.* **39,** 513–535
Miller, P. L. (1964). Factors altering spiracle control in adult dragonflies: water balance. *J. exp. Biol.* **41,** 331–343
Mittler, T. E. (1958). The excretion of honey-dew by *Tuberolachnus salignus* (Gmelin) (Homoptera: Aphididae). *Proc. Roy. Entomol. Soc. London* (*A*) **33,** 49–55
Monteith, J. L. (1960). Micrometeorology in relation to plant and animal life. *Proc. Linn. Soc. London* **171,** 71–82
Monteith, J. L. (1972). *Survey of Instruments for Micrometeorology.* Blackwell Sci. Publs., Oxford and Edinburgh
Monteith, J. L. (1973). *Principles of Environmental Physics.* Edward Arnold, London
Mordue, W. (1969). Hormonal control of Malpighian tubule and rectal function in the desert locust, *Schistocerca gregaria. J. Ins. Physiol.* **15,** 273–285
Morgan, N. O. (1964). Autecology of the adult horn fly, *Haematobia irritans* (L.) (Diptera: Muscidae). *Ecology* **45,** 728–736
Mosebach-Pukowski, E. (1937). Über die Raupengesellschaften von *Vanessa io* und *Vanessa urticae. Zeit. für Morphol. und Ökol. der Tiere* **33,** 358–380
Nagy, K. A. (1975). Water and energy budgets of free-living animals: measurement using isotopically labelled water. In: *Environmental Physiology of Desert Organisms* (N. F. Hadley, ed.) pp. 227–245. Dowden, Hutchinson and Ross, London, Stroudsburg, Penn.
Nicolson, S. W. (1980). Water balance and osmoregulation in *Onymacris plana*, a tenebrionid beetle from the Namib desert. *J. Ins. Physiol.* **26,** 315–320
Noble-Nesbitt, J. (1976). Active transport of water vapour. In: *Transport of Ions and Water in Animals* (B. L. Gupta, R. Moreton, J. Oschman and B. J. Wall, eds) pp. 571–597. Academic Press, London.
Noble-Nesbitt, J. (1978). Absorption of water vapour by *Thermobia domestica* and other insects. In: *Comparative Physiology: Water, Ions and Fluid Mechanics* (K. Schmidt-Nielsen, L. Bolis and S. H. P. Maddrell, eds) pp. 53–66. Cambridge University Press, Cambridge
Norgaard, E. (1956). Environment and behaviour of *Theridion saxatile. Oikos* **7,** 159–192
O'Donnell, M. J. (1978). The site of water vapour absorption in *Arenivaga investigata.* In: *Comparative Physiology: Water, Ions and Fluid Mechanics* (K. Schmidt-Nielsen, L. Bolis and S. H. P. Maddrell, eds) pp. 115–121. Cambridge University Press, Cambridge
O'Farrell, A. F. (1963). Temperature-controlled physiological colour change in some Australian damselflies. *Aust. J. Sci.* **25,** 437–438
Parry, D. A. (1951). Factors determining the temperature of terrestrial arthropods in sunlight. *J. exp. Biol.* **28,** 445–462
Paul, R. G. (1975). Honeydew panting in *Tuberolachnus salignus* (Gmelin) (Hemiptera: Aphididae). *Entomol. Gaz.* **26,** 70
Pepper, J. H. and Hastings, H. (1952). The effect of solar radiation on grasshopper temperature and activities. *Ecology* **33,** 96–103
Phillips, J. E. (1980). Epithelial transport and control in recta of terrestrial insects. In: *Insect Biology in the Future* (M. Locke and D. S. Smith, eds) pp. 145–177. Academic Press, New York and London
Pierre, F. (1958). Écologie et peuplement entomologique des sables vifs du Sahara Nord-Occidental. Sér. Biol. I. 332 pp. Publications Centre Recherches Sahariennes, Paris

Pimentel, D. (1958). Alteration of microclimate imposed by populations of flour beetles (*Tribolium*). *Ecology* **39,** 239–246

Platt, R. B., Love, G. J. and Williams, E. L. (1958). A positive correlation between relative humidity and the distribution and abundance of *Aedes vexans*. *Ecology* **39,** 167–169

Plowright, R. C. and Owen, R. E. (1980). The evolutionary significance of bumble bee colour patterns: a mimetic interpretation. *Evolution* **34,** 622–637

Pond, C. M. (1973). Initiation of flight and preflight behaviour of anisopterous dragonflies, *Aeshna* spp. *J. Ins. Physiol.* **19,** 2225–2229

Porter, W. P. and Gates, D. M. (1969). Thermodynamic equilibria of animals with environment. *Ecol. Monographs* **39,** 227–244

Porter, W. P., Mitchell, J. W., Bechman, W. A. and deWitt, C. B. (1973). Behavioural implications of mechanistic ecology (thermal and behavioural modelling of desert ectotherms and their microenvironment). *Oecologia* **13,** 1–54

Prince, G. J. and Parsons, P. A. (1977). Adaptive behaviour of *Drosophila* adults in relation to temperature and humidity. *Aust. J. Zool.* **25,** 285–290

Rainey, R. C., Waloff, Z. and Burnett, G. F. (1957). The behaviour of the red locust (*Nomadacris semifasciata* Serville) in relation to topography, meteorology and vegetation of the Rukwa Rift Valley, Tanganyika. *Anti-Locust Bull.* **26,** 1–96

Ramsay, J. A., Butler, C. G. and Sang, J. H. (1938). The humidity gradient at the surface of a transpiring leaf. *J. exp. Biol.* **15,** 255–265

Raschke, K. (1956). Über die physikalischen Beziehungen zwischen Wärmeübergangszahl, Strahlungsaustausch, Temperatur und Transpiration eines Blattes. *Planta* **48,** 200–238

Rayah, El Amin El. (1970). Humidity responses of two desert beetles, *Adesmia antiqua* and *Pimelia grandis* (Coleoptera: Tenebrionidae). *Ent. exp. Appl.* **13,** 438–447

Richards, A. G. (1958). Temperature in relation to activity of single and multiple physiological systems in insects. *Proc. Int. Congr. Entomol. 10*, **2,** 67–72

Riddle, W. A., Crawford, C. S. and Zeitone, A. M. (1976). Patterns of haemolymph osmoregulation in 3 desert arthropods. *J. comp. Physiol.* **112,** 295–305

Rosenberg, N. J. (1974). *Microclimate: The Biological Environment.* Wiley Interscience, New York and London

Roth, L. M. and Willis, E. R. (1951). The effects of desiccation and starvation on the humidity behaviour and water balance of *Tribolium confusum* and *Tribolium castaneum*. *J. exp. Zool.* **118,** 337–361

Rowell, C. H. F. (1971). The variable coloration of the acridoid grasshoppers. *Adv. Insect Physiol.* **8,** 146–199

Rücker, F. (1933). Die Farben der Insekten und ihre Bedeutung für den Wärmehaushalt. *Arch. ges. Physiol.* **231,** 729–741

Salt, R. W. (1969). The survival of insects at low temperatures. *Symp. Soc. exp. Biol.* **23,** 331–350

Schaffer, W. M., Jensen, D. B., Hobbs, D. E., Gurevitch, J., Todd, J. R. and Schaffer, M. V. (1979). Competition, foraging energetics and the cost of sociality in 3 species of bees. *Ecology* **60,** 976–987

Schimitschek, E. (1931). Forstentomologische Untersuchungen aus dem Gebiet von Lunz. I. Standortsklima und Kleinklima in Beziehung zum Entwicklungsablauf und zur Mortalität von Insekten. *Z. angew. Entomol.* **18,** 460–491

Schlising, R. A. (1970). Sequence and timing of bee foraging in flowers of *Ipomoea* and *Aniseia* (Convolvulaceae). *Ecology* **51,** 1061–1067

Schoener, T. W. and Janzen, D. H. (1968). Notes on environmental determinants of tropical versus temperate insect size patterns. *Amer. Nat.* **102,** 207–224
Schwerdtpfeger, P. (1976). *Physical Principles of Micro-Meteorological Measurements*. Elsevier, Amsterdam, Oxford and New York
Seely, M. K. (1979). Irregular fog as a water source for desert dune beetles. *Oecologia* **42,** 213–227
Seely, M. K. and Hamilton, W. J. (1976). Fog catchment sand trenches constructed by tenebrionid beetles, *Lepidochora*, from the Namib desert. *Science* **193,** 484–486
Seymour, R. S. (1974). Convective and evaporative cooling in sawfly larvae. *J. Ins. Physiol.* **20,** 2447–2457
Seymour, R. S. and Vinegar, A. (1973). Thermal relations, water loss and oxygen consumption of a north American tarantula. *Comp. Biochem. Physiol.* **44A,** 83–96
Shanks, R. E. (1956). Altitudinal and microclimatic relationships of soil temperature under natural vegetation. *Ecology* **37,** 1–7
Shepherd, R. F. (1958). Factors controlling the internal temperature of spruce budworm larvae, *Choristoneura fumiferana* (Clem). *Can. J. Zool.* **36,** 779–786
Sheriff, D. W. (1973). An infra-red psychrometer for detecting changes in the humidity of leaf boundary layers. *J. Exp. Botany* **24,** 641–647
Sherman, P. W. and Watt, W. B. (1973). The thermal ecology of some *Colias* butterfly larvae. *J. Comp. Physiol.* **83,** 25–40
Sinclair, J. G. (1922). Temperatures of the soil and air in a desert. *Mon. Weath. Rev.* **50,** 142–144
Slobodchikoff, C. N. and Wismann, K. (1981). A function of the subelytral chamber of tenebrionid beetles. *J. exp. Biol.* **90,** 109–114
Smith, R. F. (1954). The importance of the microenvironment in insect ecology. *J. Econ. Entomol.* **47,** 205–210
Somme, L. and Ostbye, E. (1969). Cold hardiness in some winter active insects. *Nor. Entomol. Tidsskr.* **16,** 45–48
Sotavalta, O. (1947). The flight-tone (wing-stroke frequency) of insects. *Acta Entomol. Fenn.* **4,** 1–117
Stiles, E. W. (1979). Evolution of colour pattern and pubescence characteristics in male bumblebees: automimicry vs thermoregulation. *Evolution* **33,** 941–957
Stobbart, R. M. and Shaw, J. (1974). Salt and water balance: excretion. In: *The Physiology of Insecta* (M. Rockstein, ed.) Vol. 5, pp. 361–446. Academic Press, New York and London
Stoutjesdijk, P. (1977). High surface temperatures of trees and pine litter in the winter and their biological importance. *Int. J. Biometeor.* **21,** 325–331
Stower, W. J. and Griffiths, J. E. (1966). The body temperature of the desert locust (*Schistocerca gregaria*). *Ent. exp. Appl.* **9,** 127–178
Sullivan, C. R. and Wellington, W. G. (1953). The light reactions of larvae of the tent caterpillars, *Malacasoma disstria* Hbn., *M. americanum* (Fab.) and *M. pluviale* (Dyar) (Lepidoptera; Lasiocampidae) *Can. Ent.* **85,** 297–310
Sutton, O. G. (1953). *Micrometeorology. A Study of Physical Processes in the lowest Layers of the Earth's Atmosphere*. McGraw Hill, London
Tahvanainen, J. O. (1972). Phenology and microhabitat selection of some flea beetles (Coleoptera: Chrysomelidae) on wild and cultivated crucifers in central New York. *Ent. Scand.* **3,** 120–138
Theodor, O. (1936). On the relation of *Phlebotomus papatasii* to the temperature and humidity of the environment. *Bull. Ent. Res.* **27,** 653–671

Toolson, E. C. (1978). Diffusion of water through the arthropod cuticle: thermodynamic consideration of the transition phenomenon. *J. Thermal. Biol.* **3,** 69–73

Treherne, J. E. and Willmer, P. G. (1975). Hormonal control of integumentary water loss; evidence for a novel neuroendocrine system in an insect. *J. exp. Biol.* **63,** 143–159

Tyndale-Briscoe, M., Wallace, M. M. H. and Walker, J. M. (1981). An ecological study of an Australian dung beetle, *Orthophagus granulatus* Boheman (Coleoptera; Scarabaeidae) using physiological age-grading techniques. *Bull. Ent. Res.* **71,** 137–152

Unwin, D. M. (1978). Simple techniques for microclimate measurement. *J. Biol. Ed.* **12,** 179–189

Unwin, D. M. (1980). *Microclimate Measurement for Ecologists.* Academic Press, London

Unwin, D. M. and Ellington, C. P. (1979). An optical tachometer for measurement of the wing-beat frequency of free-flying insects. *J. exp. Biol.* **82,** 377–378

Unwin, D. M. and Willmer, P. G. (1978). A simple field cryoscope-osmometer for freezing point determination with small fluid samples. *Physiol. Entomol.* **3,** 341–345

Uvarov, B. P. (1931). Insects and climate. *Trans. Ent. Soc. London* **79,** 1–232

Uvarov, B. P. (1977). *Grasshoppers and Locusts. A Handbook of General Acridology,* Vol. 2. C.O.P.R., London

Verhoef, M. A. and Witteveen, J. (1980). Water balance in Collembola and its relation to habitat selection: cuticular water loss and water uptake. *J. Ins. Physiol.* **26,** 201–208

Veron, J. E. N. (1973). Physiological control of chromatophores of *Austrolestes annulosus. J. Ins. Physiol.* **19,** 1689–1703

Vielmetter, W. (1958). Physiologie des Verhaltens zur Sonnenstrahlung bei dem Tagfalter *Argynnis paphia* L. Untersuchungen im Freiland. *J. Ins. Physiol.* **2,** 13–37

Volkonsky, M. A. (1939). Sur la photoakinèse des acridiens. *Arch. Inst. Pasteur Algérie* **17,** 194–220

Waloff, Z. (1963). Field studies on solitary and transiens desert locusts in the Red Sea area. *Anti-Locust Bull.* **40,** 1–93

Wasserthal, L. T. (1975). The role of butterfly wings in regulation of body temperature. *J. Ins. Physiol.* **21,** 1921–1930.

Waterhouse, F. L. (1950). Humidity and temperature in grass microclimates with reference to insolation. *Nature* **166,** 232–233

Waterhouse, F. L. (1955). Microclimatological profiles in grass cover in relation to biological problems. *Quart. J.R. Met. Soc.* **81,** 63–71

Watt, W. B. (1968). Adaptive significance of pigment polymorphisms in *Colias* butterflies. I Variation of melanin pigment in relation to thermoregulation. *Evolution* **22,** 437–458

Watt, W. B. (1969). Adaptive significance of pigment polymorphisms in *Colias* butterflies. II. Thermoregulation and photoperiodically controlled melanin variation in *Colias eurytheme. Proc. Nat. Acad. Sci. USA* **63,** 767–774

Weis-Fogh, T. (1967). Respiration and tracheal ventilation in locusts and other flying insects. *J. exp. Biol.* **47,** 561–587

Wellington, W. G. (1949). The effects of temperature and moisture upon the behaviour of the spruce budworm, *Choristoneura fumiferana* Clemens (Lepidoptera: Tortricidae). *Sci. Agr.* **29,** 201–215

Wellington, W. G. (1950). Effects of radiation on the temperature of insectan habitats. *Sci. Agr.* **30,** 209–234

Wellington, W G. (1957). The synoptic approach to studies of insects and climate. *Ann. Rev. Ent.* **2,** 143–162

Wellington, W. G., Sullivan, C. R. and Green, G. W. (1966). Biometeorological research in Canadian forest entomology: A review. *Int. J. Biometeor.* **10,** 3–15

Wharton, G. W. and Richards, A. G. (1978). Water vapour exchange kinetics in insects and acarines. *Ann. Rev. Ent.* **23,** 309–328

Wigglesworth, V. B. (1941). The sensory physiology of the human louse, *Pediculus humanus corporis* De Geer. (Anoplura). *Parasitology* **33,** 67–109

Willmer, P. G. (1980a). The effects of a fluctuating environment on the water relations of larval Lepidoptera. *Ecol. Entomol.* **5,** 271–292

Willmer, P. G. (1980b). The effect of insect visitors on nectar constituents in temperate plants. *Oecologia* **47,** 270–277

Willmer, P. G. (1982). Hygrothermal determinants of insect activity patterns: the Diptera of water-lily leaves. *Ecol. Entomol.* (In press)

Willmer, P. G. and Corbet, S. A. (1981). Temporal and microclimatic partitioning of the floral resources of *Justicia aurea* amongst a concourse of pollen vectors and nectar robbers. *Oecologia* **51,** 67–78

Willmer, P. G. and Unwin, D. M. (1981). Field analyses of insect heat budgets: reflectance, size and heating rates. *Oecologia* **50,** 250–255

Wilson, E. O. (1971). *The Insect Societies.* Belknap Press, Harvard, Cambridge, Mass.

Winston, P. W. and Beament, J. W. L. (1969). An active reduction of water level in insect cuticle. *J. exp. Biol.* **50,** 541–546

Wojtowski, F. (1963). Studies on heat and water economy in bumblebee nests (Bombinae). *Zoologica Pol.* **13,** 19–36

Control of Food Intake

E. A. Bernays and S. J. Simpson

Centre for Overseas Pest Research, London, UK

1 Introduction

The regulation of food intake has been investigated in a number of insects, most notably blowflies and locusts. Results have tended to suggest that the mechanisms involved vary considerably between groups, between species within groups, and even within a species under different conditions. Since the last major review of insect feeding (Barton Browne, 1975) a number of key studies have started to bridge the gaps between such apparently diverse situations. Also, it has become more apparent that results must be interpreted strictly in the context of the experimental methods used. If this is done it is possible to find a number of important features in the regulation of food intake which are shared by diverse insect groups, and to show the complex balance of regulatory mechanisms used by any one species. The aim of the present review is to draw such generalisations and attempt to provide an overall synthesis of feeding in the insects. In doing so greatest emphasis has inevitably been placed on the blowflies and locusts but where possible data from other insect groups have been incorporated.

It is intended to concentrate on the regulation of the quantity of food ingested: what initiates feeding, what determines how much is eaten in one meal and what determines the frequency with which meals are taken. Food selection and the control of the quality of food ingested are not considered except in so far as they affect the amounts consumed.

2 Initiation of ingestion

Before ingestion begins insects exhibit a pattern of behaviour which brings the head and mouthparts to a suitable position for feeding and enables them to monitor food quality critically. If the food is suitable feeding follows. Although the steps in the behaviour pattern may occur rapidly and thus be difficult to distinguish, each step requires particular sensory inputs, often involving sensilla remote from the mouthparts, and the sequence may be interrupted by unsuitable stimulants at each stage. In this section, the control of the preliminary steps in the sequence of behaviour is considered, together with aspects of chemoreception which influence the early stages of ingestion.

Most insects will not start to feed unless they receive an appropriate stimulus from the substrate which indicates to them that it is of a suitable nature. Although there are many statements to the contrary, more precise observations indicate that a lack of selectivity is associated with extreme deprivation of food. In a well-fed insect with continuous access to food some positive phagostimulation is necessary to initiate feeding.

The variety of ingestion mechanisms among insects is far greater than in any other class of animals. The mouthpart structures associated with them are taxonomic characters at the ordinal level, and there may be large differences even within families. Partly as a result of this, the preliminary aspects of feeding behaviour also vary between groups. Comprehensive studies of both behaviour and sensory physiology within a group are rare and the best known cases are discussed in this section.

2.1 BITING AND CHEWING INSECTS

2.1.1 *The beginning of feeding*

The initiation of ingestion has been studied in most detail in locusts. Although olfaction can play a major role in location of food, tarsal chemoreceptors are involved in arresting movement in *Schistocerca gregaria* (Kendall, 1971). If the information from the large number of chemo- and mechanoreceptors at the tips of the labial and maxillary palps indicates that the food is acceptable, then the insect locates a suitable starting place, perhaps a leaf edge, with the help of its palps and then lowers its head (Mordue, 1974, 1979). At this point additional information may be received by contact of external labral receptors against the surface of the leaf (Sinoir, 1969). Only then does biting follow. The first bite is often very small but the leaf tissue is crushed and because of the hydrophilic nature of the inner surface of the mouthparts, fluid from within the leaf spreads over the various groups of chemoreceptors in the cibarial cavity (Cook, 1977).

The palpation which precedes biting in acridids is a rapid vibration of the maxillary and labial palps, which brings the sensilla at their tips into intermittent contact with the surface of the leaf (Blaney and Chapman, 1970; Blaney and Duckett, 1975). The information from up to 2000 sensilla determines whether or not head lowering and biting follow (Blaney and Chapman, 1970; Bernays and Chapman, 1970, 1974). The terminal sensilla of the palps of *Locusta migratoria* respond to many different chemicals including sugars (Blaney, 1974) but Blaney and Chapman (1970) concluded that during the palpation phase of the initiation of feeding, they normally respond to long chain components of the plant surface waxes.

The conspicuous head lowering described by Sinoir (1969) is only readily distinguished on an artifical substrate where the exploratory behaviour appears to be slower than on a plant. Similarly a small test bite is often only noticeable on a plant which is relatively unpalatable.

Following the initial bite the sensilla on the inside of the clypeolabrum are stimulated by chemicals from the plant sap. These sensilla respond to a wide range of chemicals (Haskell and Schoonhoven, 1969), including phagostimulants and deterrents, the response varying in different species (for a full discussion see Bernays and Chapman, 1978, and section 3). Very large numbers of chemoreceptors are present in the cibarial cavity, mostly arranged in groups (Bitsch and Denis, 1973; Chapman and Thomas, 1978; Cook, 1977; Louveaux 1972, 1975; Viscuso *et al.*, 1978). The possible significance of the very large numbers is discussed by Dethier (1976), Chapman and Thomas (1978) and Chapman (1980, and see this volume).

In a normally feeding insect the sensilla which initiate feeding become effective in sequence; tarsi, palps, cibarial cavity. Sinoir (1970) suggests that continuous feeding requires stimulation of all the groups of chemoreceptors within the cibarial cavity, as well as the palp tips, and gives a description of probable stages of chemoreception at the beginning of a meal. Ablation experiments have also suggested that the major groups of chemoreceptors are all involved in the regulation of ingestion, since in most cases their removal was shown to decrease amounts eaten (Haskell and Mordue, 1969). However, studies involving cutting of nerves to these groups of sensilla showed that after an interval normal intake was often resumed (Louveaux, 1976). The insect can partly compensate for such loss of some receptor input by changes in feeding behaviour. For instance, after palpectomy, greater use is made of the anterior surface of the labrum in monitoring the food prior to ingestion (Sinoir, 1969). A group of chemoreceptors near the tip of the labrum appears to be particularly important in monitoring the presence of feeding deterrents, and their ablation resulted in larger amounts of a previously unacceptable food being ingested (Haskell and Mordue, 1969). All the chemoreceptors tested neurophysiologically appear to respond to many phagostimulants and deterrents and the system appears to be both complex and plastic (Blaney, 1975, 1980; Haskell and Schoonhoven, 1969). Very similar mouthparts and arrangements of sensilla in cockroaches and crickets suggest similar regulatory processes (Klein, 1979; Malz and Hintze-Podufal, 1979, 1979a; Petryszak, 1975; Rościszewska and Fudalewicz-Niemezyk, 1974; Urvoy *et al.*, 1978).

Phagostimulation with liquids can result in ingestion without movement of the mouthparts (Barton Browne *et al.*, 1975). This suggests an ability to produce suction through the closed mouthparts and such suction could play a part in normal feeding by facilitating the spread of fluids over groups of chemoreceptors, or in the actual ingestion of very wet food.

The initiation of feeding and the ingestion process have not been studied in detail in caterpillars, but it is assumed that it is broadly similar to the processes occurring in grasshoppers although the mouthparts and the number of sensilla are much reduced. There is a behavioural sequence of events which resembles that of grasshoppers, but the controlling chemoreceptor mechanisms are simpler, being associated more with single chemicals (Schoonhoven, 1973) and with surprisingly simple correlations between impulse patterns from receptors and feeding activity, at least in the oligophagous *Pieris* spp. (Blom, 1978, 1978a). Olfaction has been shown to play an important role in feeding (Dethier, 1937) and in some cases non-host plants are ingested if the antennae and maxillae are removed (Schoonhoven, 1968). These structures bear olfactory receptors as well as contact chemoreceptors. Single sensilla or small groups of them occur on the mouthparts (Henig, 1930) and the roles of these different sensilla are described by Dethier (1937), Ma (1972) and Schoonhoven (1968, 1973). As in acridids, phagostimulants are required to initiate and maintain feeding, but the process is achieved with relatively little sensory input: there are approximately 24 chemoreceptors on the mouthparts of caterpillars compared with nearly 2000 in a grasshopper of similar size (Chapman, 1974, 1980).

2.1.2 *Aspects of chemoreception*

Chemoreception is of primary importance in initiating food intake. Some chemicals, termed phagostimulants, are essential for feeding to start at all; others, termed deterrents, may inhibit feeding even in the presence of phagostimulants (see section 3). Certain chemicals may affect only some steps in the feeding sequence. In the larva of *Bombyx mori*, β-sitosterol, isoquercitrin and morin induce biting but do not lead to ingestion when presented alone, and separate swallowing factors were identified as cellulose, sugar, inositol, silica and potassium phosphate (Hamamura, 1970). Maximum amounts of food are only ingested in the presence of both biting and swallowing factors. In acridids surface chemicals, perceived by the palp receptors, may stimulate biting but then play little or no role in continued feeding. These too might be regarded as biting factors or "incitants" but mostly these categories cannot be distinguished, since many chemicals including sucrose stimulate all the chemoreceptor groups and thus elicit all the behavioural steps.

Acridid chemosensilla are responsive to a very wide range of chemicals and appear not to be specialised for detecting particular compounds. Those which have been examined have neurones responding to all classes of chemical and may be termed generalists (Blaney, 1974, 1975; Winstanley and Blaney, 1978). There are some compounds for which the deterrent threshold is extremely low, however, as with azadirachtin on palp and cibarial receptors of *Schistocerca gregaria* (Winstanley and Blaney, 1978).

Lepidopteran larvae on the other hand possess sensory neurones which are more or less specific for different types of chemicals such as water, glucose, fructose, salts and deterrents (Schoonhoven, 1973). In some cases a highly specific cell responding to particular phagostimulants may be present. The best known example of such a cell is the glycoside receptor in *P. brassicae*, which responds particularly to glucosinolates from the cruciferous host plants (Schoonhoven, 1967). These compounds will greatly enhance the amounts of a suboptimal diet (less than 0.2 M sucrose on filter paper) which is ingested (Ma, 1972; Schoonhoven, 1977) although they are only weakly stimulating when presented alone. Thus, the glucosinolates have been called incitants for *P. brassicae*, although it is possible that the effect is due to a synergistic action of the two chemicals at a central level (see section 3).

Chemoreceptors start to adapt very quickly (e.g. Blaney, 1974), and some degree of adaptation will certainly have occurred before ingestion begins, but the sequence of receptor input, palpation, and the discontinuous nature of the stimulus when feeding on solid material probably all contribute to minimising adaptation, and presumably ensure that adaptation to a threshold level is prevented. Its possible significance in determining meal length is discussed in section 4.

During the initiation of feeding on acceptable food "central excitation" is rapidly raised to a high level (see sections 4 and 5). The effect has been shown to be the result of chemoreception (Dethier *et al.*, 1965) and one of its functions may be to enhance the intensity of the feeding processes so that they will continue quickly and efficiently. The importance of such a mechanism has been discussed in relation to vertebrates where there are known to be positive feedbacks following the initiation of feeding (e.g. Wiepkema, 1971). Another function of a heightened excitatory state may be to increase the likelihood of relocating the food if contact with it is lost (see section 2.2.1). To some extent it will offset adaptation of the chemoreceptors, and its possible role in determining meal size is discussed in section 4.

Finally, the chemoreceptor input may influence the rate of feeding. Less acceptable foods are eaten more slowly, primarily due to pauses during the course of ingestion (Bernays and Chapman, 1972). Such pauses may be due to a direct effect of low palatability or to a reduced action of the chemoreceptor input in maintaining the excitatory state.

2.2 FLUID FEEDERS

2.2.1 *Blowflies*

In order to ingest food, the fly must first extend its proboscis. This generally occurs after phagostimulation of receptors remote from the mouthparts. Tarsal chemoreceptor hairs are normally involved at this stage. In *Phormia regina* there are about 250 chemoreceptors round the edge of the labellum which, if stimulated, also cause proboscis extension, but in the usual sequence of events, their stimulation occurs after proboscis extension and results in spreading of the labellar lobes (Pollack, 1977). Details of the nature and specificity of tarsal and labellar chemoreceptors are summarised by Dethier (1976). Following spreading of the labellar lobes, an array of over 100 small chemoreceptor papillae (the interpseudotracheal papillae) may come into contact with the substrate. Contact of these papillae with phagostimulants leads to the initiation and driving of sucking by the pharyngeal pump (see Dethier, 1976). Further details for *Calliphora erythrocephala* may be found in the work of Rice (1973). Food quality is monitored by cibarial chemoreceptor sensilla, whose structure and distribution are described by Rice

(1973), while Evans and Barton Browne (1960) review all the earlier work.

If during ingestion the sensory input declines below a certain threshold level or if negative feedback from stretch receptors associated with the gut leads to termination of the meal, the proboscis slowly retracts. If feeding deterrents are encountered, however, there may be fast retraction. The actions and control of the proboscis are described by Dethier (1959), while more recently van der Starre and Ruigrok (1980) have shown that "both extension and retraction are direct muscular actions, extension not being caused by a pneumatic mechanism" as was previously thought. The distribution and structure of sensilla on the proboscis are described by Felt and Vande Berg (1976), Maes and den Otter (1975), Maes and Vedder (1978) and Wilczek (1967) while the relationship between phagostimulant action on chemoreceptors and the consequent motor patterns has been demonstrated by Getting (1971).

Stimulation of single chemoreceptors can elicit the appropriate behaviour patterns at each stage of feeding. For example, stimulation of one labellar receptor hair of *P. regina* with 0.4 M sucrose will result in proboscis extension. Stimulation of two receptors simultaneously with 0.06 M sucrose will have the same effect, while stimulation of the whole labellum with only 0.016 M sucrose is sufficient to cause extension (Arab, 1959). Thus summation occurs and must normally be important. Also, stimulation of hairs on one side of the labellum causes extension in that direction, and there are suggestions that the receptor fields are represented centrally in spatially separated regions (Dethier, 1976).

Although sugars stimulate all the chemoreceptor hairs, there are some chemicals which stimulate proboscis extension, but reduce the amount of sugar ingested (Evans and Barton Browne, 1960), so that presumably decisions are made at each step in the initiation of feeding (Evans and Barton Browne, 1960).

If the fly is less "hungry" only a high frequency of impulses will cause feeding. The numbers or frequency of impulses required to produce proboscis extension for different deprivation states has received a lot of attention, and the work is fully reviewed by Barton Browne (1975) and Dethier (1976). Sensory adaptation occurs during feeding and it is more rapid with less stimulating materials. Within a second of stimulation with 1.0 M sucrose the impulse frequency is reduced by half, but reduction of impulse frequency for lower concentrations is much faster (Dethier and Hanson, 1965; McCutchan, 1969). At maximally stimulating concentrations, and in most natural situations, where the food source is chemically more complex, the role of adaptation is probably minimal (Thomson and Holling, 1974). Its possible role in determining meal size is discussed further in section 4.

To some extent sensory adaptation may be offset by phagostimulants producing a state of arousal or a heightened "central excitatory state" which may last for over a minute and lower the behavioural threshold for responsiveness to less stimulating materials over this time (Dethier *et al.*, 1965; Nelson, 1977). Proboscis extension following stimulation with water is more likely to occur if the same or other sensilla (for example on the tarsi) have been stimulated with sucrose. The effect lasts longer with increasing sucrose concentration. Thus, several chemosensory components play a role in the regulation of food intake by the fly.

The motor output causing sucking is a centrally generated rhythmic pattern, and mechanoreceptor inputs from the cibarium are necessary to both maintain continuity of rhythmicity of the motor output and to alter output as a function of load in various ways (Dethier, 1976; Rice, 1970; Rice and Finlayson, 1972).

2.2.2 *Blood feeders*

Within the blood feeding insects there is a wide variety of feeding mechanisms and sensory receptor systems, each adapted to particular host characteristics. Details of mouthpart morphology are provided by Askew (1971) and feeding within the group has been reviewed by Friend and Smith (1972, 1977) and Langley (1976).

After host location there is a basic sequence of events during feeding which is common to most blood feeders: probing, piercing and penetrating the host surface, locating blood, ingesting blood and passing it into part or parts of the gut (Friend and Smith, 1977; Galun, 1975). Details of the behavioural sequences and their chemosensory control in a number of groups have been reviewed by Friend and Smith (1977).

A temperature gradient or differential is the major stimulus required to elicit probing in all blood feeders studied to date. Other factors which are known to affect probing and piercing include CO_2, visual, mechanical and chemical stimuli. Dietary composition, including the nature and concentrations of nucleotides, sugars and electrolytes, osmotic pressure and pH may determine whether ingestion and gorging occur. Receptors responding to temperature have been found on the antennae of *Rhodnius prolixus* and *Aedes aegypti* (Wigglesworth and Gillet, 1934; Davis and Sokolove, 1975) and on the antennae and foretarsi of *Glossina morsitans* (Dethier, 1954; Reinouts van Haga and Mitchell, 1975). Stimulation of these receptors results in proboscis extension in *G. morsitans*. Receptors responding to water vapour and CO_2 have been described on the antennae of *A. aegypti* (Kellogg, 1970). The role of mechanoreceptors in the tsetse fly has been discussed by Rice (1975).

Nucleotides are the major phagostimulants in the blood feeders and receptors responding to them have been demonstrated in the tsetse fly *G. morsitans* (Mitchell, 1976, 1976a) and postulated in *Rhodnius prolixus* (Smith and Friend, 1976). Little is known of the nature of receptors which respond to blood in most species. Rice *et al.* (1973, 1973a) have catalogued the chemoreceptors on the mouthparts of the tsetse fly. There are two groups: sixteen on the labellum which are involved in probing and piercing and four in the cibarium which control pumping and ingestion. In mosquitoes and *R. prolixus* the only gustatory receptors which have been found to respond to blood are in the cibarium.

2.2.3 *Aphids*

In aphids and other hemipteroids it is likely that there is a distinct behavioural sequence leading to the initiation of feeding. For example, Klingauf *et al.* (1971) showed that surface extracts of bean leaves induced more probing attempts in *Acyrthosiphon pisum*, and Dagmar (1979) showed that simple compounds occurring on leaf surfaces can cause more probing in *Aphis fabae*. Fier and Beck (1963) describe orientation and testing of the seed surface by *Oncopeltus fasciatus*. This insect distinctly places the area of labial receptors onto the surface of the seed before stylet penetration, and ethanolic extracts from the surface of milkweed seeds induce penetration. Amino acids and starches were more important thereafter for influencing the continuation of sucking.

Aphids ingest large amounts of fluid generally from the phloem sieve tubes during feeding on an appropriate host plant, as evidenced by the copious production of honey dew which is the residue from which nutrients have been selectively removed (Auclair, 1963, 1969). McLean and Kinsey (1968), using electronic apparatus, found that *Acyrthosiphon pisum* on its favoured host plant fed for 22 hours out of 24. The rate of uptake is regulated according to apparent need as shown by Mittler (1958) who found that *Tuberolachnus salignus* fed more slowly when the nitrogen level of the sap was high. Also it has been found that *Aphis fabae* feeds fastest when ants are in attendance (Banks and Nixon, 1958). Practically nothing is known about the manner in which these regulatory mechanisms operate and nothing is known concerning the action of chemoreceptors, or whether they adapt in such a continuously stimulating environment. Contact chemoreceptors are apparently absent on the labium and stylets of aphids (Tjallingii, 1978) so that monitoring of plant sap must occur by sampling into the food canal in the proximal mouthpart region to epipharyngeal receptors (the epipharyngeal organ), (Wensler and Filshie, 1969; Tjallingii, 1980). References to mouthpart sensilla arrangement and function in Heteroptera, are given by Avé *et al.* (1978) and Cobben (1978).

3 Continuation of feeding

Continued feeding depends, in general, on continued phagostimulation and an inadequate concentraton of phagostimulants leads to early cessation of feeding (Barton Browne *et al.*, 1975a; Bernays and Chapman, 1974, Cook, 1976). This indicates that the pattern of feeding is not one which is simply switched on and then continues until repletion; it requires continued positive feedback, an important part of which is usually chemical feedback from the food. This is not true however of aquatic filter feeders which rely largely on mechanoreception to maintain the filtering mechanism, although phagostimulants do alter the rate of feeding by mosquito larvae (Dadd, 1970). Nor is it true in carnivorous insects in which feeding is often independent of phagostimulants. Mantids and robber flies (Asilidae) catch and attempt to eat anything of suitable size moving in an appropriate manner. In some cases, as with mantids, the insects appear to be completely catholic, eating anything which is caught; in other cases however, predators exhibit specific requirements which may be regulated by gustatory information following initial non-specific catching (Brues, 1946; Dadd, 1970).

3.1 PHAGOSTIMULANTS

Sugars, and in particular, sucrose are the best known and most widespread phagostimulants. Many studies of phytophagous insects including grasshoppers, beetles, caterpillars, aphids and planthoppers have now been carried out and in all cases a range of sugars including sucrose stimulates feeding (e.g. Abushama, 1968; Akeson *et al.*, 1970; Auclair, 1969; Banks 1965; Dadd, 1960; 1970a; Hsaio, 1972; Ma, 1972; Mittler, 1967; Schoonhoven and Dethier, 1966; Thorsteinson, 1958, 1960). Bees, wasps, butterflies, moths and many Diptera, including blood sucking species, are also known to feed on sugary solutions. In *Phormia regina*, maltose, fructose and sucrose are all very effective phagostimulants, while various other sugars initiate and maintain feeding but to a lesser extent (Dethier *et al.*, 1956; Dethier and Rhoades, 1954; Hassett *et al.*, 1950). The effectiveness measured in different species, as the amount ingested in one meal or over an extended period, increases as the concentration increases up to a maximum above which the amounts ingested may be reduced (Barton Browne *et al.*, 1975a; Cook, 1976; and see Dethier, 1976 for a full discussion of flies). The concentration causing maximum intake generally lies between 0.1 M and 1.0 M. Moreover, in *Pieris brassicae* at least, there is a strong correlation between impulse frequencies from the sugar receptors and the amounts ingested (Blom, 1978). The comparative stimulating effectiveness of different sugars for six phytophagous insects is shown in Table 1. Interestingly, in *Oncopeltus fasciatus*, starches are more strongly stimulating than sucrose (Fier and Beck, 1963).

TABLE 1 Comparative stimulating effectiveness of different sugars for different phytophagous insects. (+++++, highly stimulating; +, weakly stimulating; –, no effect; •, not tested)

	Locusts		Beetles		Caterpillars	
	Locusta migratoria[1]	*Schistocerca gregaria*[2]	*Hypera postica*[3]	*Leptinotarsa decemlineata*[3,4]	*Pieris brassicae*[5]	*Spodoptera* spp.[6,7,8]
Pentoses						
L-arabinose	+	•	•	–	–	–
L-rhamnose	–	–	•	–	–	–
D-ribose	–	–	•	–	–	•
D-xylose	–	–	•	–	–	–
Hexoses						
D-fructose	+++++	+++++	++++	+	–	+++++
D-galactose	++	+	•	+	–	++
D-glucose	+++	++++	+	+	++	++
D-mannose	–	+	++	–	–	+
L-sorbose	+	+	•	–	–	–
Disaccharides						
D-cellobiose	–	+	•	–	–	–
D-lactose	+	+	•	–	–	+
D-maltose	+++++	++++	++	–	–	+++
D-melibiose	+++	+++	•	–	–	++
D-sucrose	+++++	+++++	+++++	+++++	+++++	+++++
D-trehalose	+	++	+	+	–	–
Trisaccharides						
D-melizitose	++++	+	+++	++	–	++
D-raffinose	++++	+++	•	–	–	+++
Alcohols						
Inositol	+	•	•	–	–	–
Sorbitol	+	+	•	–	–	–
Mannitol	+	+	•	–	•	–

[1] Cook (1977)
[2] Dadd (1960)
[3] Hsiao (1969)
[4] Hsiao and Fraenkel (1968)
[5] Ma (1972)
[6] Ma (1976)
[7] Meisner *et al.* (1972)
[8] Khalifa *et al.* (1974)

Apart from sugars, in most phytophagous insects it is common for a selection of different nutrients normally present in the host plants to serve as phagostimulants (e.g. Dadd, 1963; Dethier, 1970; Ma, 1972; Mitchell, 1974; Schoonhoven, 1973; Schoonhoven and Dethier, 1966; Thorsteinson, 1958; Wensler and Dudzinski, 1972). Among these are amino acids, although their effects vary with the species. Thus in acridids certain amino acids are weakly stimulating (Cook, 1977a; Thorsteinson, 1958), while in aphids, *Oncopeltus fasciatus* and *Leptinotarsa decemlineata*, a number of amino acids are strongly stimulating. The differences in response to amino acids by several phytophagous insects is shown in Table 2. Other nutrients which may stimulate feeding include phospholipids, certain fatty acids and sterols, ascorbic acid and various salts (e.g. Cook, 1977a; Dadd, 1970a).

TABLE 2 Comparative stimulating effectiveness of different amino acids on different phytophagous insects. (++, stimulating; —, no effect; •, not tested; (+), stimulating only in the presence of sucrose)

	Locusta migratoria[1]	*Leptinotarsa decemlineata*[2]	*Hypera postica*[3]	*Pieris brassicae*[4]	*Aphis fabae*[5]	*Acyrthosiphon pisum*[6]	*Oncopeltus fasciatus*[7]
Glycine	—	+	•	—	•	•	++
D-alanine	—	•	•	•	•	•	++
L-alanine	•	++	—	•	++	•	++
L-serine	+	++	—	—	++	•	+
γ-aminobutyric acid	—	++	—	—	•	•	+
L-valine	+	+	—	•	•	++	•
L-threonine	+	—	•	—	•	•	+
L-leucine	—	+	•	—	•	++	•
L-isoleucine	—	—	•	•	•	—	•
L-cysteine	—	+	•	•	—	•	•
L-cystine	—	—	•	•	•	•	•
L-methionine	—	—	•	(+)	++	++	++
L-aspartic acid	—	—	•	•	•	•	•
L-asparagine	—	+	•	•	•	—	•
L-glutamic acid	—	—	•	—	•	•	+
L-glutamine	—	—	•	—	•	•	•
L-lysine	—	—	•	—	•	—	•
L-arginine	—	—	•	•	•	—	+
L-histidine	•	—	•	—	++	—	+
L-phenylalanine	—	—	•	—	—	++	+
L-tyrosine	—	—	•	•	—	•	•
L-tryptophan	—	—	•	•	•	++	•
L-proline	+	++	—	(+)	++	•	•
L-hydroxyproline	•	+	•	•	•	•	•

[1] Cook (1977a)
[2] Hsaio and Fraenkel (1968)
[3] Hsaio (1969)
[4] Ma (1972)
[5] Leckstein and Llewellyn (1974)
[6] Srivastava and Auclair (1974)
[7] Fier and Beck (1963)

Caution is required in the interpretation of experiments in which single nutrient chemicals are presented to test their phagostimulatory properties, since when they are combined the response of the insect may be different. With some combinations of amino acids, the response is greater than the sum of the responses to each amino acid separately (Bernays and Chapman, 1978; Thorsteinson, 1958). In some instances an amino acid or secondary plant chemical stimulates feeding only in the presence of sucrose (Ma, 1972). In such cases where the behavioural response to two or more chemicals in terms of amount eaten is greater than the sum of the amounts eaten when the individual compounds are presented singly, then the chemicals are synergistic in their action (Dethier and Kuch, 1971). In some cases the combinations are additive or less than additive (Bernays and Chapman, 1978). Other examples of these effects are given by Dadd (1970a).

For fluid feeders, water is commonly a phagostimulant but this is a chemical in a special category, since it is universally required (see Dethier, 1976 for discussion). Normally an adequate amount of water is ingested with the food but the case of desiccating conditions when it becomes a drinking stimulus is discussed in section 6.

Chemicals have sometimes been shown to be phagostimulants even though they are neither required nor utilised. A well-known example is the sugar fucose, which is strongly stimulating to flies and causes increased feeding yet is not used metabolically (Hassett *et al.*, 1950). However, the significance of this is not known, since fucose is probably rarely encountered by a fly in the course of its normal activity.

Non-nutrient materials which are specific to the host plants stimulate feeding in some oligophagous species, and appear to function as sign stimuli indicating suitable food. For example a number of species feed on cruciferous plants containing glucosinolates: such compounds are commonly phagostimulants for these insects and increase food intake (van Emden, 1972; Ma, 1972; Nayer and Thorsteinson, 1963; Nielson, 1978; Schoonhoven, 1968, 1972). The results of Ma (1972), however, on larvae of *Pieris brassicae*, indicate that if sucrose levels are greater than 0.05 M, no other stimulants are required (see also section 3). Similarly for insects feeding on solanaceous plants the alkaloidal glycosides have a key role in phagostimulation (Hsiao, 1974). A number of examples of oligophagy or monophagy among phytophagous insects appear to depend on the stimulating properties of some secondary plant compounds which are of chemotaxonomic significance (Dadd, 1970a; van Drongelen, 1979; Nielsen *et al.*, 1979; Norris, 1970; Rees, 1969; Schoonhoven, 1972). In polyphagous species such as the desert locust *Schistocerca gregaria*, many secondary plant chemicals of no known value to the insect enhance food intake, but the interpretation of this can only be speculative (Chapman and Bernays, 1977).

Although many blood sucking insects feed on sugar solutions and sucrose is a phagostimulant, levels which occur normally in the host blood do not approach the sensory threshold (Altman and Dittmer, 1972). Such insects are commonly stimulated to gorge by the presence of certain nucleotides such as ATP (Table 3) but they may not always encounter optimal concentrations (Friend and Smith, 1972, 1977; Galun and Rice, 1971; Hocking, 1971;

Hosoi, 1959). Thus it is possible that food intake may be limited by low levels of such chemicals, although this is unlikely to be a major reason for the occasional ingestion of small meals, since the host is chosen before probing and the nucleotides will inevitably be present in the blood, and will stimulate continuous ingestion after the choice is made (Hocking, 1971).

Although many different compounds can stimulate feeding, usually only a few provide the major phagostimulatory input. The relative importance of the different phagostimulants clearly varies with the species. Among most phytophagous insects, sucrose and in some cases fructose are by far the most important materials, even though optimal food intake may require some other particular chemicals or a complex of them. The same is true for flies except when they exhibit a "protein hunger" associated with reproduction (see section 7.2). At such times protein or chemicals associated with it is ingested preferentially although the nature of the mechanism driving this behaviour is as yet unknown.

TABLE 3 Stimulating effectiveness of adenine and some nucleotides for various insects which have been tested (++, highly stimulating; –, no effect; •, not tested: A, adult; L, larva)

	Adenine	AMP	ADP	ATP	Other nucleotides	Ref.
Phytophagous insects						
Locusta migratoria	–	–	–	–	•	1
Schistocerca gregaria	–	–	•	–	•	1
Hypera postica (L)	+++	+	+	+	+	2
Sitona cylindricollis (A)	+	+	•	+	–	3
Leptinotarsa decemlineata (L)	–	–	–	–	–	2
Pieris brassicae (L)	–	•	•	•	•	4
Spodoptera exempta (L)	++	+	+	+	–	5
Blood sucking insects						
Rhodnius prolixus	–	+	+	++	+	6
Culex pipiens (A)	•	+	+	++	–	7
Aedes aegypti (A)	•	+	+	++	–	7
Glossina austeni (A)	–	+	+	++	–	8
Xenopsylla cheopis	–	–	–	++	–	9
Simulium venustum (A)	•	+	++	+	•	10
Miscellaneous						
Culex pipiens (L)	–	–	–	+	•	11
Musca domestica (A)	•	–	–	–	+	12

[1] Bernays, unpublished
[2] Hsaio (1969a)
[3] Beland *et al.* (1973)
[4] Ma (1972)
[5] Ma and Kubo (1977); Ma (1977)
[6] Friend and Smith (1972)
[7] Hosoi (1959)
[8] Galun and Margalit (1969, 1970)
[9] Galun (1966)
[10] Sutcliffe and McIver (1975, 1979)
[11] Dadd (1970)
[12] Robbins *et al.* (1965)

Further recent references may be found in Huang Guo-Cheng (1981). Progress in the studies on the phagostimulants of haematophagous arthropods. *Acta Ent. Sin.* **24**, 344–348

Studies with artificial diets have shown that the balance of nutrients often affects the growth and development of insects. In most cases, however, natural foods and normal food selection behaviour ensure that the balance of nutrients is appropriate and only in a few groups of phytophagous insects is there any suggestion that balance *per se* is important in governing food intake. In aphids much emphasis has been placed on the importance of the balance of nutrients. There is no doubt that selection and performance are greatly affected by imbalance (Mittler, 1972) but the evidence for its action on growth via regulation of food intake is scanty. Single nutrients in excess can greatly reduce food intake, and more complex imbalances of amino acids may reduce food uptake in aphids (van Emden and Bashford, 1971; Mittler, 1967, 1972). There are examples of lepidopterous larvae having reduced food intake as a result of nutrient imbalance and a few examples among phytophagous larval Diptera where reduced food intake is one of the causes of the retarded growth found on certain artificial diets (House, 1965, 1967, 1971; Ishikawa *et al.*, 1969).

Specific requirements for individual nutrients or classes of compound, resulting from deprivation on the one hand or specific needs on the other, may affect feeding activity in some insects. Such specific effects are discussed in section 5.

3.2 FEEDING DETERRENTS

Evidence is accumulating that within the possible range of foods available to phytophagous insects in particular, deterrent chemicals play a large part in determining both the initial choice made and the amounts eaten (e.g. Beck, 1965; van Emden, 1972; Fraenkel, 1969; Hsaio, 1974; Jermy, 1966; Schoonhoven, 1972). There is a very wide range of secondary metabolites in plants which deter feeding in various insects. These chemicals include amines, alkaloids, terpenoids, phenolics and numerous types of glycosides. The deterrent effects are concentration-dependent, and may completely prevent feeding when concentrations are high. Oligophagous and monophagous species are more sensitive to the presence of deterrent compounds than are polyphagous species and the majority of non-host plants, or those eaten in small amounts, contain chemicals which are feeding deterrents (Jermy, 1961).

A relatively simple case is that of *Locusta migratoria* which is graminivorous: its restriction to grasses is determined by a wide range of feeding deterrents (Bernays and Chapman, 1977). Plants such as seedling grasses, which are eaten in small amounts, contain deterrent compounds which reduce the amount eaten (Bernays and Chapman, 1974, 1976; Bernays *et al.*, 1974). The same is probably true of other graminivorous grasshoppers (Bernays and Chapman, 1977). Amongst polyhpagous grasshoppers such as *Schistocerca gregaria*, there are many phagostimulants which tend to increase amounts ingested. Here too, however, deterrents play an important part, particularly in plants which are only eaten to a very small extent (Chapman and Bernays, 1977).

The balance of phagostimulants and deterrents is probably the final determinant of the palatability of a material. In acridids it has been shown that low levels of deterrents will be ignored if the level of phagostimulants is very high (Bernays and Chapman, 1977). In flea beetles which feed on Cruciferae, the amounts eaten depend largely on the combinations and amounts of phagostimulatory and deterrent secondary chemicals including various glucosinolates and flavonoids (Nielson, 1978, 1978a). Other examples are known from the Lepidoptera (Beck and Hanec, 1958; Ma, 1972; Schoonhoven. 1973). Moreover Blom (1978, 1978a) has shown that the amount eaten is a reflection of the balance of inputs from sugar and glucoside receptors (phagostimulants) on the one hand and the deterrent receptor (deterrents) on the other, in *Pieris brassicae* larvae.

3.3 PHYSICAL FACTORS

Food intake is also regulated by the physical nature of the food. Hardness and toughness are parameters of the food which are of particular relevance to phytophagous and wood-boring insects. There are examples in a variety of species where hardness has been shown to either reduce feeding or to increase the time taken to ingest a given amount of food (e.g. Agarwal, 1969; Baines *et al.*, 1973; Bernays and Chapman, 1970; Feeny, 1970; Pathak, 1969; Tanton, 1962; Williams, 1954). Seed feeders may be unable to feed on particular seeds or they may eat less on the harder specimens (Smith, 1973). Other examples for certain crop pests are given by Beck (1965). The effect of hardness may not necessarily be simply a matter of reduced food intake, but may involve other disadvantages such as inadequate nutrient intake for the effort required in ingestion. This, however, is difficult to separate from nutrient deficiencies related to an increased proportion of fibre (Agarwal, 1969).

In some situations where physical factors may be limiting, ingestion of food is greatly affected by the presence of other insects. The larvae of the sawfly *Neodiprion pratti* feed on certain pine trees, but when they hatch they have difficulty cutting into the tough cuticle. If one succeeds, the smell attracts others, which thereafter are able to feed in a group (Ghent, 1960). Similarly, a group of young sheep blowfly maggots is required to penetrate the skin of the sheep and create a lesion for feeding (Norris, 1959).

Physical form can also affect the ease of food intake, and the amounts ingested. Thus particle size affects ingestion rate in filter feeders such as certain mosquito larvae (Dadd, 1971). In phytophagous chewing insects.

leaf shape may have an effect, since some of these insects prefer or require an edge and rolled leaves make feeding difficult (Bernays and Chapman, 1970; Mordue (Luntz), 1979).

Many minor physical factors may affect food intake, but it is probably true that where insects are adapted for particular diets, physical factors play a minor role. Thus grasshoppers which are graminivorous have mandibular structures adapted for dealing with this hard material, and it is unlikely that the food intake is regulated by this alone. On the other hand, grasshoppers not adapted for grass feeding are more commonly affected by hardness (Mulkern, 1967). In some situations physical features of the host such as hairiness, have an indirect effect by interfering with behaviour prior to feeding. Such phenomena are reviewed by Chapman (1977).

In the case of fluid feeders, viscosity may reduce food intake. For example the rate of feeding on high concentrations of sugars by flies is relatively low due to the high viscosity (Dethier *et al.*, 1956). Insects which make use of host fluids under pressure, may be affected by changes in the pressure of such fluids. For example, mosquitoes which insert their stylets into the capillaries of their vertebrate hosts feed faster than those which feed from small local haemorrhages which they cause in the host tissues. The same individual may feed in either manner, but larger meals are ingested if feeding is directly from a capillary (Clements, 1963). Other physical factors influencing probing and ingestion in blood feeding insects are temperature gradients, humidity, the thickness and texture and hairiness of the host skin, the osmotic pressure and pH of the food medium and host tissues (Friend, 1978; Friend and Smith, 1977).

Availability of food and the mutual disturbance by abundant conspecifics are obvious physical interactions, but they have been investigated thoroughly only in carnivorous insects. It is usual for the feeding rate to increase markedly with prey density although if the prey density is very high feeding rate may again fall (Hassell *et al.*, 1976; Holling, 1961; Mori and Chant, 1966; Tostowaryk, 1972). On the other hand, predator density may affect food intake, since encounters between predators cause mutual disturbance and searching stops temporarily (Tostowaryk, 1972). Such examples are part of a multitude of minor interactions with the physical environment which can temporarily influence food intake.

3.4 INFORMATION PROCESSING

During feeding the central nervous system of the insect receives information from what may be a large number of sensilla, each containing a number of physiologically different receptor cells. The nature of this information is potentially extremely complex. The spectrum of responses of different sensilla and of individual receptors can vary considerably, from those which respond to a wide range of stimuli to others which are extremely specific (Blaney, 1975, 1980; Boeckh, 1980; Dethier, 1967, 1974; Dethier and Kuch, 1971; Winstanley and Blaney, 1978). Such highly specific chemoreceptors and their central connections have been termed "labelled lines" (Perkel and Bullock, 1968). In addition, sensitivity to a single stimulus may vary both between similar receptors, and in an individual receptor with time (Dethier, 1976; Schoonhoven, 1976). In some insects such as lepidopteran larvae there is a basal firing rate or "noise level" in unstimulated olfactory cells, which also varies. In such cases, a chemical stimulus may result in either an increase or a decrease in the basal firing rate, the same stimulus having opposite effects in different receptors (Dethier and Schoonhoven, 1969). Further, differences in the latencies and rates of adaptation of receptors provide more potential information to the CNS (Dethier, 1976; Dethier and Schoonhoven, 1969).

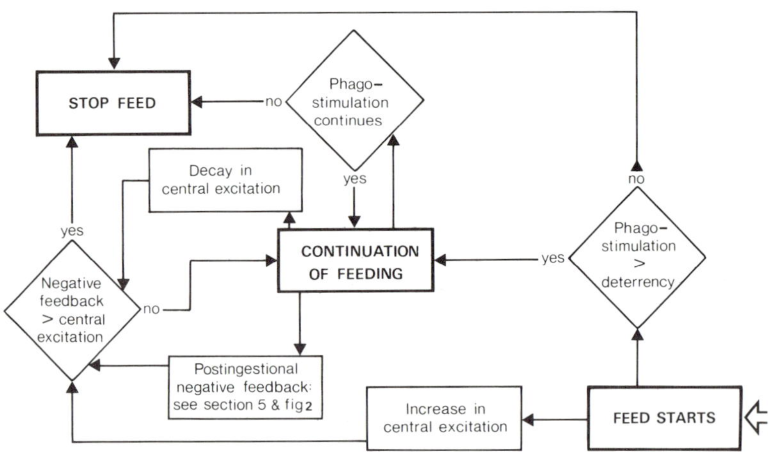

Fig. 1 A model of the processes involved in the continuation of feeding by an insect such as the locust or fly

Ultimately, the CNS integrates incoming information with input from other sensory systems to give the appropriate motor output. Some of this integration may occur peripherally, depending on the degree of electrical coupling between inputs from individual receptors (Dethier, 1976). Generally there will be considerable convergence of peripheral inputs to second and higher order neurones in the CNS, possibly in conjunction with selective filtering (Boeckh, 1980; Boeckh *et al.*, 1975). In attempting to describe the manner in which the CNS may recognise a particular food, the concept of "across-fibre patterning" has been developed. Each material making gustatory contact with the mouthparts, will produce a different net effect or response profile. Discrimination by across-fibre patterning involves the ability of the central nervous system to distinguish between such profiles. The degree to which the CNS uses labelled lines in analysing peripheral input has been a matter for conjecture. The concepts of labelled lines and across-fibre patterning have been discussed by Blaney (1975, 1980), Boeckh (1980, 1980a), Dethier (1973, 1974, 1976), van Drongelen *et al.* (1978) and Winstanley and Blaney (1978).

Such work on information processing concerns the chemical discrimination of appropriate food, and the manner in which food quality is monitored. No attempt has yet been made to understand the complex neural regulation of the whole feeding process but a model of the behavioural sequences and the factors which appear to be relevant for an insect such as a grasshopper is given in Fig. 1.

4 Termination of feeding: the control of meal size

Most insects eat their food in discrete meals with relatively long periods between them, and in the extreme case of some blood sucking insects there may be only one meal in each instar. What regulates the amount taken in? The possible mechanisms which have been suggested to date are (*a*) chemosensory adaptation, (*b*) decay of an excitatory state, (*c*) volumetric factors detected by mechanoreceptors, and (*d*) negative feedback from nutrient, humoral and other factors associated with the haemolymph.

In this section these various possibilities are discussed; it is assumed that the insect is ready to feed, has ample stimulating food available, and that the process of ingestion has begun. The importance of food quality on meal size is considered in section 3.

4.1 ACRIDIDS

Adaptation of chemoreceptors on the mouthparts has been suggested as a regulatory mechanism in the termination of a meal by *Chortoicetes terminifera* (Barton Browne *et al.*, 1975). When drops of sucrose were placed on the mouthparts of restrained insects, more were imbibed if they were alternated with drops of water than if they were given in an unbroken sequence. An interpretation of this in terms of sensory adaptation is reasonable, but a single chemical, presented in this way bears little resemblance to a natural situation. It is unlikely that complete chemoreceptor adaptation occurs during a normal meal on a normal solid food, because the natural feeding process is such that a complex and continuously changing series of stimuli is presented to the mouthpart sensilla (Bernays and Chapman, 1974). For example adaptation of the palp receptors is reduced by palpation (Blaney and Duckett, 1975; see also section 3). On balance, it must be concluded, that while locust chemoreceptors do adapt to phagostimulants (Blaney, 1975), and sensory adaptation may play a role in terminating feeding when a single chemical is presented almost continuously, the evidence from naturally feeding insects which can feed continuously for twenty minutes, suggests that this cannot be a major factor in the regulation of meal size on normal foods.

The level of excitation associated with the start of feeding (Blaney and Chapman, 1970) is enhanced by phagostimulation and this effect is probably more extreme after a period of food deprivation (see section 5). In *C. terminifera* the enhancement was measured in terms of increased readiness to imbibe a relatively unstimulating drop of fluid following contact with a highly phagostimulatory concentration of sucrose (Barton Browne *et al.*, 1975a). In *Locusta migratoria* it was measured as increased food-searching behaviour after temporary loss of contact with the food (Bernays and Chapman, 1974). A high excitatory state may be necessary to maintain feeding activity long enough to ensure optimum intake, and if so, then its level could be considered as a factor involved in regulating meal size. With constant access to a suitably stimulating food the initial level should be close to optimal, but it may be higher after a period of deprivation (Blaney *et al.*, 1973), and allow slightly larger meals because negative feedbacks take longer to offset it. Also artificially stimulating the cibarial chemoreceptors of *L. migratoria* with grass juice before feeding on blades of grass leads to slightly larger meals (Bernays and Chapman, 1974). These experiments show that heightened excitation may increase meal size. It is also possible that towards the end of a meal the decay of the excitatory state is a cause of terminating that meal. The apparently enhanced state of arousal associated with feeding, indicated by palpation and movement, continues for a short period after the completion of a meal (Blaney *et al.*, 1973; Simpson, 1981). This suggests that a fall in an excitatory state is not a dominant reason for the termination of feeding. This may be because a heightened level of excitation is maintained during a meal by continued optimal phagostimulation. Once such phagostimulation ceases the excitatory state returns to a baseline level.

Locusts feeding without imposed periods of food deprivation ingest their food in discrete meals, each of several minutes duration, separated by longer periods, averaging about an hour, when no feeding occurs (Blaney *et al.*, 1973; Simpson, 1981) If insects are deprived of food for long enough to empty the foregut, about 5 hours, and then allowed to feed, the size of the meal taken by nymphs of *L. migratoria* is remarkably constant (Bernays and Chapman, 1972). This suggests a sharp cut-off of feeding.

In nymphs of *L. migratoria*, the backward movement of solid food such as tough mature grass to the midgut in the course of a meal is slower than the ingestion rate, so that the foregut is filled from the posterior end forwards (Bernays and Chapman, 1973). Hyperphagia results if the posterior pharyngeal nerves, at the front of the foregut, are cut; the insects attempt to feed continuously but are physically incapable of ingesting much more. Sectioning other nerves to the foregut does not have this effect, but Rowell (1963) observed hyperphagia in *Schistocerca gregaria* after sectioning the recurrent nerve. This operation had no effect on meal size in *L. migratoria*, but it is very easy to damage the posterior pharyngeal nerves while sectioning the recurrent nerve, because, at least in *L. migratoria*, their positions are very variable and they often branch from the recurrent nerve itself. It may well be that the observations made by Rowell resulted from damage to the posterior pharyngeal nerves. These nerves supply the anterior part of the foregut, the last part to fill during a meal. Over this region is a network of nerves containing approximately ten large cells (Bernays, unpublished) and these are probably the perikarya of stretch receptors. It is suggested that input from these cells, indicating complete fullness of the foregut, is the signal for cessation of feeding.

When, however, the interval between meals is less than four hours, feeding often stops before the foregut is fully distended (Bernays and Chapman, 1972). This is most obvious if the food consists of soft lush vegetation which passes to the midgut faster than tough and relatively dry food (Simpson, unpublished), so that some other mechanism may be involved. Barton Browne *et al.* (1976), working on adults of *C. terminifera*, found that meal size was correlated with weight loss during the previous period of food deprivation. However, their food source was water or water with sucrose and, since the muscles at the posterior end of the foregut do not prevent backward movement of fluid, such material cannot provide adequate pressure to stimulate the stretch receptors in the way that normal solid food does. Further, the insects were in a state of negative water balance, and drinking responses have been shown to be affected by a reduction in body volume (Bernays, 1977, and see section 6).

In adult female *C. terminifera* it was also suggested that total body volume may provide a negative feedback regulating meal size (Moorhouse *et al.*, 1976). This was not apparent in nymphs of *L. migratoria* deprived for five hours and then fed on a tough mature grass; body volume did not alter because the full foregut displaced air from thoracic air sacs (Bernays and Chapman, 1973). It is possible that, in the *C. terminifera* experiments, the manner of liquid food presentation, by drops placed on the mouthparts, provided excessive stimulation and, in the absence of a normal foregut-mediated cut-off, an unusual increase in body volume occurred. Perhaps a body volume component is important in some situations, particularly where the insect has continuous access to a soft lush food having a high water content, because such food passes back to the midgut relatively quickly, and preliminary observations indicate that sometimes the crop is only half filled at the end of the meal. Further investigations are required to establish the exact role of volumetric factors other than foregut distension.

Finally there is the possibility that haemolymph factors limit meal size. In *L. migratoria* nymphs, altering nutrient concentrations and osmotic pressure during the course of a meal via a chronically implanted cannula had no effect on the size of that meal (Bernays and Chapman, 1974a). Likewise, corpus cardiacum extracts were without effect. On the other hand, if the haemolymph osmotic pressure was raised without a significant volume change about twenty minutes before food was presented, meal size was reduced. Such high levels of haemolymph osmotic pressure occur only at certain times such as late in the instar, and may contribute to the reduction in meal size which occurs then (Bernays and Chapman, 1974a). In general, however, the effect is relatively small compared with volumetric factors, and the role of the haemolymph is more significant in determining the length of the period between meals (see section 5).

4.2 BLOWFLIES

Chemosensory adaptation has been given a lot of attention in the blowfly *Phormia regina*. When feeding on sugar solutions, blowflies seldom feed for more than two minutes, and then may not feed again for two hours or more. The meal taken is not continuous, a relatively long initial bout of feeding being followed by successively shorter ones (Gelperin and Dethier, 1967). The first bout is lengthened if the feeding stimulus is stronger (except at very high concentration ranges). The sensory adaptation to a threshold level is faster when the stimulating solution is in low concentration (Dethier and Hanson, 1965; McCutchan, 1969; see also section 2), so that this adaptation of the chemoreceptors may play a part in regulating the length of the bout (Dethier, 1976; Dethier *et al.*, 1956; Thomson and Holling, 1974). Since, however, time between bouts allows for at least some disadaptation, it is ultimately the number of bouts which will relate to meal size. This is unlikely to be a function of chemosensory adaptation.

As with locusts, a "central excitatory state" is generated by contact with phagostimulants, and this temporarily lowers the behavioural threshold for responsiveness to less stimulating materials (Dethier *et al.*, 1965). Loss of contact with the food during the excitatory state leads to the so-called fly dance which increases the chances of locating food again (Nelson, 1977). As with locusts, the main effect of such a heightened state of excitation is to maximise feeding processes when food is in the vicinity, although decay of such excitation during the meal could be a significant factor in terminating feeding (Dethier *et al.*, 1965; Barton Browne, 1975). This is because more highly phagostimulatory solutions are ingested in larger amounts, and such solutions induce a higher level of excitation which takes longer to decay. The relative importance of phagostimulation acting directly on motor systems as compared with phagostimulation enhancing excitation and indirectly stimulating feeding is considered further in section 3.

The exact role of the decay of the excitatory state in the termination of feeding on an optimal food is difficult to assess, but it probably interacts with direct negative feedbacks such that these become dominant when the level of excitation falls. Attempts at feeding actively continue for a long time when volumetric feedbacks inhibiting feeding are removed, so that a fall in the excitatory state as a result of continuing contact with food is unlikely itself to lead to the cessation of feeding. It may be, however, that the inhibitory inputs themselves influence the rate of decay of the excitatory state as well as directly terminating feeding.

During the first part of a meal on sugar solution, some fluid passes directly into the midgut and some into the crop which is a diverticulum of the oesophagus. Towards the end of a meal food is directed entirely into the crop, and finally, at the end of a meal, the crop is closed by a valve (for a detailed account see Knight, 1962, and Green, 1964). The crop is in a more or less central position in the abdomen so that its filling also enlarges the abdomen. The first reports that cutting the ventral nerve cord produced hyperphagia in *Phormia regina* and related flies, suggested that receptors in the abdominal body wall were involved (Dethier and Gelperin, 1967; Nuñez, 1964). It has now been shown that branches of the abdominal nerve form a loose basketwork over the crop (Gelperin, 1971), and several stretch receptors within the basketwork have an increased firing rate when the nerves are extended, a situation which arises towards the end of a meal when the crop is distended. The input from these receptors probably provides the brain with the information required to inhibit further feeding (Gelperin, 1971a).

A second negative feedback mechanism occurs in flies. At the end of feeding on a sugar solution when the midgut and crop are replete, there is a certain amount of food in the oesophagus. At intervals, small quantities of food from the crop are returned to the oesophagus and then passed into the midgut. Stretch receptors in a branch of the recurrent nerve which innervates the foregut are stimulated by oesophageal expansion and cutting the recurrent nerve anterior to this branch results in overfeeding (Dethier and Bodenstein, 1958; Dethier and Gelperin, 1967; Gelperin, 1966). Thus it is believed that input from the stretch receptors here also inhibits further feeding. A fuller account is given by Gelperin (1971a). Both mechanisms depend on a volumetric component rather than feedbacks involving nutrients or energy reserves.

Protein-deprived flies which take a full sugar meal will then take a protein meal if given the opportunity. (The size of this meal depends upon the amount of sugar solution previously ingested. Even if the fly is severely bloated after ingesting 2 M sucrose, some protein is ingested subsequently, Belzer, 1970.) The separate control of protein and sugar meals in *Phormia regina* has been studied by Belzer (1978, 1978a, b, 1979). Frontal ganglionectomy (and the presumed removal of a negative feedback via the recurrent nerve) led to sugar or protein hyperphagia in both protein-deprived and gravid females, when either food source was provided. Operated, protein-deprived flies, given a choice of foods, ingested more sugar than protein. Unoperated flies however, preferred protein. These results have been interpreted in the following way: negative feedback via the recurrent nerve exerts some control on both protein and sugar meals, with the effect being stronger for sugar.

Sectioning the abdominal nerve led to protein hyperphagia or some elevation of sugar ingestion in protein-deprived flies when either food source was presented. In gravid flies the operation led only to a slight increase in protein ingestion, suggesting that abdominal feedback is most important in regulating protein ingestion, with sugar ingestion being regulated primarily via the recurrent nerve feedback. Belzer (1979) demonstrated that the presence of mature oocytes reduced protein intake because of feedback from abdominal stretch receptors. By having recurrent nerve feedback as the primary regulator of sugar ingestion, a gravid female can still take the sugar meals necessary for survival despite a high level of abdominal nerve feedback due to the presence of the oocytes. Similarly, by having two different volumetric controls, a protein-hungry fly which has just fed to repletion on a sugar solution can then take a protein meal if it is available. Alternatively, a fly which has just taken a protein meal may ingest a sugar solution immediately afterwards. By integrating incoming chemosensory information on the nature of the food with volumetric information from the oesophagus and abdomen, a fly can make the best use of available resources.

4.3 MISCELLANEOUS FLUID FEEDERS

4.3.1 *Rhodnius*

Rhodnius prolixus takes just one blood meal of about six times its own weight in each nymphal instar (Buxton, 1930). Severe hyperphagia results from cutting the ventral nerve cord. Cutting a hole in the abdomen and midgut, such that the ingested blood can leak out, also greatly increases the duration of feeding and the volume ingested (Maddrell, 1963). This suggests that negative feedback from stretch receptors is a probable control mechanism. Appropriate stretch receptors in each complete abdominal segment have been described, and their adapted discharge frequency increases with the intensity of stretching (Anwyl, 1972). This indicates that the theory of Bennet-Clark (1963), that back pressure prevents the proper functioning of the cibarial pump and this inability to feed is the cause of termination of the meal, is now untenable. No other experiments have been carried out to examine the possibility of other control mechanisms.

4.3.2 *Mosquitoes*

In several species of mosquito, sectioning of the ventral nerve cord anterior to the second abdominal ganglion resulted in about a four-fold increase in the volume of blood consumed (Gwadz, 1969). In *Aedes aegypti* meal size depended on the point at which the cord was sectioned: the more anterior the cut, the greater was the size and duration of the meal taken. It is apparent that inhibitory inputs resulting from abdominal distention bring about the termination of feeding, and that there are probably a number of receptors involved which are segmentally arranged. There has been no investigation of carbohydrate meals, which may be separately controlled as in blowflies, particularly as the partitioning of these into the crop, with blood going into the midgut, is usual in this group (Hosoi, 1954).

4.3.3 *Milkweed bug*

In *Oncopeltus fasciatus* it has been suggested that sensory adaptation is a likely cause of the cessation of feeding (Fier and Beck, 1963). The argument was based on the fact that the length of a probe (meal) was greater on a substrate of cellulose or starch than on sucrose. There are however, other interpretations of this phenomenon. There is a suggestion that adaptation may play some part in aphids since, when given artificial diets alternating with water, they ingest more than the same aphids given the artificial diet continuously (Cull and van Emden, 1977). Results of McClain and Fier (1973), give evidence for the presence of a volumetric feedback in *O. fasciatus*. When 4 μl of distilled water was injected into the haemolymph less of both liquid and solid diets was ingested than after injection of only 1 μl. On the other hand, a number of amino acids had different effects on the lengths of subsequent meals on both liquid and solid diets after injection into the haemolymph. For example glycine reduced meal length on milkweed seeds while histidine increased it. These results are difficult to interpret since the final haemolymph concentrations were probably higher than those occurring naturally, but they do suggest that the effect is not a simple osmotic one.

Finally it has been suggested that in *O. fasciatus*, the depletion of salivary secretions is a cause of cessation of feeding (Miles, 1959). At the end of a meal, if a drink is taken, feeding can again be initiated (Fier and Beck, 1963).

4.4 SYNTHESIS

The regulation of meal size is a means of preventing physical damage due to overeating and probably of ensuring that the most effective amount for optimal digestion and conservation of effort is eaten. In all cases examined so far, the major component in this short term regulation of feeding is a volumetric factor mediated by stretch receptors which are variously located to suit the particular morphology of the insect. It must be assumed that increased input from stretch receptors provides a negative or inhibitory feedback, which overrides the chemosensory input from the partly adapted mouthpart receptors and the positive effects of what may remain of the central nervous excitation. In the case of flies the two negative feedback mechanisms are partly additive, and certainly there is spatial summation between chemoreceptors associated with feeding (Arab, 1959; Dethier, 1953; Gelperin, 1971a), but the details of the central connections remain to be discovered.

5 Length of time between meals

The timing of feeding depends on the interactions which occur physiologically within the insect and behaviourally in relation to the environment. Fullness of the gut is important, and physiological and environmental factors influencing the rate at which the gut empties are thus important in determining the frequency at which meals can be taken. Other factors may affect chemoreception and so influence the amount of peripheral information entering the central nervous system. Changes in the responsiveness and activity of the insect also affect the pattern of feeding.

5.1 THE EFFECT OF FOOD INTAKE

The intake of food has a variety of effects on the insect which persist for some time after feeding, and which reduce the chances of another feed occurring. In *Locusta migratoria*, for example, large feeds tend to be followed by long interfeeds, and *vice versa*, though no precise correlation exists (Blaney *et al.*, 1973; Simpson, 1981). Details as to how the effects of food intake are integrated into an overall control system can only be a matter for speculation, and only for acridids and flies are there many data.

5.1.1 *Acridids*

Stretching the crop not only switches off feeding, it also causes the release of one or more hormones from the storage lobes of the corpora cardiaca.

One such hormone is the diuretic hormone which presumably has the effect of offsetting any increase in haemolymph volume as a result of the absorption of water from the food (Mordue, 1969). A hormone also causes the pores on the terminal sensilla of the palps to close so that the sensilla are no longer functional. This effect, as measured by the changes in electrical resistance across the tips of the palps, persists for an hour or more in insects which have had a large meal following a period of deprivation. By two hours after feeding the sensilla are again fully functional (Bernays *et al.*, 1972; Bernays and Chapman, 1972a). Injections of homogenates of the storage lobes of the corpora cardiaca indicate that the change is concentration-dependent (Bernays and Mordue, 1973). Pore closure may reduce the probability of further feeding for a period after the meal by reducing sensory input, although direct proof is difficult to obtain.

There are other changes following feeding which result from hormone release. For example, locomotor activity is reduced. The same effect is produced in insects deprived of food for four hours, by filling the foregut with agar or by injecting homogenates of the storage lobes of the corpora cardiaca into the haemocoel. This effect is most marked if the donors are themselves deprived of food, and it can also be mimicked by haemolymph transfusions from newly fed insects (Bernays, 1980). Injection of haemolymph from newly fed insects has been shown to delay the beginning of the next meal, although the size of the meal is not affected (Bernays and Chapman, 1974). Cazal (1969) has shown that homogenates of the storage lobes of the corpora cardiaca of *Locusta migratoria* enhance active movements of the foregut. This may increase the rate of foregut emptying immediately after a meal (Bernays and Chapman, 1974). Most other studies have involved the use of whole corpora cardiaca from donors of various ill-defined states and the results, which have been very variable, probably tell us little of the events which are naturally occurring, since different hormones with perhaps opposite effects are known to occur in the glandular and storage lobes of the corpora cardiaca.

The composition of the haemolymph changes after feeding. Best known is the increase in osmotic pressure which was found to be the result of a net removal of water into the gut during the course of the meal, with the result that ion, amino acid, carbohydrate and lipid concentrations in the haemolymph all increased by similar amounts (Bernays and Chapman, 1974a). The osmotic pressure rose by 6% during a meal in insects with free access to food consisting of tough mature grass blades, and by 12% during a meal following five hours of food deprivation. Osmotic pressure in turn affects gut movements (see section 5.2). It has also been found that increased haemolymph osmotic pressure slightly decreases the meal size (Bernays and Chapman, 1974b).

Other changes occur following feeding whose effects are quite unknown. For example, there may be a considerable increase in weight and also specific gravity of the insect since air sac space is replaced by enlargement of the crop and midgut (Bernays and Chapman, 1973). Whether such factors have roles in the regulation of feeding is not known, although their multiplicity may help to explain the extreme variability between individuals.

Figure 2 illustrates the network of processes occurring when a locust feeds and which are known to reduce the probability of further feeding.

5.1.2 *Flies*

Hormonal effects following feeding have not been studied in the blowfly, although it has been suggested that hormones have similar roles to those described in acridids (Green, 1964a). Davis and Takahashi (1980) have shown that chemoreceptor sensitivity to host odours is reduced in the mosquito *Aedes aegypti* after a blood meal and that this is caused by humoral agents in the haemolymph.

Changes in concentration of sugars in the haemolymph following feeding have been studied in the blowfly. If a high molarity sugar meal is taken, the level of haemolymph sugars is quickly increased, and this in turn reduces the rate of crop emptying (Gelperin, 1966a). Alternatively, if the molarity is low, digested and absorbed sugars may be removed to the fat body very quickly and the haemolymph osmotic pressure remains low, with the result

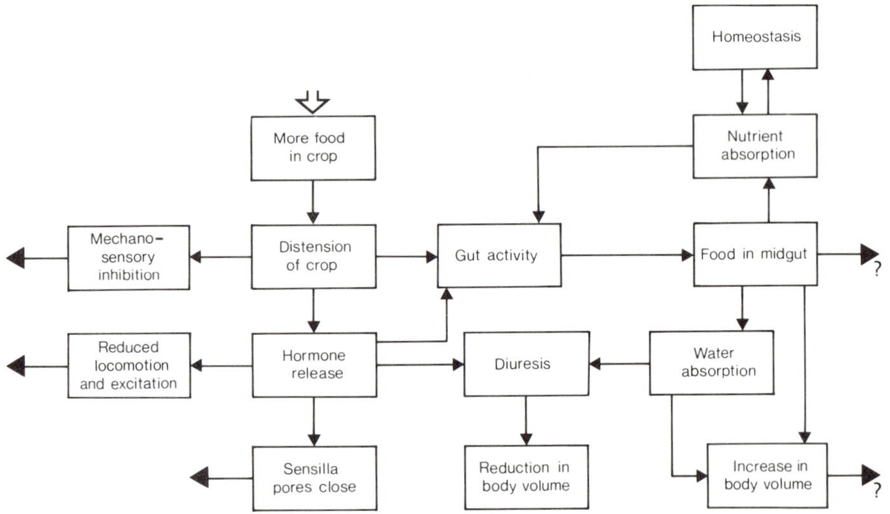

Fig. 2 A diagram of the network of interacting physiological processes which occur when a locust feeds and which reduce the probability of further feeding. Larger arrows pointing outwards indicate the possible negative feedbacks which may be involved in terminating the meal

that the crop empties quickly, and feeding can occur sooner (Gelperin, 1971a). Gut emptying is discussed further in section 5.2.2.

Barton Browne and Evans (1960) showed that increased levels of sugars in the haemolymph reduced locomotor activity in *Phormia regina*. This will in turn certainly reduce the chances of feeding (Bowdan, 1981), so that active digestion of a meal rich in sugars, can be expected to have an effect on the timing of the next meal via its effect on haemolymph sugars. The possible interactions are discussed more fully by Barton Browne (1975).

Work with the blood feeding tsetse fly *Glossina morsitans* suggests that other mechanisms could be involved in the reduced responsiveness following a meal. Brady (1975) has shown that in this insect, abdominal weight loss is correlated with readiness to feed. By implication therefore, increased abdominal weight as a result of feeding adds either directly or indirectly to the inhibitory factors reducing the chances of taking another meal.

5.2 GUT EMPTYING

Since fullness of the whole gut or part of it results in inhibitions of feeding via gut or body wall receptors (see section 4), then the pattern and rate of emptying must play an important role in determining when feeding occurs again.

5.2.1 *Acridids and cockroaches*

In fifth instar nymphs of *L. migratoria* fed in a standard manner at 30°C on a relatively fibrous mature grass such as *Agropyron repens*, the amount of food in the foregut begins to decrease slowly after the termination of feeding. The decrease is slight at first and probably due to digestion alone, but subsequently waves of movement carry the food through to the midgut, and the foregut is empty in five hours (Baines *et al.*, 1973; Bernays and Chapman, 1972). The distension of that region of the crop from which inhibitory input is derived, is noticeably reduced at one hour, so that inhibition of feeding will be reduced. Insects commonly feed again after about one hour if food is available, and rarely is the interfeed as long as two hours at least during daylight (Blaney *et al.*, 1973). These times may vary, particularly with temperature and different types of food. For example a softer, lusher food such as seedling wheat, passes back into the midgut during the course of a meal and the crop is commonly empty within two hours (Simpson, unpublished).

The rate of crop emptying in locusts can be drastically reduced by injection of nutrients into the haemolymph at the end of a meal and movements of the crop which cause emptying are decreased by the addition of nutrients to the bathing medium of *in vitro* preparations (Baines, 1979). This has also been demonstrated in the cockroach *Blatella germanica* (Gordon, 1968). Trehalose injections, which were designed to increase the haemolymph osmotic pressure of fifth instar nymphs of *L. migratoria* by 200 mOsm, almost completely prevented crop emptying during the first hour following a feed. The same trends were shown to a lesser extent with glycine and NaCl injection. Glycerol was as effective as trehalose, while injection of water increased the rate of crop emptying (Bernays, unpublished). Because the haemolymph concentrations of individual nutrients may be regulated at different rates it is not possible to distinguish between a general osmotic effect and differential effects of specific nutrients, but it must be concluded that one or both of these factors are important in regulation of crop emptying. It is thus conceivable, but unproven, that diets with high nutrient levels may provide, at least temporarily, higher levels of solutes in the haemolymph, which in turn reduce the rate of crop emptying. This would provide an explanation for the various reports that diets which have reduced nutrient levels, are ingested by various orthopterous insects in greater quantity over a period of time, compared with nutrient-rich diets (Dadd, 1960; Gordon, 1968; van Herrewege, 1974; Le Berre and Mainguet, 1973; McGinnis and Kasting, 1967). Further, nymphs of *L. migratoria* which had been deprived of food for over 24 hours and then fed, were able to ingest significantly larger amounts at subsequent meals compared with insects which had initially been deprived for only five hours (Bernays, unpublished). The differences must have been due to variation in rate of emptying of the crop, since the size of the first meal taken after deprivation is not altered by deprivation periods of between five and 24 hours (Bernays and Chapman, 1972). It is possible that the faster crop emptying in the more deprived insect was mediated by some reduction in haemolymph nutrient levels. Louveaux (1977) showed that over a period, the amount which would have been eaten during a 24 hour deprivation is almost entirely made up by extra feeding on subsequent days, but not necessarily the next day. This indicates that there are probably other factors involved in compensation following a period of food deprivation.

The role of foregut stretch receptors in the inhibition of feeding in cockroaches is unknown since the crop, in *Periplaneta americana* at least, is always kept relatively full of air which is simply displaced during feeding (Davey and Treherne, 1963). There is, however, a 20% increase in volume during feeding, so that a system analogous to that in locusts is feasible. In any case emptying the crop of solid material is a prerequisite for further feeding. Treherne (1957) showed that the rate of crop emptying in *P. americana* is linearly related to the concentration of ingested sugar. Increased concentration resulted in a decreased rate of emptying. Furthermore, feeding the insects with equiosmolar concentrations of different substances, resulted in similar rates of emptying, so that osmotic pressure of the crop lumen was indicated as the important factor. It was suggested that receptors in the pharynx directly measure osmotic pressure and thereby regulate crop emptying (Davey and Treherne, 1963a). Detailed histological investigations failed however to reveal the proposed osmoreceptors (Moulins, 1974). None of the work on cockroaches described above rules out the possibility of an effect from the haemolymph, the composition of which could be altered within minutes of ingestion. This is perhaps the most likely feedback route as, in nymphs of *L. migratoria*, radioactively labelled sugar applied to the food could be detected in the haemolymph within three minutes of the start of feeding (Bernays, unpublished).

Larger meals result in greater crop activity in both locusts and cockroaches, with a fast rate of emptying initially (Baines *et al.*, 1973; McGinnis and Kasting, 1967; Treherne, 1957). In locusts at least the stomatogastric system has a role to play but little is known of how feedback mechanisms function (Clarke and Grenville, 1960; Highnam *et al.*, 1966). In the cockroach *Leucophaea maderae*, the neural control of crop emptying may be different from that in *P. americana* in that stretch itself and consistency of the food seem to be the major factors influencing rate of crop emptying (Engelmann, 1968), although no investigation of nutrient or osmotic factors was made. Further work is needed to establish whether the different factors, shown in different species, are actually common to many species.

As mentioned in section 5.1.1 hormones released as a result of feeding may affect gut movements and have some effect on the rate of crop emptying (Cazal, 1969).

5.2.2 *Flies*

The most detailed studies on control of crop emptying are in blowflies (Gelperin, 1966a; Thomson, 1975, 1975a; Thomson and Holling, 1974, 1975, 1975a, 1976, 1976a, 1977). As in cockroaches and locusts, larger meal sizes result in faster crop emptying, at least initially. However, with larger meals, the crop takes longer to empty during the second half of the emptying time (Gelperin, 1966a). Also feeding on more concentrated sugar solutions leads to slower crop emptying. An accumulating inhibitory effect of digested and absorbed sugar on the rate of emptying has been found, the relevant factor being the haemolymph osmotic pressure (Thomson and Holling, 1977). Simple diffusion of sugars from the midgut to the haemolymph increases the haemolymph osmotic pressure, which in turn causes a decrease in the rate of crop emptying. Thus sucrose at 1 M and glucose at 2 M when fed to the fly, empty from the crop at the same rate. Injection into the haemolymph of a range of different solutes including fucose (which cannot be utilised by the fly) retard crop emptying. Increasing concentrations of sugars injected into the haemolymph, increasingly reduce the rate of crop emptying. In a general way the mechanism is adequate to control nutrient intake, since the stimulating sugars naturally encountered are of greatest nutrient value. Exactly how osmotic pressure affects crop activity is not known. Locomotor activity such as prolonged flight increases the rate of crop emptying, and this is probably through the same mechanism (Gelperin, 1972; Thomson and Holling, 1977).

In the haematophagous fly *Stomoxys calcitrans*, variation in crop emptying rates was also found after meals on different concentrations of sucrose (Venkatesh and Morrison, 1980), and it may also be true for mosquitoes (Jones and Madhukar, 1976). The situation with blood meals however is complicated. Such food usually passes directly to the midgut (see section 4) but its destination varies to some extent with age. When blood does pass partly into the crop, it is quickly removed to the midgut (Venkatesh and Morrison, 1980). Its fate in the midgut has not been studied. Similarly in *Glossina brevipalpis*, blood entering the crop is passed into the midgut shortly after the cessation of feeding (Moloo, 1971). Although in *G. morsitans* fed on aqueous salt solutions containing ATP, crop emptying is retarded by high concentrations of potassium (Langley and Pimley, 1973), the roles of specific components from a natural blood meal, or the overall osmotic pressure, are as yet unknown.

In many mosquitoes a blood meal is taken prior to vitellogenesis (see section 7). Rapid diuresis during the first hour after feeding concentrates the protein (Nijhout and Carrow, 1978), and unless the volume or quality of the meal is insufficient to allow egg maturation, this is retained in the midgut until shortly before oviposition some forty eight hours later. If egg maturation cannot proceed the meal is voided after twenty four to thirty six hours. Up to this time retention is probably due to the cohesive nature of the meal or the peritrophic membrane. There may even be some physical barrier between the mid- and hindgut preventing food passage (Freyvogel and Staubli, 1965). Cole and Gillett (1979) and Rosenberg (1980) have demonstrated in two species that prolonged retention is due to hormonal feedback from the ovaries. This allows continued absorption of nutrients throughout vitello-genesis.

In a few other insects, notably the larvae of *Celerio euphorbiae* and both the larvae and adults of *Pieris brassicae* (David and Gardiner, 1961; House, 1965; Ma, 1972), it has been shown that dilution of the nutrient components of an artificial diet, or a period of prior deprivation, increases the total intake over a period of days. Nothing is known of the mechanism of compensation, but there is reason to suppose that rate of crop emptying via an osmotic effect or a nutrient feedback from the gut or haemolymph has a part to play.

In the system so far proposed, with the haemolymph composition playing an important role in controlling crop emptying, a further set of regulating factors must be involved. These factors concern the manner in which levels of haemolymph constituents are themselves regulated by the fat body. Removal of haemolymph sugar for growth and energy purposes is met by rapid release from the crop, and also by hormonally controlled synthesis and end point inhibited release of sugar (trehalose) from the fat body (Friedman, 1967; Steele, 1976). These homeostatic mechanisms are usually directly linked to the gut function and daily food intake. They are discussed further by Gelperin (1971a).

5.3 ACTIVITY LEVELS

Food intake leads to a reduced probability of further feeding. An important component of this is a reduction in the level of locomotor activity following a meal (see section 5.1). The length of an interfeed period is determined firstly by the time between the termination of one meal and the increase in locomotor activity associated with location of food for the next meal (the "primary" phase) and, secondly, by the time taken to actually find the food once activity has commenced (the "activated" phase). In the case of insects living in close association with the food source the latter may be only a small part of the interfeed period, while in other insects it may be much more important. Factors which influence activity levels directly affect the length of both the primary and the activated components of an interfeed period.

The basis of the reduction in activity which determines the length of the primary interfeed phase are unknown, although hormonal factors certainly play a part (Bernays, 1980, and see section 5.1). A reduction of incoming stimuli could be important. For example, the peripheral effect of hormones on chemosensilla of locusts (section 5.1.1) could reduce incoming stimuli by preventing the arrival of stimulants at the receptors. Omand (1971) found a reduction in impulse frequency in sensilla of replete flies when phagostimulants were offered, but recent work shows that this is not usual (Hall, 1980; Rachman, 1979). It is possible that haemolymph osmotic pressure plays a part also, since increases induced artificially in locust nymphs can cause a reduction in activity (Bernays, unpublished). Decreases in osmotic pressure have not been shown in locusts to increase activity or responsiveness although no studies have been made with changes comparable to those occurring naturally (Bernays and Chapman, 1974; Moorhouse, 1971). Increased levels of haemolymph potassium were proposed as a cause of decreased activity by Ellis and Hoyle (1954). The balance of evidence now suggests that this is not the case, as Chapman (1958) and Moorhouse (1969) found no correlation between changes in haemolymph potassium concentration and feeding activity. No other controlling agents are known, but a full discussion of the possibilities is given by Barton Browne (1975) and Brady (1975).

It is probable that hormones and haemolymph osmotic pressure act via the central nervous system, possibly influencing the activity of some excitatory centre (Huber, 1967). This may also be affected rhythmically or by peripheral inputs resulting from exogenous stimuli. It is also possible that there is variation between individuals in the rate of firing in such a centre, accounting for some of the enormous variation in respect of apparently spontaneous locomotor activity.

Circadian changes in central excitability have been discussed by Brady (1974, 1975a) who notes their qualitative similarity to other changes in excitation. Hall (1980a) demonstrated circadian rhythmicity in the proboscis extension response of the blowfly *Protophormia terraenovae*. A central control mechanism was inferred as no daily change in the response of chemoreceptors was found. Recently Simpson (1981) has presented evidence for the presence of a shorter term oscillation underlying feeding behaviour and other activities in fifth instar nymphs of *Locusta migratoria* reared with constant access to food under a 12 hour light: 12 hour dark photoregime. The period of the oscillation differs slightly between insects, the range found being 12 minutes to 16.5 minutes, but is constant for each insect. If the oscillation, for convenience, is represented as a sine wave, then feeding does not occur during every cycle, but when it does occur, it usually begins near an oscillation peak. Other behaviours which occur more frequently, such as the initiation of locomotion and a number on non-locomotory behaviours, occur on peaks when there is no feeding. Once feeding or locomotion has commenced, and during the short period of settling afterwards, rhythmicity in the nonlocomotory behaviours is lost. The nature of the oscillator is not known, but it is endogenous and is not reset during a twelve hour light period. The presence of such an intermediate length oscillator has important implications in the organisation of complex behaviour, as well as in determining the precise timing of feeding, at least when there is a low level of exogenous stimulation.

Exogenous stimuli can affect activity and thus the duration of the primary interfeed period. In locusts it has has been well demonstrated by Moorhouse (1971) and Kennedy and Moorhouse (1969) that newly imposed stimuli have a general arousal effect, while Blaney *et al*., (1973) showed that interfeed lengths are shorter if regular changes of light intensity occur. Factors which reduce locomotor activity also reduce the likelihood of feeding. Thus, primary interfeed lengths for locust nymphs are longer in darkness than in light (Blaney *et al*., 1973). Where a stimulus which promotes activity is lacking, activation presumably eventually results from the expression of central nervous endogenous activity. The degree to which the responsiveness to external stimuli varies with the oscillator described above is not yet known.

Once the activated phase of the interfeed period has commenced there is a gradual increase in the time spent in locomotor activity with continued food deprivation. This is true of locusts, various flies, bugs, caterpillars and beetles (Barton Browne and Evans, 1960; Blaney and Chapman, 1970; Brady, 1973, 1975; Dingle, 1968; Green, 1964, 1964a; Hans and Thorsteinson, 1961; Leonard, 1970). The subject is fully discussed by Barton Browne (1975). Extended deprivation will eventually reverse the change. In normal circumstances the increase in activity ultimately leads to feeding.

Responsiveness to food-related stimuli also increases with deprivation. In tsetse flies, where visual cues are important in food finding, there is a progressive increase in the responsiveness of variously aged male and female flies to a moving visual stimulus over four days of deprivation (Brady, 1972). With locusts, an anemotactic response to food odour reaches its maximum gradually over several hours of deprivation (Moorhouse, 1971). Increases in general locomotor activity of *Phormia regina* have been shown to occur after exposing the flies to various food odours. The activity has no directional component and odours relating to protein cause a more dramatic change in flies fed only on sugar than in flies fed on both sugar and protein (Bowdan, 1981). Probing responses of blowflies to sugar also increase (Evans and Barton Browne, 1960) as do probing responses of tsetse flies (Brady, 1973). The responsiveness of mosquitoes to food-related stimuli increases more quickly after small than large meals (Klowden and Lea, 1978). A number of other such studies with different insects are discussed by Barton Browne (1975).

5.4 THE INFLUENCE OF THE ENVIRONMENT

In the biology of an insect species, particular behaviour patterns may take precedence over feeding activities during certain times, and at least briefly interrupt food intake. As discussed in section 5.3 diurnal rhythms are certainly important since many insects are restricted in the time of day at which they feed, even under conditions of constant temperature. Mosquitoes for example, exhibit marked rhythms, with different times of feeding in different species (Clements, 1963). Many caterpillars feed primarily at night while the larvae of *Barathra brassicae* feed only during the hours of darkness. *B. brassicae* loses a considerable amount of weight during the day and during periods of increased daylength the amount of food ingested in severely limited (Danilevskii, 1965).

Temperature affects the length of the interfeed. Crop emptying is faster at higher temperatures (Baines *et al.*, 1973) and it is likely that other post-ingestional factors change more rapidly at higher temperature, although this is not known. Temperature may also influence the feeding pattern indirectly. For example the need to regulate body temperature may interfere with the normal feeding responses. Thus, during the very hot periods of the day tropical grasshoppers often climb high in the vegetation, keeping only in the shade and not feeding (Chapman, 1959). Again, the need to control loss of water under hot and dry conditions may prevent insects from feeding on dry food which is normally quite acceptable in moderate amounts (Chapman, 1959; Loveridge, 1974; Saxena, 1967, and see section 6).

Many other examples could be cited, but in general it must be assumed that the insect is in fact adapted to obtain adequate food in the niche in which it has evolved. Ranges of environmental parameters outside the norm will generally be the ones which impose real restrictions at any particular time.

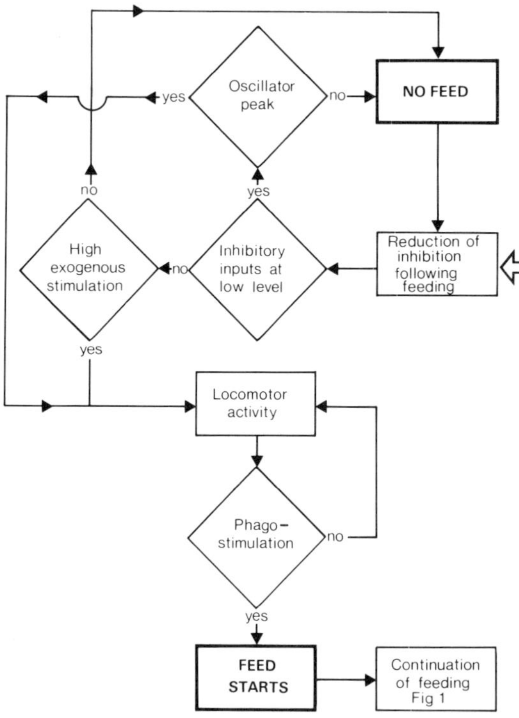

Fig. 3 A model of the interacting factors associated with the initiation of feeding in *Locusta migratoria*

5.5 CONCLUSION

Excluding environmental variables, a system for the patterning of feeding may be proposed: feeding induced inhibitory inputs, often from the gut, are reduced with time after a meal. When a subsequent heightened arousal state develops, movement and contact with the food follows. Feeding proceeds until inhibition from the gut intervenes. The behavioural pattern should be a simple alternation of feeding and non-feeding, as the balance shifts from dominance of excitation to dominance of inhibition, leading to regular meals. The extremely complex pattern that actually occurs results from the modification of this simple model by a multitude of other factors. The possible interrelationships between such factors in the locust are schematically represented in Fig. 3.

Any model probably gives a picture of the average behaviour, with predictability for any one individual being low. In attempting, however, to understand or explain variations in behaviour and ultimately what is responsible for the timing of the next meal, it is important not to lose sight of the fact that in spite of the variability, the normally healthy individuals in a population, in a suitable environment for feeding, *do* satisfy their requirement for food. An intrinsically less active insect, which rests for a long period, is more likely then to feed for a longer period, so that one returns to the basic model: ultimately, in spite of all the intervening variables regulating behaviour, food intake depends primarily on needs, acting through inputs derived from the gut.

6 Drinking

Most insects ingest water with the food, and normally it is in adequate amounts. On the other hand most insects which have been examined have the capacity to drink if, for one reason or another, they are suffering from a water deficit (Barton Browne, 1964; Edney, 1977; Mellanby and French, 1958). Only in acridids and flies have the regulatory mechanisms been investigated.

Acridids readily drink and respond positively to water, but only if they have been deprived of it for a period, or have fed solely on relatively dry food (Loveridge, 1974). Water satiated insects actively move away from an area of wet filter paper whereas insects with a deficit move actively on to the wet surface and make biting movements on it (Kendall and Seddon, 1975). This behaviour is probably mediated through tarsal receptors though an antennal response cannot be ruled out in these experiments.

In *Chortoicetes terminifera* Barton Browne and van Gerwen (1976) showed that the amount of water ingested was correlated with haemolymph osmotic pressure, but it is possible that there were other factors. In *L. migratoria*

Bernays (1977) has shown that the readiness to drink, measured as a positive response when freely moving insects make contact with water, is correlated with reduction in the abdominal volume. The amount ingested after initiation of the positive response is related to the reduction in haemolymph osmotic pressure. Thus initiation and termination of the drink have different controlling mechanisms (Fig. 4). It is possible that the same is true of *C. terminifera*, since only amounts ingested by restrained insects were monitored in the experiments described.

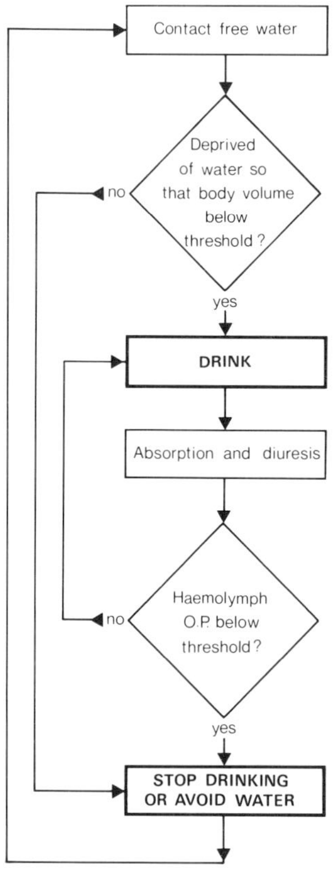

Fig. 4 A model of the regulation of drinking in *Locusta migratoria*

In the case of *Phormia regina*, a volumetric or pressure factor has been shown to affect responsiveness to water, since injections into the haemolymph of over 2 μl of a variety of solutions, including concentrated saline and glucose, prevented the response (Dethier and Evans, 1961; Evans, 1961). Also, removal of haemolymph caused the responsiveness to develop. In *Lucilia cuprina* no correlation was found between drinking behaviour and volumetric factors

(Barton Browne, 1968; Barton Browne and Dudzinski, 1968). Osmotic pressure was found to have some effect, but the best correlation was found between amounts of water drunk and increasing concentration of chloride ions in the haemolymph. The separation of positive responses and amount ingested may eventually lead to an explanation of a more general nature, but for the moment, it appears that different species of fly have different mechanisms regulating ingestion of water. Some change in the haemolymph reflecting the state of water balance, be it volume or solute concentration, seems likely to be a factor in regulation. How the haemolymph factors interact with those involved in feeding is not known.

7 Changes during the life history

7.1 CHANGES WITH DEVELOPMENT

During larval development, the intake of food is always less for a time before and after moulting. The change is well documented for a variety of insects (e.g. Beck *et al.*, 1958; Davey, 1954; Hill and Goldsworthy, 1968; Saxena, 1967; Shrihari, 1970), although the physiological factors involved are still obscure. In *Locusta migratoria* nymphs, the meal size of individual insects which were previously deprived of food for five hours, became progressively larger for the first few days of the instar, while for the last third of the instar they decreased progressively (Bernays and Chapman, 1972). This suggests that there may be a shift in some level of central excitation which requires differing degrees of inhibition from volumetric factors to offset it. Insects kept under a 12 hour light: 12 hour dark photoregime with constant access to food show an increase in daily intake up until mid-instar, due in most insects to increases in average meal size (Simpson, 1982, and unpublished). Blaney *et al.* (1973) however, found that under constant light the increase in food intake during the first half of the instar was due primarily to a decrease in length of the interfeed in most cases. Support is given to the idea of a central nervous change by the fact that responsiveness to other stimuli alters through the instar in the same way, and that locomotor activity also reaches a peak (Chapman, 1954; Ellis, 1951; Moorhouse, 1971). This pattern was found to be less marked when insects were isolated and had continuous access to highly nutritious food (Simpson, unpublished). Changes in the peripheral system during the instar seem unlikely since chemoreceptor responses appear to remain consistent over most of the period (Blaney, unpublished). Stoffolano and Bernays (1980) found that the post-ecdysial fast of fifth instar *L. migratoria* nymphs could be considerably shortened by providing stimuli causing arousal: again suggesting that central arousal state is an important factor influencing food intake.

During the period of somatic growth in adult hemimetabolous insects, food intake is particularly high, and again most work has been with acridids. In these insects there is a very large weight increase from the final moult to sexual maturity and an increase in oxygen consumption (Walker *et al.*, 1970). The increase in food intake necessary to account for these changes has been measured for *S. gregaria* (Hill *et al.*, 1968) and *L. migratoria* (McCaffery, 1975). Mordue (Luntz) and Hill (1970) showed that adult female locusts ingest relatively more bran than lettuce during the period of somatic growth when provided with both foods. During the ovarian growth phase, total amounts eaten are much less and no distinct choice is made between bran and lettuce. It was suggested that bran is chosen during somatic growth as it provides readily available carbohydrates without the problem of water loading. There is probably a specific nutrient feedback since young adult *L. migratoria* feeding on albumen-coated wheat leaves ate less than those having plain wheat leaves, although the final assimilation of nitrogen was similar in both treatments (Bernays, unpublished). This is in contrast to the situation with nymphs where such an effect was found with sugar but not protein. In adults, the protein levels in the haemolymph reach a plateau at the time when somatic growth is completed, and however the control is achieved, it is likely to be mediated through the haemolymph (Tobe and Loughton, 1967).

In holometabolous insects, larval feeding through the instar probably follows a similar pattern in general to that in Hemimetabola (Grosse, 1974; Shrihari, 1970). Feeding by the larva must be such as to provide for growth and build up of reserves for the nonfeeding pupal stage. To do so the fat body must build up reserves and maintain the low levels of haemolymph nutrients presumed necessary for increased food intake. Very often larval feeding must also provide nutrient for maturation, and even, for example in some Lepidoptera, for the whole of the adult life including egg production. Larval feeding in certain Hemimetabola, for example mayflies, must also sustain the adult for the length of its life. In none of these cases is there a study of the controlling mechanisms governing the food intake.

7.2 CHANGES IN RELATION TO REPRODUCTION

The food intake of insects is commonly increased or altered for the production of eggs. In acridids, the females with good quality food have no particular requiremints for extra protein, but they do increase the amount consumed, in a cyclic manner, with a peak of feeding betwen the production of successive egg pods, and little or no feeding for about a day prior to oviposition (Hill *et al.*, 1968; McCaffery, 1975).

In holometabolous insects, many adult forms feed on nectar and other sugary solutions and the protein for egg production must either be ingested separately or obtained from stores lasting over from the larval stages. In some insects exogenous protein is only required to initiate the secretion of hormones necessary for egg production (Braken and Nair, 1967; Johansson, 1955, 1958; Larsen and Bodenstein, 1959), while in others sufficient protein is required to both trigger hormonal secretion and to subsequently synthesise yolk.

Many Hymenoptera and Coleoptera eat pollen but nothing at all is known of how this is regulated. Blood sucking is another expedient for obtaining protein and occurs in many Diptera. Decomposing vegetation or animal matter is a source of protein-rich food sought out by many Coleoptera and Diptera. The regulation of feeding on this specialised diet by blowflies is the only situation where detailed studies have been made.

Blowflies and their relatives require a separate source of protein, with a regular requirement related to egg production. Protein is not usually required for previtellogenic growth unless the larva was subject to food deficiency (Trepte, 1980). One of the interesting features of such protein ingestion is its restriction to time of need, and at such times blowflies are attracted by odours which emanate from decaying proteinaceous material (Pospíšil, 1958), even though they may be well fed on carbohydrates; in choice situations high protein diets are ingested preferentially (Belzer, 1978; Dethier, 1961, Strangways Dixon, 1961). This temporarily restricted diet selection is also of interest because protein alone is quite inadequate for survival for any length of time.

Belzer (1970, 1978, 1978a, b, 1979) has proposed a model for the control of protein ingestion which is superimposed on the pattern existing in relation to feeding on a sugar solution. The volumetric regulatory mechanisms have been discussed in section 4. An additional inhibition of feeding is postulated, namely satiety resulting from relieving a deficit of some reserve material related to protein synthesis. Shortage of available body protein, due to the synthesis of yolk protein, leads to a deficit of the proposed material, and protein is again ingested. Very high levels of inhibition from the foregut and abdominal stretch receptors are required to terminate protein ingestion in one meal, so that even if the crop is filled with sugar solutions, protein solutions will be taken up and into the migdut directly, until, at a higher level, stretch receptor inhibition comes into play. The products of digestion of the protein meal are readily available, and the postulated extra inhibition which they induce reduces further feeding on protein.

Other aspects of feeding behaviour are consonant with Belzer's theory. Thus gravid females have a protein deficit, but because the egg masses cause massive abdominal expansion, feeding is decreased. After oviposition however, when all three inhibitory factors are low, very long protein drinks are taken. This is a suitable behaviour pattern since the oviposition site is usually a source of decaying proteinaceous material. Treatment of flies with cycloheximide, a specific inhibitor of protein synthesis, prevented the expected rise in protein "hunger" prior to vitellogenesis (Belzer, 1978a; Price, 1969). On the other hand treatment after the normal development of protein "hunger" did not reduce it.

The results of Rachman (1980) suggest that Belzer's "reserve material related to protein synthesis" which determines the effectiveness of a protein source as a feeding stimulus, is in fact the fly's level of carbohydrate reserves. Rachman proposed that a protein source only becomes an effective feeding stimulus when carbohydrate reserves are above a certain threshold level. Reduction of carbohydrate reserves below this level, for example during the protein synthesis associated with vitellogenesis, or as a result of extended starvation or prolonged flight, leads to a loss of effectiveness of protein as a feeding stimulant. Also flies given prior access to high concentrations of sucrose develop "protein hunger" earlier than flies given lower concentrations. The effect of treatments which interfere with protein metabolism, for example cycloheximide poisoning, may be explained in terms of interference with carbohydrate metabolism. Thus the relationship between protein hunger and the need for protein may be indirect, being mediated by the demands of protein metabolism on energy reserves.

Strangways Dixon (1961) put forward a different hypothesis concerning the mediation of the protein "hunger". He suggested that the neural mechanisms that mediate ingestive behaviour are under hormonal control, and that the humoral changes occurring in relation to the ovarian cycle, also affect feeding. Further discussion of these theories and their various drawbacks is given by Dethier (1976).

7.3 DIAPAUSE

Many insects survive periods of environmental stress by entering into a state of reduced or arrested development called diapause. The capacity to do this is genetically determined, and is usually triggered by environmental signals that reliably precede the actual stress (Lees, 1955). When diapause occurs during a stage which is normally actively feeding, feeding behaviour is greatly altered. Usually there is increased feeding prior to entering diapause, and always there is reduced feeding during diapause.

In adult grasshoppers and locusts, diapause takes the form of a delay in maturation which may last six months or more. Feeding does occur during diapause but at a reduced rate (Uvarov, 1966). In *Nomadacris septemfasciata* feeding was found to increase during early adult life in April and May.

but was then greatly reduced until November when maturation processes began (Chapman, 1957). The elimination of the normal high level of feeding in the morning was characteristic while the large evening feed before roosting occurred as normal. The difference must lie in some central mechanism, since food was available and conditions were appropriate for feeding in the mornings, and the crops of the insects were empty.

In many cases, there is no feeding during diapause. Mosquitoes entering diapause will not bite when given the opportunity (Washino, 1970). Various Coleoptera are said not to feed during diapause, although *Galleruca taneceta* feeds intermittently. It is possible that thorough examination of insects throughout diapause will indicate that there is feeding, but at a very low rate (Guerra and Bishop, 1972; Hodek, 1967; Siew, 1966).

Only in the blowfly has a detailed study of feeding behaviour during diapause been made. *Musca domestica* and *Phormia regina* have been found to feed little during diapause, and an analysis of responsiveness showed that in the latter species, with insects of similar age, more of the nondiapausing insects respond to 1 M sucrose than the diapausing insects (Stoffolano, 1974). Also, nondiapausing insects take much longer meals, with an intake of four times the volume of a meal taken by diapausing insects. The difference was less extreme in *Protophormia terraenovae* (Greenberg and Stoffolano, 1977). One factor possibly affecting intake is a change in chemoreceptors: as with aging insects, more sensilla appear to be nonfunctional during diapause (Stoffolano, 1973). It is not known whether these sensilla regain their sensitivity at the termination of diapause, so the causal relationship must remain in doubt.

It is more likely that some central inhibition occurs during diapause, and it has been suggested that accumulated stores in the form of an enlarged fat body provide some inhibition from abdominal stretch receptors (Stoffolano, 1968). While this may be true, experimental proof is lacking and can hardly account for the continued reduced responsiveness towards the end of diapause when fat body reserves are greatly reduced. A more attractive hypothesis would relate hormonal events occurring during diapause to a central nervous inhibition of feeding activity or lowered "feeding drive". Such a hypothesis may then be applicable to the various insect groups and life history stages where diapause occurs.

7.4 AGING

The daily intake of food is affected by age *per se* at least in flies (Gelperin and Dethier, 1967). A contributory factor in these insects lies in the changes in the chemoreceptor sensilla (Stoffolano, 1973). In males the frequency of firing in sensilla contacting particular solutions was reduced with age, and in both sexes, the percentage of inoperative sensilla increased by up to 50% with age. The readiness with which flies respond to sucrose solutions is reduced by approximately 50% with age and it is possible that the changes in sensilla may account for it.

Rees (1970) and Stoffolano *et al.* (1978) showed that the number of inoperative salt and water receptors in labellar chemoreceptors of *Phormia* spp. increased with age, so that by thirty days after emergence almost all receptors were inoperative.

8 Concluding remarks

It must be acknowledged that while chemosensory and behavioural factors involved in feeding, and the control of meal size in two insect groups have been well examined, there are some major gaps in the story for any one group. Flies have received most attention, but even here the final mystery is untouched: what is the central mechanism controlling feeding and which must integrate the incoming information? What are the hormonal and other causal factors influencing the likelihood of feeding? What complex interplay is occurring at any one time which on the one hand ensures that food requirements are satisfied, and on the other hand allows enormous variation between individuals?

Among the mass of detailed experiment on a wide range of insects, three major facts seem to be emerging. Firstly, continued feeding generally requires continued phagostimulation the nature of which is specialised in different ways to suit particular individual feeding habits. Secondly, the amount of food ingested in one meal depends to a large extent on volumetric factors. Finally there is some regulation of periods between meals which depends on haemolymph composition. The central nervous factors which ultimately govern feeding must now be investigated.

Acknowledgements

We would like to thank the following for advice and criticism: Drs L. Barton Browne, W. M. Blaney, R. F. Chapman, V. G. Dethier and T. H. Hsiao.

References

Abushama, F. F. (1968). Food plant selection by *Poecilocerus hieroglyphicus* (Klug.) (Orthoptera: Pyrgomorphidae) and some of the receptors involved. *Proc. Roy. Entomol. Soc. London A* **43,** 96–104

Agarwal, R. A. (1969). Morphological characteristics of sugar cane and insect resistance. *Ent. exp. Appl.* **12,** 767–776

Akeson, W. R., Gorz, H. J. and Haskins, F. A. (1970). Sweet clover weevil (*Sitona cylindricollis*: Col., Curculionidae) feed stimulants: variation in levels of glucose, fructose and sucrose in *Melilotus* leaves. *Crop Sci.* **10,** 477–479

Altman, P. L. and Dittmer, D. S. (eds) (1972). Blood and other body fluids. In: *Biological Handbooks*. American Societies for Experimental Biology, Washington

Anwyl, R. (1972). The structure and properties of an abdominal stretch receptor in *Rhodnius prolixus*. *J. Ins. Physiol.* **18,** 2143–2153

Arab, Y. M. (1959). Some chemosensory mechanisms in the blowfly, *Bull. Coll. Arts Sci. Baghdad* **4,** 77–85

Askew, R. F. (1971). *Parasitic Insects*. Heinemann, London

Auclair, J. L. (1963). Aphid feeding and nutrition. *Ann. Rev. Ent.* **8,** 439–470

Auclair, J. L. (1969). Nutrition of plant sucking insects on chemically defined diets. *Ent. exp. Appl.* **12,** 623–641

Avé, D., Frazier, J. L. and Hatfield, L. D. (1978). Contact chemoreception in the tarnished plant bug *Lygus lineolaris*. *Ent. exp. Appl.* **24,** 217–227

Baines, D. M. (1979). Studies of weight changes and movements of dyes in the caeca and midgut of fifth-instar *Locusta migratoria migratorioides* (R. & F.), in relation to feeding and food deprivation. *Acrida* **8,** 95–105

Baines, D. M., Bernays, E. A. and Leather, E. M. (1973). Movement of food through the gut of *Locusta migratoria migratorioides*. *Acrida* **2,** 319–332

Banks, C. J. (1965). Aphid nutrition and reproduction. *Rep. Rothamsted exp. Stn.* 1964, 299–309

Banks, C. J. and Nixon, H. L. (1958). Effects of the ant *Lasius niger* L. on feeding and excretion of the bean aphid, *Aphis fabae* Scop. *J. exp. Biol.* **35,** 703–711

Barton Browne, L. (1964). Water regulation in insects. *Ann. Rev. Ent.* **9,** 63–82

Barton Browne, L. (1968). Effects of altering the composition and volume of the haemolymph on water ingestion of the blowfly, *Lucilia cuprina*. *J. Ins. Physiol.* **14,** 1603–1620

Barton Browne, L. (1975). Regulatory mechanisms in insect feeding. *Adv. Insect Physiol.* **11,** 1–116

Barton Browne, L. and Dudzinski, A. (1968). Some changes resulting from water deprivation in the blowfly, *Lucilia cuprina*. *J. Ins. Physiol.* **14,** 1423–1434

Barton Browne, L. and Evans, D. R. (1960). Locomotor activity of the blowfly as a function of feeding and starvation. *J. Ins. Physiol.* **4,** 27–37

Barton Browne, L. and Gerwen, A. C. M. van (1976). Regulation of water ingestion by the locust, *Chortoicetes terminifera*: the effect of injections into the haemolymph. *Physiol. Ent.* **1,** 159–167

Barton Browne, L., Moorhouse, J. E. and Gerwen, A. C. M. van (1975). Sensory adaptation and the regulation of meal size in the Australian plague locust, *Chortoicetes terminifera*. *J. Ins. Physiol.* **21,** 1633–1639

Barton Browne, L., Moorhouse, J. E. and Gerwen, A. C. M. van (1975a). An excitatory state generated during feeding in the locust, *Chortoicetes terminifera*. *J. Ins. Physiol.* **21,** 1731–1735

Barton Browne, L., Moorhouse, J. E. and Gerwen, A. C. M. van (1976). A relationship between weight loss during a period of food deprivation and the sizes of subsequent meals taken by the Australian plague locust, *Chortoicetes terminifera*. *J. Ins. Physiol.* **22,** 89–94

Beck, S. D. (1965). Resistance of plants to insects. *Ann. Rev. Ent.* **10,** 207–232

Beck, S. D., Edwards, C. A. and Medler, J. T. (1958). Feeding and nutrition of the milkweed bug, *Oncopeltus fasciatus* (Dallas). *Ann. Ent. Soc. Am.* **51,** 283–288

Beck, S. D. and Hanec, W. (1958). Effect of amino acids on feeding behaviour of the European corn borer, *Pyrausta nubilalis* (Hubn). *J. Ins. Physiol.* **2,** 85–96

Beland, G. L., Haskins, F. A., Manglitz, G. R. and Gorz, H. J. (1973). Sweet clover weevil: adenosine as a feeding stimulant. *J. econ. Ent.* **66,** 1037–1039

Belzer, W. R. (1970). The control of protein ingestion in the black blowfly *Phormia regina* (Meigen). Ph.D. thesis, University of Pennsylvania, Philadelphia

Belzer, W. R. (1978). Patterns of selective protein ingestion by the blowfly *Phormia regina. Physiol. Ent.* **3,** 169–175

Belzer, W. R. (1978a). Factors conducive to increased protein feeding by the blowfly *Phormia regina. Physiol. Ent.* **3,** 251–257

Belzer, W. R. (1978b). Recurrent nerve inhibition of protein feeding in the blowfly *Phormia regina. Physiol. Ent.* **3,** 259–263

Belzer, W. R. (1979). Abdominal stretch in the regulation of protein ingestion by the black blowfly, *Phormia regina. Physiol. Ent.* **4,** 7–13

Bennet-Clark, H. C. (1963). The control of meal size in the blood-sucking bug *Rhodnius prolixus. J. exp. Biol.* **40,** 741–750

Bernays, E. A. (1977). The physiological control of drinking behaviour in *Locusta migratoria. Physiol. Ent.* **2,** 261–273

Bernays, E. A. (1980). The post-prandial rest in *Locusta migratoria* nymphs and its hormonal regulation. *J. Ins. Physiol.* **26,** 119–123

Bernays, E. A. and Chapman, R. F. (1970). Experiments to determine the basis of food selection by *Chorthippus parallelus* (Zetterstedt) (Orthoptera: Acrididae) in the field. *J. Anim. Ecol.* **39,** 761–776

Bernays, E. A. and Chapman, R. F. (1972). Meal size in nymphs of *Locusta migratoria. Ent. exp. Appl.* **15,** 399–410

Bernays, E. A. and Chapman, R. F. (1972a). The control of changes in peripheral sensilla associated with feeding in *Locusta migratoria. J. exp. Biol.* **57,** 755–763

Bernays, E. A. and Chapman, R. F. (1973). The regulation of feeding in *Locusta migratoria.* Internal inhibitory mechanisms. *Ent. exp. Appl.* **16,** 329–342

Bernays, E. A. and Chapman, R. F. (1974). The regulation of food intake by acridids. In: *Experimental Analysis of Insect Behaviour* (L. Barton Browne, ed.) pp. 48–59. Springer Verlag, Berlin

Bernays, E. A. and Chapman, R. F. (1974a). Changes in haemolymph osmotic pressure in *Locusta migratoria* in relation to feeding. *J. Ent. A.* **48,** 149–155

Bernays, E. A. and Chapman, R. F. (1974b). The effects of haemolymph osmotic on the meal size of nymphs of *Locusta migratoria* L. *J. exp. Biol.* **61,** 473–480

Bernays, E. A. and Chapman, R. F. (1976). Antifeedants in seedling grasses. In: *The Host Plant in Relation to Insect Behaviour and Reproduction* (T. Jermy, ed.) pp. 41–46. Plenum, New York

Bernays, E. A. and Chapman, R. F. (1977). Deterrent chemicals as a basis of oligophagy in *Locusta migratoria. Ecol. Ent.* **2,** 1–18

Bernays, E. A. and Chapman, R. F. (1978). Plant chemistry and acridoid feeding behaviour. In: *Biochemical Aspects of Plant and Animal Coevolution* (J. B. Harborne, ed.) pp. 99–141. Academic Press, London

Bernays, E. A. and Mordue, A. J. (1973). Changes in palp tip sensilla of *Locusta migratoria* in relation to feeding: the effects of different levels of hormone. *Comp. Biochem. Physiol.* **45A,** 451–454

Bernays, E. A., Blaney, W. M. and Chapman, R. F. (1972). Changes in chemoreceptor sensilla on the maxillary palps of *Locusta migratoria* in relation to feeding. *J. exp. Biol.* **57,** 745–753

Bernays, E. A., Chapman, R. F., Horsey, J. and Leather, E. M. (1974). Inhibitory effects of seedling grasses on feeding and survival of acridids. *Bull. ent. Res.* **64,** 413–420

Bitsch, J. and Denis, J. R. (1973). La tête. In: *Traité de Zoologie* (P. P. Grasse, ed.). Tome VIII, fasc. 1. Masson et Cie, Paris

Blaney, W. M. (1974). Electrophysiological responses of the terminal sensilla on the maxillary palps of *Locusta migratoria* (L.) to some electrolytes and non-electrolytes. *J. exp. Biol.* **60,** 275–293

Blaney, W. M. (1975). Behavioural and electrophysiological studies of taste discrimination by the maxillary palps of larvae of *Locusta migratoria* (L.). *J. exp. Biol.* **62,** 555–569

Blaney, W. M. (1980). Chemoreception and food selection by locusts. In: *Olfaction and Taste VII* (H. van der Starre, ed.) pp. 127–130. IRL Press, London

Blaney, W. M. and Chapman, R. F. (1970). The functions of the maxillary palps of Acrididae (Orthoptera). *Ent. exp. Appl.* **13,** 363–376

Blaney, W. M., Chapman, R. F. and Wilson, A. (1973). The pattern of feeding of *Locusta migratoria* (Orthoptera, Acrididae). *Acrida* **2,** 119–137

Blaney, W. M. and Duckett, A. M. (1975). The significance of palpation by the maxillary palps of *Locusta migratoria* (L.): an electrophysiological and behavioural study. *J. exp. Biol.* **63,** 701–712

Blom, F. (1978). Sensory activity and food intake: a study of input-output relationships in two phytophagous insects. *Neth. J. Zool.* **28,** 277–340

Blom, F. (1978a). Sensory input behavioural output relationships in the feeding activity of some lepidopterous larvae. *Ent. exp. Appl.* **24,** 258–263

Boeckh, J. (1980). Neural mechanisms of odour detection and odour discrimination in the insect olfactory pathway. In: *Insect Neurobiology and Pesticide Action,* pp. 367–374

Boeckh, J. (1980a). Ways of nervous coding of chemosensory quality at the input level. In: *Olfaction and Taste VII* (H. van der Starre, ed.) pp. 113–122. IRL Press, London

Boeckh, J., Ernst, K. D., Sass, H. and Waldow, U. (1975). Coding of odour quality in the insect olfactory pathway. In: *Olfaction and Taste V* (D. A. Denton and J. P. Coghlan, eds) pp. 239–246. Academic Press, New York

Bowdan, E. (1981). Activity of female blowflies, (*Phormia regina*), in response to novel odours. *Ent. exp. Appl.* **29,** 297–304

Brady, J. (1972). The visual responsiveness of the tsetse fly *Glossina morsitans* Westw. (Glossinidae) to moving objects: the effects of hunger, sex, host odour and stimulus characteristics. *Bull. ent. Res.* **62,** 257–279

Brady, J. (1973). Changes in the probing responsiveness of starving tsetse flies (*Glossina morsitans* Westw.) (Diptera, Glossinidae). *Bull. ent. Res.* **63,** 247–255

Brady, J. (1974). The physiology of insect circadian rhythms. *Adv. Insect Physiol.* **10,** 1–116

Brady, J. (1975). "Hunger" in the tsetse fly: the nutritional correlates of behaviour. *J. Ins. Physiol.* **21,** 807–830

Brady, J. (1975a). Circadian changes in central excitability–the origin of behavioural rhythms in tsetse flies and other animals? *J. Ent. A.* **50,** 79–96

Braken, G. K. and Nair, K. K. (1967). Stimulation of yolk deposition in an ichneumonid by feeding synthetic juvenile hormone. *Nature* **216,** 483–484

Brues, C. T. (1946). *Insect Dietary.* Harvard University Press, Massachusetts

Buxton, P. A. (1930). The biology of the blood-sucking bug *Rhodnius prolixus. Trans. Roy. Entomol. Soc. London* **78,** 227–236

Cazal, M. (1969). Actions d'extraits de corpora cardiaca sur le peristaltisme intestinal de *Locusta migratoria*. *Archs Zool. exp. gén.* **110,** 83–89

Chapman, R. F. (1954). Responses of *Locusta migratoria migratorioides* (R. & F.) to light in the laboratory. *Brit. J. anim. Behav.* **2,** 146–152

Chapman, R. F. (1957). Observations on the feeding behaviour of adult Red Locusts (*Nomadacris septemfasciata* (Serville)). *Brit. J. anim. Behav.* **5,** 60–75

Chapman, R. F. (1958). A field study of the potassium concentration in the blood of the red locust *Nomadacris septemfasciata* (Serv.), in relation to its activity. *Anim. Behav.* **6,** 60–67

Chapman, R. F. (1959). Field observations on the behaviour of hoppers of the Red Locust. *Anti-Locust Bull.* No. 33, 54pp.

Chapman, R. F. (1974). *Feeding in Leaf Eating Insects.* Oxford University Press, London

Chapman, R. F. (1977). The role of the leaf surface in food selection by acridids and other insects. *Colloq. Internal. CNRS* **265,** 133–149

Chapman, R. F. (1980). The evolution of insect chemosensory systems and food selection behaviour. In: *Olfaction and Taste VII* (H. van der Starre, ed.) pp. 131–134. IRL Press, London

Chapman, R. F. and Bernays, E. A. (1977). Chemical resistance of plants to insect attack. *Scr. varia pont. Acad. Sci.* **41,** 603–633

Chapman, R. F. and Thomas, J. G. (1978). The numbers and distribution of sensilla on the mouthparts of Acridoidea. *Acrida* **7,** 115–148

Clarke, K. U. and Grenville, H. (1960). Nervous control of movements in the foregut of *Schistocerca gregaria* Forsk. *Nature* **186,** 98–99

Clements, A. H. (1963). *The Physiology of Mosquitoes.* Pergamon Press, Oxford

Cobben, R. H. (1978). Evolutionary trends in Heteroptera. Part II. Mouthpart structures and feeding strategies. *Meded. Landbouwhoges, Wageningen* 78–5

Cole, S. J. and Gillett, J. D. (1979). The influence of the brain hormone on retention of blood in the midgut of the mosquito *Aedes aegypti* (L.) (Diptera Culicidae). II. The involvement of the ovaries and ecdysone. *Proc. Roy. Soc. B* **205,** 411–422

Cook, A. G. (1976). A critical review of the methodology and interpretation of experiments designed to assay the phagostimulatory activity of chemicals to phytophagous insects. *Symp. Biol. Hung.* **16,** 47–54

Cook, A. G. (1977). The anatomy of the clypeo-labrum of *Locusta migratoria* (L.) (Orthoptera: Acrididae) *Acrida* **6,** 287–306

Cook, A. G. (1977a). Nutrient chemicals as phagostimulants for *Locusta migratoria. Ecol. Ent.* **2,** 113–121

Cull, D. C. and Emden, H. F. van (1977). The effect on *Aphis fabae* of diet changes in their food quality. *Physiol. Ent.* **2,** 109–115

Dadd, R. H. (1960). Observations on the palatability and utilisation of food by locusts, with particular reference to the interpretation of performances in growth-trials using synthetic diets. *Ent. exp. Appl.* **3,** 283–304

Dadd, R. H. (1963). Feeding behaviour and nutrition in grasshoppers and locusts. *Adv. Insect Physiol.* **1,** 47–109

Dadd, R. H. (1970). Comparison of rates of ingestion of particulate solids by *Culex pipiens* larvae: phagostimulant effect of water soluble yeast extract. *Ent. exp. Appl.* **13,** 407–419

Dadd, R. H. (1970a). Arthropod nutrition. In: *Chemical Zoology VA* (M. Florkin and B. T. Scheer, eds) pp. 35–95. Academic Press, New York

Dadd, R. H. (1971). Effects of size and concentration of particles on rates of ingestion of latex particles by mosquito larvae. *Ann. Ent. Soc. Am.* **64,** 687–692

Dagmar, J. R. (1979). The role of phenolic substances for host selection behaviour of the black bean aphid. *Ent. exp. Appl.* **26,** 49–54

Danilevskii, A. S. (1965). *Photoperiodism and Seasonal Development of Insects.* Oliver and Boyd, London

Davey, P. M. (1954). Quantities of food eaten by the desert locust, *Schistocerca gregaria* (Forsk.), in relation to growth. *Bull. ent. Res.* **45,** 539–551

Davey, K. G. and Treherne, J. E. (1963). Studies on crop function in the cockroach (*Periplaneta americana* L.) I. The mechanism of crop emptying. *J. exp. Biol.* **40,** 763–777

Davey, K. G. and Treherne, J. E. (1963a). Studies on crop function in the cockroach (*Periplaneta americana* L.) II. The nervous control of crop emptying. *J. exp. Biol.* **40,** 775–780

David, W. A. L. and Gardiner, B. O. C. (1961). Feeding behaviour of the adults of *Pieris brassicae* (L.) in a laboratory culture. *Bull. ent. Res.* **52,** 741–762

Davis, E. E. and Sokolove, P. G. (1975). Temperature responses of antennal receptors of the mosquito *Aedes aegypti. J. Comp. Physiol.* **96,** 223–236

Davis, E. E. and Takahashi, F. T. (1980). Humoral alterations of chemoreceptor sensitivity in the mosquito. In: *Olfaction and Taste VII* (H. van der Starr, ed.) pp. 139–142. IRL Press, London

Dethier, V. G. (1937). Gustation and olfaction in lepidopterous larvae. *Biol. Bull.* **72,** 7–23

Dethier, V. G. (1953). Summation and inhibition following contralateral stimulation of tarsal chemoreceptors of the blowfly. *Biol. Bull.* **105,** 257–268

Dethier, V. G. (1954). Notes on the biting response of tsetse flies. *Am. J. trop. Med. Hyg.* **3,** 160–171

Dethier, V. G. (1959). The nerves and muscles of the proboscis of the blowfly *Phormia regina* Meigen in relation to feeding responses. *Smithson. misc. Collns.* **137,** 157–174

Dethier, V. G. (1961). Behavioural aspects of protein ingestion by the blowfly *Phormia regina* Meigen. *Biol. Bull.* **121,** 456–470

Dethier, V. G. (1967). Feeding and drinking behaviour of invertebrates. In: *Handbook of Physiology, Section B, Alimentary Canal* (C. F. Code, ed.) pp. 79–96. *Amer. Physiol. Soc.* Vol. 1, Washington, D.C.

Dethier, V. G. (1970). Some general considerations of insects' responses to the chemicals in food plants. In: *Control of Insect Behaviour by Natural Products* (D. L. Wood, R. M. Silverstein and M. Nakajima, eds) pp. 21–28. Academic Press, New York

Dethier, V. G. (1973). Electrophysiological studies of gustation in lepidopterous larvae. II. Taste spectra in relation to food-plant discrimination. *J. Comp. Physiol.* **82,** 103–134

Dethier, V. G. (1974). The specificity of the labellar chemoreceptors of the blowfly and the response to natural foods. *J. Ins. Physiol.* **20,** 1859–1869

Dethier, V. G. (1976). *The Hungry Fly.* Harvard University Press, Cambridge, Massachusetts

Dethier, V. G. and Bodenstein, D. (1958). Hunger in the blowfly. *Z. Tierpsychol.* **15,** 129–140

Dethier, V. G. and Evans, D. R. (1961). Physiological control of water ingestion in the blowfly. *Biol. Bull.* **121,** 108–116

Dethier, V. G. and Gelperin, A. (1967). Hyperphagia in the blowfly. *J. exp. Biol.* **47,** 191–200

Dethier, V. G. and Hanson, F. E. (1965). Taste papillae of the blowfly. *J. cell. comp. Physiol.* **65,** 93–100

Dethier, V. G. and Kuch, J. H. (1971). Electrophysiological studies of gustation in lepidopterous larvae I. Comparative sensitivity to sugars, amino acids and glycosides. *Z. vergl. Physiol.* **72,** 343–363

Dethier, V. G. and Rhoades, M. V. (1954). Sugar preference-aversion functions for the blowfly. *J. exp. Zool.* **126,** 177–204

Dethier, V. G. and Schoonhoven, L. M. (1969). Olfactory coding by lepidopterous larvae. *Ent. exp. Appl.* **12,** 535–543

Dethier, V. G., Evans, D. R. and Rhoades, M. V. (1956). Some factors controlling the ingestion of carbohydrates by the blowfly. *Biol. Bull.* **111,** 204–222

Dethier, V. G., Solomon, R. L. and Turner, L. H. (1965). Sensory input and central excitation and inhibition in the blowfly. *J. Comp. Physiol.* **60,** 303–313

Dingle, H. (1968). The influence of environment and heredity on flight activity in the milkweed bug, *Oncopeltus. J. exp. Biol.* **48,** 175–184

Drongelen, W. van (1979). Contact chemoreception of host plant specific chemicals in larvae of various *Yponomeuta* species (Lepidoptera). *J. Comp. Physiol.* **134,** 265–279

Drongelen, W. van, Holley, A., and Døving, K. B. (1978). Convergence in the olfactory system: quantitative aspects of odour sensitivity. *J. theor. Biol.* **71,** 39–48

Edney, E. B. (1977). *Water Balance in Land Arthropods.* Springer Verlag, Berlin

Ellis, P. E. (1951). The marching behaviour of hoppers of the African migratory locust in the laboratory. *Anti-Locust Bull.* No. 7, 48 pp.

Ellis, P. E. and Hoyle, G. (1954). A physiological interpretation of the marching of hoppers of the African migratory locust (*Locusta migratoria migratorioides* R. and F.). *J. exp. Biol.* **31,** 271–279

Emden, H. F. van (1972). Aphids as phytochemists. In: *Phytochemical Ecology* (J. B. Harborne, ed.) pp. 25–43. Academic Press, London

Emden, H. F. van and Bashford, M. A. (1971). The performance of *Brevicoryne brassicae* and *Myzus persicae* in relation to plant age and leaf amino acids. *Ent. exp. Appl.* **14,** 349–360

Engelmann, F. (1968). Feeding and crop emptying in the cockroach *Leucophaea maderae. J. Ins. Physiol.* **14,** 1525–1532

Evans, D. R. (1961). Control of the responsiveness of the blowfly to water. *Nature* **190,** 1132–1133

Evans, D. R. and Barton Browne, L. (1960). The physiology of hunger in the blowfly. *Am. Midl. Nat.* **64,** 282–300

Feeny, P. P. (1970). Seasonal changes in oak leaf tannins and nutrients as a cause of spring feeding by winter moth caterpillars. *Ecology* **51,** 565–581

Felt, B. T. and Vande Berg, J. S. (1976). Ultrastructure of the blowfly chemoreceptor sensillum (*Phormia regina*). *J. Morph.* **150,** 763–784

Fier, D. and Beck, S. D. (1963). Feeding behaviour of the large milkweed bug *Oncopeltus fasciatus. Ann. Ent. Soc. Am.* **56,** 224–229

Fraenkel, G. (1969). Evaluation of our thoughts on secondary plant substances. *Ent. exp. Appl.* **12,** 473–486

Freyvogel, T. A. and Staubli, W. (1965). The formation of the peritrophic membrane in Culicidae. *Acta Trop.* **22,** 119–147

Friedman, S. (1967). The control of trehalose synthesis in the blowfly, *Phormia regina. J. Ins. Physiol.* **13,** 397–405

Friend, W. G. (1978). Physical factors affecting the feeding responses of *Culiseta inornata* to ATP, sucrose and blood. *Ann. Ent. Soc. Am.* **71,** 935–940

Friend, W. G. and Smith, J. J. B. (1972). Feeding stimuli and techniques for studying the feeding of haematophagous arthropods under artificial conditions, with special reference to *Rhodnius prolixus.* In: *Insect and Mite Nutrition* (J. G. Rodriguez, ed.) pp. 241–256. North Holland, Amsterdam

Friend, W. G. and Smith, J. J. B. (1977). Factors affecting feeding by blood-sucking insects. *Ann. Rev. Ent.* **22,** 309–331

Galun, R. (1966). Feeding stimulants of the rat flea *Xenopsylla cheopis. Life Sci.* **5,** 1335–1342

Galun, R. (1975). Behavioural aspects of chemoreception in blood-sucking invertebrates. In: *Sensory Physiology and Behaviour* (R. Galun, P. Hillman, I. Parnas and R. Werman, eds) pp. 211–221. Plenum, New York

Galun, R. and Margalit, J. (1969). Adenine nucleotides as feeding stimulants of the tsetse fly *Glossina austeni* Newst. *Nature* **222,** 583–584

Galun, R. and Margalit, J. (1970). Some properties of the ATP receptors of *Glossina austeni. Trans. Roy. Soc. trop. Med. Hyg.* **64,** 171–174

Galun, R. and Rice, M. J. (1971). Role of blood platelets in haematophagy. *Nature New Biol.* **233,** 110–111

Gelperin, A. (1966). Stretch receptors in the foregut of the blowfly. *Science* **157,** 208–210

Gelperin, A. (1966a). Control of crop emptying in the blowfly. *J. Ins. Physiol.* **12,** 331–345

Gelperin, A. (1971). Abdominal sensory neurones providing negative feedback to the feeding behaviour of the blowfly. *Z. vergl. Physiol.* **72,** 17–31

Gelperin, A. (1971a). Regulation of feeding. *Ann. Rev. Ent.* **16,** 365–378

Gelperin, A. (1972). Neural control systems underlying insect feeding behaviour. *Am. Zool.* **12,** 489–496

Gelperin, A. and Dethier, V. G. (1967). Long-term regulation of sugar intake by the blowfly. *Physiol. Zool.* **40,** 218–228

Getting, P. A. (1971). The sensory control of motor output in fly proboscis extension. *Z. vergl. Physiol.* **74,** 103–120

Ghent, A. W. (1960). A study of the group feeding behaviour of larvae of the Jack Pine Sawfly, *Neodiprion pratti banksianae* Roh. *Behaviour* **16,** 110–148

Gordon, H. T. (1968). Intake rates of various solid carbohydrates by male German cockroaches. *J. Ins. Physiol.* **14,** 41–52

Green, G. W. (1964). The control of spontaneous locomotor activity in *Phormia regina* Meigen–I. Locomotor patterns in intact flies. *J. Ins. Physiol.* **10,** 727–752

Green, G. W. (1964a). The control of spontaneous locomotor activity in *Phormia regina* Meigen–II. Experiments to determine the mechanism involved. *J. Ins. Physiol.* **10,** 711–726

Greenberg, S. L. and Stoffolano, J. G. (1977). The effect of age and diapause on the long-term intake of protein and sugar by two species of blowflies, *Phormia regina* (Meig.) and *Protophormia terraenovae* (R.D.). *Biol. Bull.* **153,** 282–298

Grosse, W. R. (1974). Nahrungsaufnahme und Frassrhythmik der Larven von *Calliphora erythrocephala* Meig. (Diptera, Calliphoridae). *Biol. Zbl.* **93,** 687–695

Guerra, A. A. and Bishop, J. L. (1962). The effect of aestivation on sexual maturation in the female alfalfa weevil (*Hypera postica*). *J. econ. Ent.* **55,** 747–749

Gwadz, R. W. (1969). Regulation of blood meal size in the mosquito. *J. Ins. Physiol.* **15,** 2039–2044

Hall, M. J. R. (1980). Central control of tarsal thresholds for proboscis extension in the blowfly. *Physiol. Ent.* **5,** 17–24

Hall, M. J. R. (1980a). Circadian rhythm of proboscis extension responsiveness in the blowfly: central control of threshold changes. *Physiol. Ent.* **5,** 223–233

Hamamura, Y. (1970). The substances that control feeding behaviour and growth of the silkworm, *Bombyx mori* L. In: *Control of Insect Behaviour by Natural Products* (D. L. Wood, R. M. Silverstein and M. Nakajima, eds) pp. 55–80. Academic Press, New York

Hans, H. and Thorsteinson, A. J. (1961). The influence of physical factors and host plant odour on the induction and termination of dispersal flights in *Sitona cylindricollis* Fahr. *Ent. exp. Appl.* **4,** 165–177

Haskell, P. T. and Mordue (Luntz) A. J. (1969). The role of mouthpart receptors in the feeding behaviour of *Schistocerca gregaria. Ent. exp. Appl.* **12,** 591–610

Haskell, P. T. and Schoonhoven, L. M. (1969). The function of certain mouthpart receptors in relation to feeding in *Schistocerca gregaria* (Forsk) and *Locusta migratoria migratorioides* (R. & F.). *Ent. exp. Appl.* **12,** 423–440

Hassell, M. P., Lawton, J. H. and Beddington, J. R. (1976). The components of arthropod predation 1. The prey death rate. *J. Anim. Ecol.* **45,** 135–164

Hassett, C. C., Dethier, V. G. and Gans, J. (1950). A comparison of nutritive values and taste thresholds for the blowfly. *Biol. Bull.* **99,** 446–453

Henig, B. (1930). O unerwieniu tak zwanych niž szych organów zymslowych gąsienic motyli. *Prace Wydz. Mat.-Przyrod. Tow. Przyj. Nauk w Wilnie* **6,** 41–81

Herrewege, C. van (1974). Regulation of food consumption under various alimentary conditions after a period of starvation by the German cockroach. *Ent. exp. Appl.* **17,** 234–244

Highnam, K. C., Hill, L. and Mordue, W. (1966). The endocrine system and oocyte growth in *Schistocerca* in relation to starvation and frontal ganglionectomy. *J. Ins. Physiol.* **12,** 977–994

Hill, L. and Goldsworthy, G. J. (1968). Growth, feeding activity and the utilisation of reserves in larvae of *Locusta. J. Ins. Physiol.* **14,** 1085–1098

Hill, L. Luntz, A. J. and Steele, P. A. (1968). The relationships between somatic growth, ovarian growth, and feeding activity in the adult desert locust. *J. Ins. Physiol.* **14,** 1–20

Hocking, B. (1971). Blood sucking behaviour of terrestrial arthropods. *Am. Rev. Ent.* **16,** 1–26

Hodek, I. (1967). Bionomics and ecology of predaceous Coccinellidae. *Ann. Rev. Ent.* **12,** 79–104

Holling, C. S. (1961). Principles of insect predation. *Ann. Rev. Ent.* **6,** 163–182

Hosoi, T. (1954). Mechanism enabling the mosquito to ingest blood into the stomach and sugary fluids into the oesophageal diverticula. *Annot. Zool. Jap.* **27,** 82–90

Hosoi, T. (1959). Identification of blood components which induce gorging of the mosquito. *J. Ins. Physiol.* **3,** 191–218

House, H. L. (1965). Effects of low levels of nutrient content of a food and of nutrient imbalance on the feeding and the nutrition of a phytophagous larva *Celerio euphorbiae* (Linnaeus) (Lepidoptera: Sphingidae). *Can. Ent.* **97,** 62–68

House, H. L. (1967). The role of the nutritional factors in food selection and preference, as related to larval nutrition of an insect *Pseudosarcophaga affinis* (Diptera: Sarcophagidae), on synthetic diets. *Can. Ent.* **99,** 1310–1321

House, H. L. (1971). Relations between dietary proportions of nutrients, growth rate, and choice of food in the fly larva *Agria affinis* (Fall.). *J. Ins. Physiol.* **17,** 1225–1238

Hsiao, T. H. (1969). Chemical basis of host selection and plant resistance in oligophagous insects. *Ent. exp. Appl.* **12,** 777–788

Hsiao, T. H. (1969a). Adenine and related substances as potent feeding stimulants for the alfalfa weevil, *Hypera postica. J. Ins. Physiol.* **15,** 1785–1790

Hsiao, T. H. (1972). Chemical feeding requirements of oligophagous insects. In: *Insect and Mite Nutrition* (J. G. Rodriguez, ed.) pp. 225–240. North Holland, Amsterdam

Hsiao, T. H. (1974). Chemical influence on feeding behaviour of *Leptinotarsa* beetles. In: *Experimental Analysis of Insect Behaviour* (L. Barton Browne, ed.) pp. 237–248. Springer Verlag, Berlin

Hsiao, T. H. and Fraenkel, G. (1968). The influence of nutrient chemicals on the feeding behaviour of the Colorado beetle, *Leptinotarsa decemlineata* (Coleoptera: Chrysomelidae). *Ann. ent. Soc. Am.* **61,** 44–54

Huber, F. (1967). Central control of movements and behaviour in invertebrates. In: *Invertebrate Nervous Systems* (C. A. G. Wiersma, ed.) pp. 333–351. University of Chicago Press, Chicago

Ishikawa, S., Hirao, T. and Arai, N. (1969). The chemosensory basis of host plant selection in the silkworm. *Ent. exp. Appl.* **12,** 544–554

Jermy, T. (1961). On the nature of the oligophagy in *Leptinotarsa decemlineata* Say (Coleoptera Chrysomelidae). *Acta Zool. Acad. Sci. Hung.* **7,** 119–132

Jermy, T. (1966). Feeding inhibitors and food preference in chewing phytophagous insects. *Ent. exp. Appl.* **9,** 1–12

Johansson, A. S. (1955). The relationship between corpora allata and reproductive organs in starved female *Leucophaea maderae* (Blattaria). *Biol. Bull.* **108,** 40–44

Johansson, A. S. (1958). Relation of nutrition to endocrine-reproductive functions in the milkweed bugs *Oncopeltus fasciatus* (Dal.). *Nytt. Mag. Zool.* **7,** 1–132

Jones, J. S. and Madhukar, B. V. (1976). Effects of sucrose on blood avidity in mosquitoes. *J. Ins. Physiol.* **22,** 357–360

Kellogg, F. E. (1970). Water vapour and carbon dioxide receptors in *Aedes aegypti. J. Ins. Physiol.* **16,** 99–108

Kendall, M. D. (1971). Studies on the tarsi of *Schistocerca gregaria* Forskål. Ph.D. Thesis, University of London

Kendall, M. D. and Seddon, A. (1975). The effect of previous access to water on the responses of locusts to wet surfaces. *Acrida* **4,** 1–7

Kennedy, J. S. and Moorhouse, J. E. (1969). Laboratory observations on locust responses to wind-borne grass odour. *Ent. exp. Appl.* **12,** 487–503

Khalifa, A., Salama, H. S., Azmy, N. and El-Sharaby, A. (1974). Taste sensitivity of the cotton leafworm, *Spodoptera littoralis*, to chemicals. *J. Ins. Physiol.* **20,** 67–76

Klein, U. (1979). Zur Funktion und Struktur der Palpen und ihres Sensillenfeldes bei *Gryllus bimaculatus* De Geer. Ph. D.Thesis, Munich

Klingauf, F., Nöcker-Wenzel, K. and Klein, W. (1971). Influence of some wax components of *Vicia faba* L. on the food plant selection behaviour of *Acyrthosiphum pisum* (in German). *Z. Planzenkrank. Planzenschutz* **76,** 641–648

Klowden, M. J. and Lea, A. O. (1978). Blood meal size as a factor affecting continued host seeking by *Aedes aegypti. Am. J. trop. Med. Hyg.* **27,** 827–831

Knight, M. R. (1962). Rhythmic activities of the alimentary canal of the black blowfly *Phormia regina. Ann. ent. Soc. Am.* **55,** 380–382

Langley, P. A. (1976). Initiation and regulation of ingestion by haematophagous arthropods. *J. med. Ent.* **13,** 121–130

Langley, P. A. and Pimley, R. W. (1973). Influence of diet composition on feeding and water excretion by the tsetse fly, *Glossina morsitans*. *J. Ins. Physiol.* **19,** 1097–1109

Larsen, J. R. and Bodenstein, D. (1959). The humoral control of egg maturation in the mosquito. *J. exp. Biol.* **140,** 343–381

Le Berre, J. R. and Mainguet, A. M. (1973). Nutrition du criquet migrateur *Locusta migratoria* L. I. Compartement alimentaire, survie et variation pondérale en fonction de quelques substances. *Rev. Comp. Anim.* **7,** 203–218

Leckstein, P. M. and Llewellyn, M. (1974). The role of amino acids in diet intake and selection and the utilisation of dipeptides by *Aphis fabae*. *J. Ins. Physiol.* **20,** 877–885

Lees, A. D. (1955). *The Physiology of Diapause in Arthropods*. Cambridge University Press, London

Leonard, D. E. (1970). Effects of starvation on behaviour, number of larval instars, and development rate of *Porthetria dispar*. *J. Ins. Physiol.* **16,** 25–31

Louveaux, A. (1972). Equipment sensoriel et systeme nerveux périphérique des pièces buccales de *Locusta migratoria* L. *Insectes soc.* **19,** 359–368

Louveaux, A. (1975). Étude de l'innervation sensorielle de l'hypopharynx de larves de *Locusta migratoria migratorioides* R. & F. (Orthoptère, Acrididae). *Insectes Soc.* **22,** 3–11

Louveaux, A. (1976). Répercussion sur la prise de nourriture désafférences sensitives de la région peribuccale de la larve de *Locusta*. *J. Ins. Physiol.* **22,** 9–18

Louveaux, A. (1977). Regulation of meal size and growth of fifth instar nymphs of *Locusta migratoria* (Orthoptera, Acrididae) in different conditions of starvation and temperature. *Annls Nutr. Aliment.* **31,** 85–103

Loveridge, J. P. (1974). Studies in water regulations of adult locusts–II. Water gain in the food and loss in the faeces. *Trans. Rhod. scient. Ass.* **56,** 1–30

Ma, W. C. (1972). Dynamics of feeding responses in *Pieris brassicae* Linn. as a function of chemosensory input: a behavioural, ultrastructural and electrophysiological study. *Meded. Landbouwhoges. Wageningen* 72–11

Ma, W. C. (1976). Mouthparts and receptors involved in feeding behaviour and sugar perception in the African armyworm, *Spodoptera exempta* (Lepidoptera, Noctuidae). *Symp. Biol. Hung.* **16,** 139–151

Ma, W. C. (1977). Electrophysiological evidence for chemosensitivity to adenosine, adenine and sugars in *Spodoptera exempta* and related species. *Experientia* **33,** 356–357

Ma, W. C. and Kubo, I. (1977). Phagostimulants for *Spodoptera exempta*: identification of adenosine from *Zea mais*. *Ent. exp. Appl.* **22,** 107–112

Maddrell, H. S. P. (1963). Control of ingestion in *Rhodnius prolixus* Stal. *Nature* **198,** 210

Maes, F. W. and den Otter, C. J. (1975). A functional arrangement of labellar taste hairs in the blowfly *Calliphora vicina*. In: *Olfaction and Taste V* (D. A. Denton and J. P. Coghlan, eds) pp. 199–202. Academic Press, New York

Maes, F. W. and Vedder, C. G. (1978). A morphological and electrophysiological inventory of labellar taste hairs of the blowfly *Calliphora vicina*. *J. Ins. Physiol.* **24,** 667–672

Malz, D. von and Hintze-Podufal, C. (1979). Bau und Verteilung von Sinnesorganen auf dem Clypeolabrum, Hypopharynx und der Mandibel der Schabe *Gromphadorhina brunneri* Butler (Blaberoidea, Oxyhaloidae). *Zool. Jb. Anat.* **102,** 249–266

Malz, D. von and Hintze-Podufal, C. (1979a). Bau und Verteilung von Sinnesorgannen auf dem Labium und der Maxille der Schabe *Gromphadorhina brunneri* Butler (Blaberoidea, Oxyhaloidae). *Zool. Beitr.* **35,** 81–100

McCaffery, A. R. (1975). Food quality and quantity in relation to egg production in *Locusta migratoria migratorioides. J. Ins. Physiol.* **21,** 1551–1558

McClain, E. and Fier, D. (1973). Role of volume and amino acid concentration in the feeding of *Oncopeltus fasciatus* on a liquid and a solid diet. *J. Ins. Physiol.* **19,** 287–292

McCutchan, M. C. (1969). Responses of tarsal chemoreceptive hairs of the blowfly, *Phormia regina. J. Ins. Physiol.* **15,** 2058–2068

McGinnis, A. J. and Kasting, R. (1967). Dietary cellulose: effect on food consumption and growth of a grasshopper. *Can. J. Zool.* **45,** 365–367

McLean, D. L. and Kinsey, M. G. (1968). Probing behaviour of the pea aphid, *Acyrthosiphon pisum* II. Comparisons of salivation and ingestion in host and non-host leaves. *Ann. ent. Soc. Am.* **61,** 730–739

Meisner, J., Ascher, K. R. S. and Flowers, H. M. (1972). The feeding response of the larvae of the Egyptian cotton leaf worm, *Spodoptera littoralis* Boisd. to sugars and related compounds–I: phagostimulatory and deterrent effects. *Comp. Biochem. Physiol.* **13,** 955–962

Mellanby, K. and French, R. A. (1958). The importance of drinking water to larval insects. *Ent. exp. Appl.* **1,** 116–124

Miles, P. W. (1959). The salivary secretions of a plant sucking bug, *Oncopeltus fasciatus* (Dall.) (Heteroptera: Lygaeidae). I. The types of secretion and their role during feeding. *J. Ins. Physiol.* **3,** 243–255

Mitchell, B. K. (1974). Behavioural and electrophysiological investigations on the responses of larvae of the Colorado potato beetle (*Leptinotarsa decemlineata*) to amino acids. *Ent. exp. Appl.* **17,** 245–264

Mitchell, B. K. (1976). ATP reception by the tsetse fly, *Glossina morsitans* West. *Experientia* **32,** 192–193

Mitchell, B. K. (1976a). Physiology of an ATP receptor in labellar sensilla of the tsetse fly *Glossina morsitans* Westw. *J. exp. Biol.* **65,** 259–271

Mittler, T. E. (1958). Studies in the feeding and nutrition of *Tuberolachnus salignus* (Gmelin) (Homoptera, Aphididae). III. The nitrogen economy. *J. exp. Biol.* **35,** 626–638

Mittler, T. E. (1967). Effect of amino acid and sugar concentrations on the food intake of the aphid *Myzus persicae. Ent. exp. Appl.* **10,** 39–51.

Mittler, T. E. (1972). Interactions between dietary components. In: *Insect and Mite Nutrition* (J. G. Rodriguez, ed.) pp. 211–224. North Holland, Amsterdam

Moloo, S. K. (1971). An artificial feeding technique for *Glossina. Parasitology* **63,** 507–512

Moorhouse, J. E. (1969). *Locomotor activity and orientation in locusts.* Ph.D. thesis, University of London

Moorhouse, J. E. (1971). Experimental analysis of the locomotor behaviour of *Schistocerca gregaria* induced by odour. *J. Ins. Physiol.* **17,** 913–920

Moorhouse, J. E., Barton Browne, L. and Gerwen, A. C. M. van (1976). Factors affecting the rate of ingestion of liquids by the locust, *Chortoicetes terminifera J. Ins. Physiol.* **22,** 259–264

Mordue, A. J. (1974). Studies on the role of the mouthpart and antennal receptors of *Schistocerca Gregaria* (Forskål). Ph.D. thesis, University of London

Mordue (Luntz), A. J. (1979). The role of the maxillary and labial palps in the feeding behaviour of *Schistocerca gregaria*. *Ent. exp. Appl.* **25,** 279–288
Mordue (Luntz), A. J. and Hill, L. (1970). Utilisation of food by the adult female desert locust, *Schistocerca gregaria*. *Ent. expl Appl.* **13,** 352–358
Mordue, W. (1969). Hormonal control of Malpighian tubule and rectal function in the desert locust, *Schistocerca gregaria*. *J. Ins. Physiol.* **15,** 273–285
Mori, H. and Chant, D. A. (1966). The influence of prey density, relative humidity, and starvation on the predaceous behaviour of *Phytoseiulus persimilis* Athias-Henriot (Acarina, Phytoseidae). *Can. J. Zool.* **44,** 483–491
Moulins, M. (1974). Récepteurs de tension de la région de la bouche chez *Blaberus craniifer* Burmeister (Dictyoptera: Blaberidae). *Int. J. Insect. Morph. Embryol.* **3,** 171–192
Mulkern, G. B. (1967). Food selection by grasshoppers. *A. Rev. Ent.* **12,** 59–78
Nayer, J. K. and Thorsteinson, A. J. (1963). Further investigations into the chemical basis of insect-host relationships in an oligophagous insect, *Plutella maculipennis* (Curtis). *Can. J. Zool.* **41,** 923–929
Nelson, M. C. (1977). The blowfly's dance: role in the regulation of food intake. *J. Ins. Physiol.* **23,** 603–612
Nielsen, J. K. (1978). Host plant selection of monophagous and oligophagous flea beetles feeding on crucifers. *Ent. exp. Appl.* **24,** 562–569
Nielsen, J. K. (1978a). Host plant discrimination within Cruciferae: Feeding responses of four leaf beetles to glucosinolates, cucurbitacins and cardenolides, *Ent. exp. Appl.* **24,** 41–54
Nielsen, J. K., Larsen, L. M. and Sørensen, H. (1979). Host plant selection of the horseradish flea beetle *Phyllotreta armora* (Coleoptera: Chrysomelidae): identification of two flavonol glycosides stimulating feeding in combination with glucosinolates. *Ent. exp. Appl.* **26,** 40–48
Nijhout, H. F. and Carrow, G. M. (1978). Diuresis after a blood meal in female *Anopheles freeborni*. *J. Ins. Physiol.* **24,** 293–298
Norris, D. M. (1970). Quinol stimulation and quinone deterrency of gustation by *Scolytus multistriatus* (Coleoptera: Scolytidae). *Ann. ent. Soc. Am.* **63,** 476–478
Norris, K. R. (1959). The ecology of sheep blowflies in Australia. In: *Biogeography and Ecology in Australia. Monographiae Biol.* Vol. 8, pp. 514–544
Nuñez, J. A. (1964). Trinktriebregelung bei Insekten. *Naturwiss.* **17,** 419
Omand, E. (1971). A peripheral sensory basis for behavioural regulation. *Comp. Biochem. Physiol. A* **38,** 265–278
Pathak, M. D. (1969). Stem borer and leaf hopper-plant hopper resistance in rice varieties. *Ent. exp. Appl.* **12,** 789–800
Perkel, D. H. and Bullock, T. H. (1968). Neural coding. *Neurosciences Res. Program. Bull.* **6,** 221–348
Petrysak, A. (1975). The sensory peripheric nervous system in *Periplaneta americana* (Blattoidea) 1. Mouthparts. *Zesz. nauk. U.J. Prace zool.* **393,** 41–83
Pollack, G. S. (1977). Labellar lobe spreading in the blowfly: regulation by taste and satiety. *J. Comp. Physiol. A* **121,** 115–134
Pospíšil, J. (1958). Some problems of the smell of saprophilic flies. *Acta Soc. Entomol. Bohem. (Čsl)* **55,** 316–334
Price, G. M. (1969). Protein synthesis and nucleic acid metabolism in the fat body of the blowfly *Calliphora erythrocephala*. *J. Ins. Physiol.* **15,** 931–944
Rachman, N. J. (1979). The sensitivity of the labellar sugar receptors of *Phormia regina* in relation to feeding. *J. Ins. Physiol.* **25,** 733–740

Rachman, N. J. (1980). Physiology of feeding preference patterns of female black blowflies (*Phormia regina* Meigen). I. The role of carbohydrate reserves. *J. comp. Physiol.* **139,** 59–66

Rees, C. J. C. (1969). Chemoreceptor specificity associated with choice of feeding site by the beetle *Chrysolina brunsvicensis* on its foodplant *Hypericum hirsutum. Ent. exp. Appl.* **12,** 565–583

Rees, C. J. C. (1970). Age dependency of response in an insect chemoreceptor sensillum. *Nature* **227,** 740–742

Reinouts van Haga, H. A. and Mitchell, B. K. (1975). Temperature receptors on the tarsi of the tsetse fly *Glossina morsitans* Westw. *Nature* **255,** 255–256

Rice, M. J. (1970). Cibarial stretch receptors in the tsetse fly (*Glossina austeni*) and the blowfly (*Calliphora erythrocephala*). *J. Ins. Physiol.* **16,** 277–289

Rice, M. J. (1973). Cibarial sense organs of the blowfly, *Calliphora erythrocephala* (Meigen) (Diptera: Calliphoridae). *Int. Insect Morph. Embryol.* **2,** 109–116

Rice, M. J. (1975). Insect mechanoreceptor mechanisms. In: *Sensory Physiology and Behaviour* (R. Galun, P. Hillman, I. Parnas and R. Werman, eds) pp. 135–165. Plenum Press, New York

Rice, M. J. and Finlayson, L. H. (1972). Response of blowfly cibarial pump receptors to sinusoidal stimulation. *J. Ins. Physiol.* **18,** 841–846

Rice, M. J., Galun, R. and Margalit, J. (1973). Mouthpart sensilla of the tsetse fly and their function II. Labial sensilla. *Ann. trop. Med. Parasitol.* **67,** 101–107

Rice, M. J., Galun, R. and Margalit, J. (1973a). Mouthpart sensilla of the tsetse fly and their function III. Labiocibarial sensilla. *Ann. trop. Med. Parasitol.* **67,** 109–116

Robbins, W. E., Thompson, M. J., Yamamoto, R. I. and Shortino, T. J. (1965). Feeding stimulants for the female house fly *Musca domestica* Linneaus. *Science* **147,** 628–630

Rościszewska, M. and Fudalewicz-Niemezyk, W. (1974). The peripheral nervous system of the larva of *Gryllus domesticus* L. (Orthoptera) Part II. Mouthparts. *Acta Biol. Crac. Ser. Zool.* **XVII,** 19–39

Rosenberg, R. (1980). Ovarian control of blood meal retention in the mosquito *Anopheles freeborni. J. Ins. Physiol.* **26,** 477–480

Rowell, C. H. F. (1963). A method of chronically implanting electrodes into the brains of locusts, and some results of stimulation. *J. exp. Biol.* **40,** 271–284

Saxena, K. N. (1967). Some factors governing olfactory and gustatory responses of insects. In: *Olfaction and Taste* (T. Hayashi, ed.) pp. 799–819. Pergamon Press, Oxford

Schoonhoven, L. M. (1967). Chemoreception of mustard oil glycosides in larvae of *Pieris brassicae. Proc. K. ned. Akad. Wet. C.* **70,** 556–568

Schoonhoven, L. M. (1968). Chemosensory bases of host plant selection. *Ann. Rev. Ent.* **13,** 115–136

Schoonhoven, L. M. (1972). Secondary plant substances and insects. *Recent Adv. Phytochem.* **4,** 197–224

Schoonhoven, L. M. (1973). Plant recognition by lepidopterous larvae. In: *Insect Plant Relationships* (H. F. van Emden, ed.) pp. 87–100. Blackwell, Oxford

Schoonhoven, L. M. (1976). On the variability of chemosensory information. *Symp. Biol. Hung.* **16,** 261–266

Schoonhoven, L. M. (1977). Feeding behaviour in phytophagous insects: on the complexity of the stimulus situation. In: *Comportement des Insectes et Milieu Trophique*. Colloques Internationaux du C.N.R.S. No. 265, pp. 391–398. C.N.R.S., Paris

Schoonhoven, L. M. and Dethier, V. G. (1966). Sensory aspects of host plant discrimination by lepidopterous larvae. *Arch. Neerl. Zool.* **16,** 497–530.

Shrihari, T. (1970). Étude quantitive de la consommation et de l'utilisation de la nourriture au cours de la croissance larvaire de *Pieris brassicae* (Lep. Pieridae). *Annls. Soc. ent. Fr.* **6,** 1003-1014.

Siew, Y. C. (1966). Some physiological aspects of adult reproductive diapause in *Galleruca taneceti* (L.) (Coleoptera: Chrysomelidae). *Trans. Roy. Entomol Soc. London* **118,** 359–374

Simpson, S. J. (1981). An oscillation underlying feeding and a number of other behaviours in fifth instar *Locusta migratoria* nymphs. *Physiol. Ent.* **6,** 315–324

Simpson, S. J. (1982). Changes in the efficiency of utilisation of food throughout the fifth instar of *Locusta migratoria* nymphs. *Entomologia exp. appl.* (in press)

Sinoir, Y. (1969). Le rôle des palpes et du labre dans le comportement de prise de nourriture chez la larve du criquet migrateur. *Ann. Nutr. Alim.* **23,** 167–194

Sinoir, Y. (1970). Quelques aspects du comportement de prise de nourriture chez la larve de *Locusta migratoria migratorioides* R. & F. *Ann. Soc. ent. Fr.* **6,** 391–405

Smith, C. C. (1973). The coevolution of plants and seed predators. In: *Coevolution of Plants and Animals* (L. E. Gilbert and P. H. Raven, eds) pp. 53–77. University of Texas Press, Austin

Smith, J. J. B. and Friend, W. G. (1976). Further studies on potencies of nucleotides as gorging stimuli during feeding in *Rhodnius prolixus*. *J. Ins. Physiol.* **22,** 607–611

Srivastava, P. N. and Auclair, J. L. (1974). Effect of amino acid composition on diet uptake by the pea aphid *Acyrthosiphon pisum* (Homoptera, Aphididae). *Can. Ent.* **106,** 149–156

Starre, H. van der, and Ruigrok, T. (1980). Proboscis extension and retraction in the blowfly *Calliphora vicina*. *Physiol. Ent.* **5,** 87–92

Steele, J. E. (1976). Hormonal control of metabolism in insects. *Adv. Insect Physiol.* **12,** 239–324

Stoffolano, J. G. (1968). The effect of diapause and age on the tarsal acceptance threshold of the fly *Musca autumnalis*. *J. Ins. Physiol.* **14,** 1205–1214

Stoffolano, J. G. (1973). Effect of age and diapause on the mean impulse frequency and failure to generate impulses in the labellar chemoreceptor sensilla of *Phormia regina*. *J. Geront.* **28,** 35–39

Stoffolano, J. G. (1974). Control of feeding and drinking in diapausing insects. In: *Experimental Analysis of Insect Behaviour* (L. Barton Browne, ed.) pp. 32–47. Springer Verlag, Berlin

Stoffolano, J. G. and Bernays, E. A. (1980). The post-ecdysial fast in nymphs of *Locusta migratoria*. *Ent. exp. Appl.* **28,** 213–221

Stoffolano, J. G., Damon, R. A. and Desch, C. E. (1978). The effect of age, sex and anatomical position on peripheral responses of taste receptors in blowflies, genus *Phormia* and *Protophormia*. *Exp. Geront.* **13,** 115–124

Strangways Dixon, J. (1961). The relationship between nutrition, hormones and reproduction in the blowfly *Calliphora erythrocephala* (Meigen) II. The effect of removing the ovaries, the corpus allatum and the neurosecretory cells upon selective feeding, and the demonstration of the corpus allatum cycle. *J. exp. Biol.* **38,** 637–646

Sutcliffe, J. F. and McIver, S. B. (1975). Artificial feeding of simuliids (*Simulium venustum*): factors associated with probing and gorging. *Experientia* **31,** 694–695

Sutcliffe, J. F. and McIver, S. B. (1979). Experiments on biting and gorging behaviour in the black fly, *Simulium venustum*. *Physiol. Ent.* **4,** 393–400

Tanton, M. T. (1962). The effect of leaf "toughness" on the feeding of larvae of the mustard beetle. *Ent. exp. Appl.* **5,** 74–78
Thomson, A. J. (1975). Synchronization of function in the foregut of the blowfly *Phormia regina* (Diptera: Calliphoridae) during the crop emptying process. *Can. Ent.* **107,** 1193–1198
Thomson, A. J. (1975a). Regulation of crop contraction in the blowfly *Phormia regina* Meigen. *Can. J. Zool.* **53,** 451–455
Thomson, A. J. and Holling, C. S. (1974). Experimental component analysis of blowfly feeding behaviour. *J. Ins. Physiol.* **20,** 1553–1563
Thomson, A. J. and Holling, C. S. (1975). Experimental component analysis of the feeding rate of the blowfly *Phormia regina* (Diptera: Calliphoridae). *Can. Ent.* **107,** 167–173
Thomson, A. J. and Holling, C. S. (1975a). A model of forgut activity in the blowfly *Phormia regina* Meigen. I. The crop contraction mechanism. *Can. J. Zool.* **53,** 1039–1046
Thomson, A. J. and Holling, C. S. (1976). A model of foregut activity in the blowfly *Phormia regina* Meigen. II. Peristalsis in the crop duct during the crop emptying process. *Can. J. Zool.* **54,** 172–179
Thomson, A. J. and Holling, C. S. (1976a). A model of foregut activity in the blowfly *Phormia regina* Meigen. III. Analysis of crop valve function during the crop emptying process. *Can. J. Zool.* **54,** 1140–1142
Thomson, A. J. and Holling, C. S. (1977). A model of carbohydrate nutrition in the blowfly *Phormia regina* (Diptera, Calliphoridae). *Can. Ent.* **109,** 1181–1198
Thorsteinson, A. J. (1958). The chemotactic influence of plant constituents on feeding by phytophagous insects. *Ent. exp. Appl.* **1,** 23–27
Thorsteinson, A. J. (1960). Host selection in phytophagous insects. *Ann. Rev. Ent.* **5,** 193–218
Tjallingii, W. F. (1978). Mechanoreceptors on the aphid labium. *Ent. exp. Appl.* **24,** 731–737
Tjallingii, W. F. (1980). The functioning of taste in the food selection by aphids. In: *Olfaction and Taste VII* (H. van der Starre, ed.) p. 200. IRL Press, London
Tobe, S. S. and Loughton, B. G. (1967). The development of blood proteins in the African migratory locust. *Can. J. Zool.* **45,** 975–984
Tostowaryk, W. (1972). The effect of prey defense on the functional response of *Podisus modastus* (Hemiptera: Pentatomidae) to densities of the sawflies *Neodiprion swainei* and *N. pratti banksianae* (Hymenoptera: Neodiprionidae). *Can. Ent.* **104,** 61–69
Treherne, J. E. (1957). Glucose absorption in the cockroach. *J. exp. Biol.* **34,** 478–485
Trepte, H.–H., (1980). Delayed vitellogenesis due to larval food deficiency in the house fly. *J. Ins. Physiol.* **26,** 801–806
Urvoy, J., Fudalewicz-Niemezyk, W. and Rościszewska, M. (1978). Contribution à l'étude des organes sensoriels de clypeo-labrum chez *Gryllus domesticus* L. (Orthoptera). *Acta biol. cracov. Ser. Zool.* **XXI**, 57–67
Uvarov, B. (1966). *Grasshoppers and Locusts*, Vol. I. Cambridge University Press, Cambridge
Venkatesh, K. and Morrison, P. E. (1980). Crop filling and crop emptying by the stable fly *Stomoxys calcitrans. Can. J. Zool.* **58,** 57–63
Viscuso, R., Longo, G. and Sottile, L. (1978). Studio comparato della lamina epifaringea degli Ortotteri e di altri ordini affini. *XI Congresso Nazionale Italiano di Entomologia* 1976, 95–103

Walker, P. R., Hill, L. and Bailey, E. (1970). Feeding activity, respiration, and lipid and carbohydrate content of the male desert locust during adult development. *J. Ins. Physiol.* **16,** 1001–1015

Washino, R. K. (1970). Physiological condition of overwintering female *Anopheles freeborni* in California (Diptera: Culicidae). *Ann. ent. Soc. Am.* **63,** 210–216

Wensler, R. J. and Dudzinski, A. E. (1972). Gustation of sugars, amino acids and lipids by larvae of the scarabeid *Sericesthis geminata* (Coleoptera). *Ent. exp. Appl.* **15,** 155–165

Wensler, R. J. and Filshie, B. K. (1969). Gustatory sense organs in the food canal of aphids. *J. Morph.* **129,** 473–492

Wiepkema, P. R. (1971). Positive feedbacks at work during feeding. *Behaviour* **39** 266–273

Wigglesworth, V. B. and Gillett, J. D. (1934). The function of antennae in *Rhodnius prolixus*: confirmatory experiments. *J. exp. Biol.* **11,** 408

Wilczek, M. (1967). The distribution and neuroanatomy of the labellar sense organs of the blowfly *Phormia regina* Meigen. *J. Morph.* **122,** 175–201

Williams, L. H. (1954). The feeding habit and food preferences of Acrididae and the factors which determine them. *Trans. Roy. Entomol. Soc. London* **105,** 423–454

Winstanley, C. and Blaney, W. M. (1978). Chemosensory mechanisms of locusts in relation to feeding. *Ent. exp. Appl.* **24,** 750–758

Biology of Eye Pigmentation in Insects

K. M. Summers,[1] A. J. Howells[1] and N. A. Pyliotis[2]

[1]*Department of Biochemistry and* [2]*the Electron Microscopy Unit, Australian National University, Canberra, Australia*

1 Introduction

The colour of insect eyes is determined largely by the nature of the screening pigments which they contain. These pigments are not involved in light detection but they function to isolate optically each facet of the compound eye from its neighbours and thus enhance contrast sensitivity and visual acuity. Investigations of the screening pigments and of eye colour mutants in a variety of insect species began almost fifty years ago. Indeed studies of the eye colour mutants of *Drosophila melanogaster* and *Ephestia kühniella* in the 1930s and 1940s, provided the foundation on which our present understanding of the relationship between genes and enzymes is based. Summaries of the early literature can be found in reviews by Ephrussi (1942) Butenandt (1952), Ziegler (1961), Ziegler and Harmsen (1969) and Linzen (1974).

This chapter concentrates primarily on developments since 1974 and draws largely on information gained from the study of dipteran species, although reference will be made to other insect species where appropriate. The first section describes ultrastructural information concerning the nature of the dipteran peripheral retina and the location of the screening pigments within the eyes. In the second and third sections, the biochemistry of the two types of compound which function as screening pigments in insect eyes, the ommochromes (derived from the amino acid tryptophan) and the pteridines (derived from guanine nucleotides) is considered. We discuss the biosynthetic pathways which lead to the production of these two classes of screening pigment and also the characteristics of those eye colour mutants, in various insect species, which lack a specific enzyme of the pathways. In the final sections, the interrelationships between the ommochrome and pteridine pathways and the possible functions of genes which affect the production of both types of screening pigment are considered in detail.

2 Ultrastructure of the peripheral retina

Since almost all insects have compound eyes, we will consider only the anatomy of the compound eye, concentrating on the peripheral retina (also known as the retina, photoreceptor layer or ommatidial layer; see Fig. 1). Reviews on the insect eye in which various aspects of the compound eye have been treated in detail include those by Bullock and Horridge (1965), Trujillo-Cenóz (1972) and Carlson and Chi (1979). Many terms have been used over the years to describe the different regions and features of the compound eye. No attempt will be made here to rationalise this nomenclature although alternative names used in the literature will be presented.

2.1 GENERAL ANATOMY OF THE COMPOUND EYE

The compound eye may for convenience be subdivided into two parts. One, the peripheral retina, is made up mainly of nonnervous tissue while the other consists mainly of nerve fibres and contains the three optic neuropile masses. These two parts are separated by a basement membrane (strictly a basal lamina).

In simple functional terms, the peripheral retina receives optical information on such parameters as movement, direction, contrast and spectral composition and transmits it electrically to the optic neuropile masses. The optic neuropile masses in their turn integrate the information and transmit it to the protocerebrum of the brain for analysis (Bullock and Horridge, 1965; Laughlin, 1975).

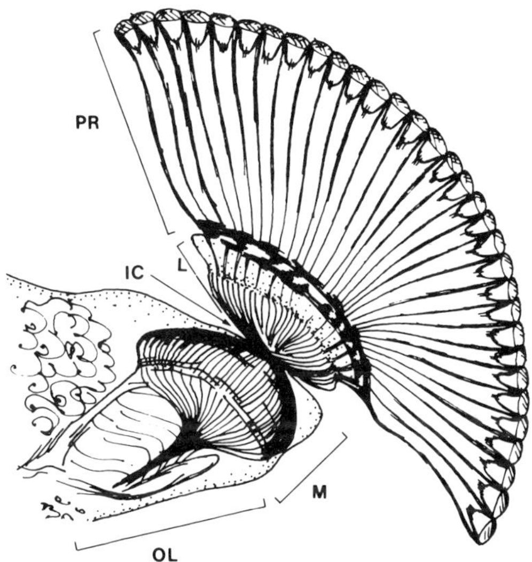

Fig. 1 Diagram of a dipteran compound eye showing the peripheral retina (PR), the lamina (L), the intermediate chiasma (IC), the medulla (M), and the optic lobe (OL)

The outermost of the three optic neuropile masses is called the lamina (also termed the intermediate retina, lamina ganglionaris, first optic or synaptic neuropile layer). The lamina is the initial site of visual integration and is the best characterised, both functionally and morphologically, of the three masses (Laughlin, 1975). The next neuropile mass is labelled as the medulla (or deep retina or second optic neuropile) and the third is known as the lobula (or optic lobe or third optic neuropile). The lamina is joined to the medulla by the optic nerve (intermediate chiasma or first optic chiasma) and the medulla in its turn is joined to the lobula via the second optic chiasma. These areas of the compound eye are illustrated diagrammatically in Fig. 1. As yet, there is no clear understanding of the functional, or for that matter morphological, integration of the neuropile masses with the proto-cerebrum (see reviews cited above and Meyerowitz and Kankel, 1978).

2.2 PERIPHERAL RETINA

The peripheral retina of a compound eye is composed of closely packed units termed facets or ommatidia. The number of ommatidia in different insect species varies considerably, from a few in more primitive groups to several tens of thousands in some moths and beetles (references cited in Trujillo-Cenóz, 1972; Carlson and Chi, 1979). Functionally, the ommatidium consists of two parts, one (the dioptric apparatus) being involved with image formation and the other (the photosensitive modified receptor neurons) with the reception of light transmitted through the dioptric apparatus and its conversion to electrical impulses. Morphologically, however, the ommatidium consists of three major components, the dioptric apparatus, the photosensitive retinula cells and the pigment cells. The organisation of these morphological components in the ommatidium is shown in Figs 2–5 and is considered in more detail below.

2.3 DIOPTRIC APPARATUS

This comprises an external corneal lens (cornea, lenslet or lens) which is a modified, transparent portion of insect exoskeleton and various subcorneal components. The nature of the subcorneal components is variable and has been used to classify compound eyes into four types; those with a eucone (most compound eyes), those with a pseudocone (the eyes of most flies), those with an acone and those with an exocone (for detailed descriptions see the reviews cited earlier). Generally, a structure called the crystalline cone (replaced by the pseudocone in most flies) is located directly beneath the corneal lens and rests on four Semper cells (also termed cone cells). The Semper cells serve to couple mechanically and optically the dioptric apparatus to the retinula cells.

2.4 RETINULA CELLS

Two major regions can be identified in the retinula cells (also called sense, sensory, visual, receptory, central or photoreceptor cells). The first, the somatic part, is associated with the peripheral retina and is surrounded by pigment cells. The second, the axon part, is contiguous with the first and is associated with the lamina. Insect ommatidia usually have eight retinula cells per ommatidium, collectively termed the retinula.

Retinula cells possess a unique microvillar arrangement of the plasma membrane forming a structure called the rhabdomere. This is viewed as the fundamental photosensitive component of the retinula cell with which the visual pigment rhodopsin is intimately associated (reviewed by Eakin, 1972; Carlson and Chi, 1979). Rhabdomeres in insect ommatidia are usually apposed, although in flies they are commonly not in contact with one another. Collectively the rhabdomeres of an ommatidium are called the rhabdom. The fly retinula has a central cavity into which the microvilli of the eight rhabdomeres project. The tip of each rhabdomere projects beyond the retinula cell into a cavity in the Semper cells and it is through this connection that the retinula cells are linked, both optically and mechanically, to the dioptric apparatus. The visual pigment rhodopsin probably does not contribute significantly to the colour of compound eyes (Langer, 1975). Retinula cells do very occasionally contain coloured granules, presumably of screening pigment (Eakin, 1972), although almost nothing is known about their function or composition.

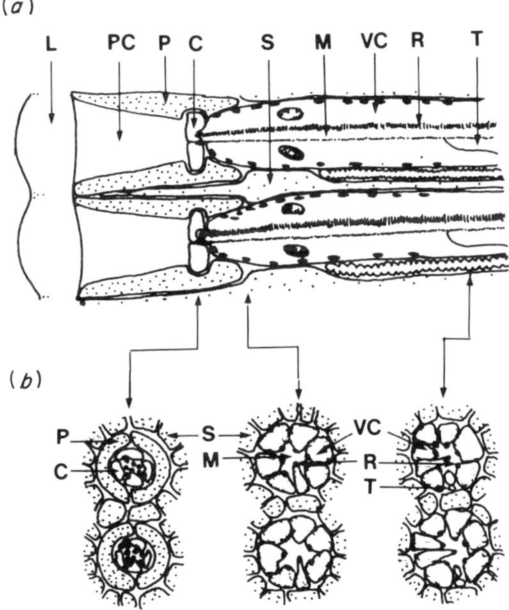

Fig. 2 Diagrams of two ommatidia from a dipteran peripheral retina. (*a*) A longitudinal section, and (*b*) cross-sections, at the different levels indicated by the arrows, showing crystalline lens (L), pseudocones (PC), primary pigment cells (P), cone cells (C), secondary pigment cells (S), central matrix (M), visual (retinula) cells (VC), rhabdomeres (R) and tracheae (T). (Redrawn from diagrams in Stavenga, 1975)

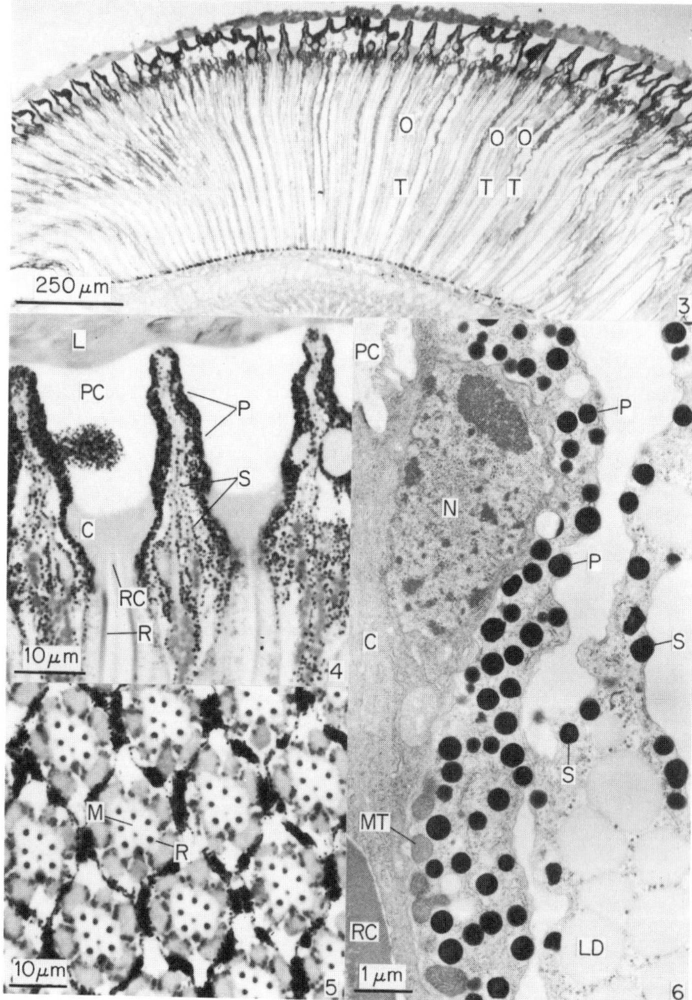

Figs 3–6 Light (Figs 3–5) and electron (Fig. 6) micrographs of structures seen in the eyes of wild type *Lucilia*. Fig. 3 represents a longitudinal section through the peripheral retina showing the columnar arrangement of ommatidia (O) and associated tracheae (T). Fig. 4 illustrates part of Fig. 3 showing the osmiophilic granules associated with the primary (P) and secondary (S) pigment cells and additionally showing the corneal lens (L), pseudocone (PC), cone cell (C), rhabdomere (R) and associated retinular cap (RC). Fig. 5 represents a cross-section through the peripheral retina showing the asymmetric arrangement of the rhabdomeres (R) around the central matrix (M) of each ommatidium. Fig. 6 illustrates part of Fig. 4 at the level of the cone cell (C) additionally resolving supposed lipid droplets (LD), mitochondria (MT) and a cone cell nucleus (N)

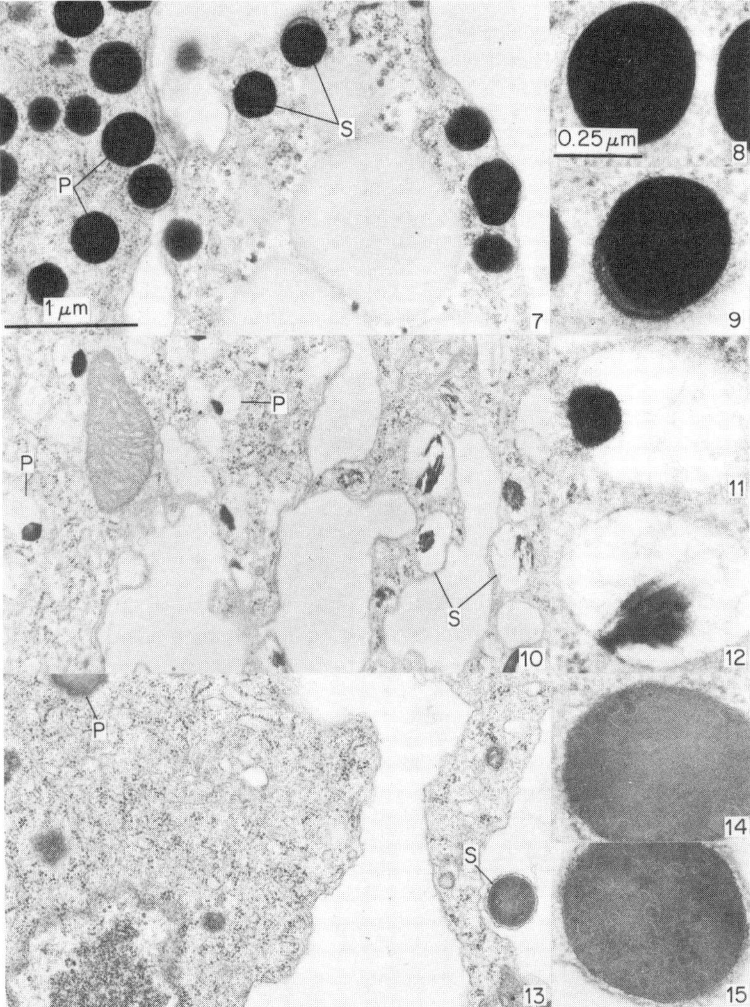

Figs 7–15 Electron micrographs showing the ultrastructure of granules in primary and secondary pigment cells in the eyes of *Lucilia*. Figs 7, 10 and 13 represent longitudinal sections, at the level of the cone cells, through the pigment cells (see Fig. 6). Granules in the primary pigment cells are designated P in these figures and are illustrated at higher magnification in Figs 8, 11 and 14. Granules in secondary pigment cells are designated S in Figs 7, 10 and 13 and are shown at higher magnification in Figs 9, 12 and 15. Figs 7–9 illustrate the eyes of wild type, with round, electron opaque, membrane-bound Type I granules in primary pigment cells and less regularly shaped, membrane-bound Type II granules (with a lighter staining region at the periphery) in secondary pigment cells. Figs 10–12 show the almost electron transparent nature of the granules found in the mutant *yellow*. The Type III granules of the primary pigment cells have a small amount of uniformly electron opaque material usually at one end, while the granules in secondary pigment cells have a more fibrous substructure. Figs 13–15 show the sparsely scattered Type IV granules with grainy substructure in both primary and secondary pigment cells of the mutant *white* of *Lucilia*. Magnification of Figs 10 and 13 as Fig. 7; magnification of Figs 9, 11, 12, 14 and 15 as Fig. 8

2.5 PIGMENT CELLS AND PIGMENT GRANULES

Two types of pigment cell (modified sheath cells) are commonly identified in insect eyes, called primary and secondary pigment cells. Primary pigment cells (also called distal or iris pigment cells, primary iris cells or corneal pigment cells) surround the dioptric apparatus. There are two such cells per ommatidium. Secondary pigment cells (proximal or retinal pigment cells, secondary iris cells, accessory cells or large pigment cells) surround both the retinula cells and the primary pigment cells. They extend from the corneal lens to the basement membrane of the peripheral retina. There are six or more secondary pigment cells per ommatidium. Most of the cytological work on screening pigments (reviewed by Langer, 1975) has centered on these two types of pigment cells. However, a third type of pigment cell, the basal pigment cell, has occasionally been described (Bullock and Horridge, 1965; Trujillo-Cenóz, 1972). The function of basal pigment cells is unclear although they also contain pigment granules which may well regulate the quantity of light reflected from the refractile tracheae at the level of the basement membrane. The known optical role of tracheae is reviewed in Carlson and Chi (1979).

Conclusions regarding the cellular and subcellular localisation of the two classes of screening pigments associated with pigment cells have been reached through the use, where practicable, of microspectrophotometric methods of examining granules *in situ*. Additional electron microscopic support for the subcellular localisation of screening pigments comes from the examination of eye colour mutants lacking (or having low levels of) one class of pigment but containing normal levels of the other (Shoup, 1966; Fuge, 1967; Langer, 1967; Schwabl and Linzen, 1972; Summers, 1979; Pyliotis and Summers, 1980). All the available information suggests that ommochromes always occur in the secondary pigment cells and are apparently usually found in membrane-bound pigment granules (reviewed in Langer 1975; see also Schneider *et al.*, 1978; Pyliotis and Summers, 1980). A similar generalised statement cannot as yet be made about the presence of ommochromes in primary pigment cells although a number of flies, including *D. melanogaster* (Shoup, 1966; Fuge, 1967), *Calliphora erythrocephala* (Langer, 1967, 1975) and *Lucilia cuprina* (Pyliotis and Summers, 1980) certainly appear to contain membrane-bound ommochrome (xanthommatin) in their primary pigment cells. These xanthommatin-containing granules are called Type I granules by Shoup (1966). Figure 6 is an electron micrograph showing the ultrastructure of the primary and secondary pigment cells in *Lucilia*. Electron-opaque, membrane-bound, ommochrome granules can be seen in both types of pigment cell. Higher magnifications of these granules are given in Figs 7–9. Whereas the ommochromes appear to be associated with granules, the pteridines usually occur in solution in the cytoplasm (see Langer, 1975). A notable exception to this general statement seems to apply in *D. melanogaster* in which the major pteridine eye pigments (the drosopterins) also appear to occur in membrane-bound granules (Shoup, 1966; Fuge, 1967). These membrane-bound drosopterin granules (called Type II granules by Shoup, 1966) were claimed to occur only in the secondary pigment cells of this insect.

2.6 PIGMENT GRANULES IN EYE COLOUR MUTANTS

Several electron microscopic studies have been conducted on pigment-deficient mutants of *D. melanogaster* (Shoup, 1966; Fuge, 1967; Schwabl and Linzen, 1972). The consistent finding in these studies has been that mutants devoid of both ommochrome and pteridine pigments (*white* and the double mutants *vermilion; brown* and *cinnabar; brown*) lack recognisable pigment granules in their pigment cells. However, in the pigment cells of mutants lacking only ommochrome (e.g. *vermilion*), there are almost empty, membrane-bound bodies which are about the same size as pigment granules. These empty granules have been called Type III granules by Shoup (1966). Similar empty granules have also been observed in the ommochrome-deficient *white* mutant of *Calliphora* (Langer, 1967).

Recent electron microscopic studies on the eye colour mutants of *Lucilia* (Summers, 1979; Pyliotis and Summers, 1980) have provided similar results. The ommochrome-deficient mutants (*yellow*, *yellowish* and *topaz*) all contain empty granules in their pigment cells, whereas the pigment cells of *white* (which are devoid of both ommochrome and pteridine pigments) have no recognisable pigment granules. Electron micrographs showing the ultrastructure of pigment cells from the *yellow* and *white* mutants of *Lucilia* are given in Figs 10–12 and 13–15 respectively. The large, irregularly shaped, sparsely-scattered structures with a granular substructure, which seem to replace the pigment granules in *white*, (Figs 13–15) have been called Type IV granules by Shoup (1966). This author has suggested that these are sites of breakdown of materials not used in pigment synthesis. Evidence that the empty granules seen in *yellow* (Figs 10–12) are those involved in ommochrome (xanthommatin) deposition has been presented by Pyliotis and Summers (1980). The empty granules become electron opaque and more normal in appearance when this mutant receives a dietary supplement of 3-hydroxykynurenine (the final intermediate in the xanthommatin pathway—see Fig. 16). Such a dietary supplement brings about the deposition of some xanthommatin in this mutant (Summers and Howells, 1978).

Although the amount of information available is still obviously limited, it does seem to be a consistent finding that the ultrastructure of the pigment cells can be altered in two different ways in ommochrome-deficient mutants, depending on the presence or absence of pteridine pigments. When pteridines are also lacking, no recognisable pigment granules are observed, however when pteridines are present empty granules are seen. This interaction between ommochrome and pteridine pigment production is apparent at the biochemical as well as at the ultrastructural level, as will be discussed in detail later in this review.

3 The ommochrome biosynthetic pathway

The main ommochrome screening pigment found in dipteran eyes is xanthommatin. The series of oxidation steps involved in xanthommatin biosynthesis is shown in Fig. 16 and the chemical structures of the compounds involved in this pathway are given in Fig. 17. Each step is catalysed by a different enzyme and, in any insect species, each enzyme is probably coded for at a distinct and different genetic locus. Hence, in the species which have been

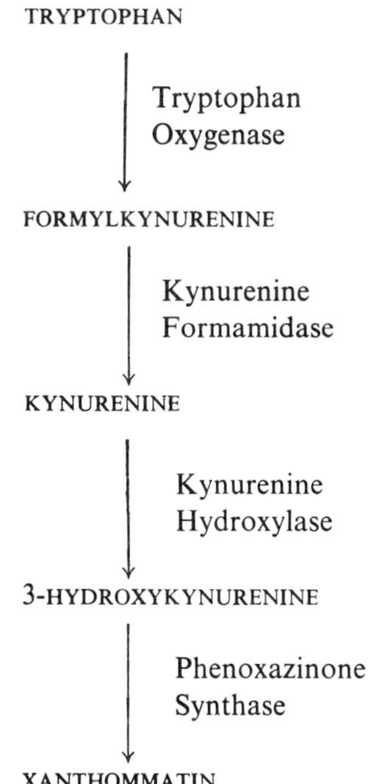

Fig. 16 The xanthommatin biosynthetic pathway in insects

well studied, mutations at several different loci have been found to cause deficiencies in xanthommatin production. Those mutants which lack the activity of a pathway enzyme frequently accumulate the substrate of the missing enzyme and, in holometabolous insects, this accumulation occurs especially in the period between pupariation and adult emergence. Intermediates accumulated during this period are often excreted in the meconium soon after emergence (Gilmour, 1961; Dustmann, 1968; Bonse, 1969; Howells and Ryall, 1975). Some conversion to other minor metabolites, such as transamination products of kynurenine and 3-hydroxykynurenine (kynurenic acid and xanthurenic acid respectively), can also occur in mutants which accumulate high levels of pathway intermediates (Danneel and Zimmerman, 1954; Leibenguth, 1970, 1971).

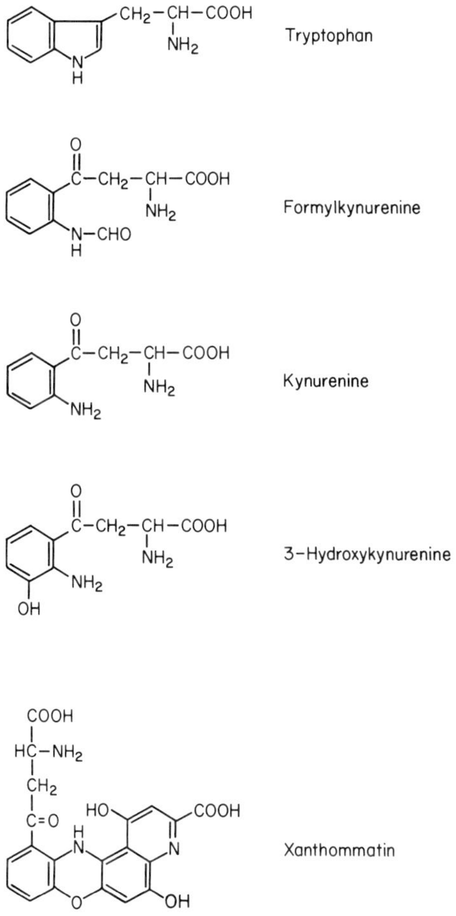

Tryptophan

Formylkynurenine

Kynurenine

3-Hydroxykynurenine

Xanthommatin

Fig. 17 Chemical structures of compounds involved in the xanthommatin biosynthetic pathway

3.1 TRYPTOPHAN OXYGENASE

The first enzyme of the xanthommatin biosynthetic pathway, tryptophan oxygenase, catalyses the conversion of tryptophan to formylkynurenine. Mutants lacking tryptophan oxygenase activity, which characteristically build up elevated levels of tryptophan and have an absence of subsequent intermediates of the pathway, include *vermilion* of *D. melanogaster* (Green, 1949; Baglioni, 1959), *green* of the house fly *Musca domestica* (Milani, 1975), *yellowish* of the blowfly *Lucilia* (Summers and Howells, 1978) and ivory of the flesh fly *Sarcophaga barbata* (Trepte, 1978) as well as *a* of the meal moth, *Ephestia* (Caspari and Gottlieb, 1975) and *snow* of the honey-bee, *Apis mellifera* (Green, 1955; Dustmann, 1968, 1975).

Vermilion (*v*) mutants of *D. melanogaster* have been shown to be able to produce brown pigment if subsequent pathway intermediates (i.e. kynurenine or 3-hydroxykynurenine) are provided to larvae by feeding (Beadle and Law, 1938; Green, 1952), injection (Ephrussi, 1942), transplantation of the developing eye into a wild type host (Beadle and Ephrussi, 1936) or genetic mosaicism (Sturtevant, 1932). The mutants are thus said to have a nonautonomous phenotype. Similarly, *yellowish* of *Lucilia* is nonautonomous, since these insects make xanthommatin when the larvae are raised on a diet containing kynurenine or 3-hydroxykynurenine (Summers and Howells, 1978). The *a* mutant of *Ephestia* is also nonautonomous (Caspari, 1933); in fact the eyes of adults acquire brown pigment even if the adult moth is injected with kynurenine after emergence (Hadorn and Kühn, 1953; Kühn and Egelhaaf, 1955). The best evidence that the structural gene for tryptophan oxygenase is located at the *v* locus in *D. melanogaster* is the finding that the activity of the enzyme is proportional to the number of v^+ alleles present in the genome (Tobler *et al.*, 1971).

Tryptophan oxygenase occurs in the soluble fraction of the cell. Its activity is widely reported to show a sigmoidal response to increasing substrate concentrations in microbial (Maeno and Feigelson, 1967), mammalian (Badaway, 1977) and insect (Baillie and Chovnick, 1971; Schartau and Linzen, 1976) preparations. In species where the pathway has been studied in detail, the activity of tryptophan oxygenase is lower *in vitro* than that of any other enzyme of the pathway, so that this enzyme probably controls the rate of flux of tryptophan through the pathway (Ryall and Howells, 1974).

Detailed developmental profiles of tryptophan oxygenase activity have been published for the blowfly *Protophormia terrae-novae* (Linzen and Schartau, 1974) and for *D. melanogaster* (Tobler, 1975; Sullivan and Kitos, 1976). It is difficult to compare the two profiles directly, since one is expressed as activity per mg protein while the other is expressed on a per insect basis. However, the general features do seem to be similar. Activity is present throughout larval life, with a peak in activity at about the time of pupariation, after which it declines until adult emergence. During adult life the activity rises again reaching a peak several days after emergence. As Sullivan and Kitos (1976) point out, the principal biological function of this enyzme is probably concerned with degrading excess tryptophan, which can either be excreted (larvae and adults), stored as kynurenine or 3-hydroxykynurenine for later excretion, or converted to xanthommatin. The peaks of activity at pupariation (when the histolysis of larval tissues and the breakdown of larval protein is occurring) and in mature adults probably coincide with periods when the level of free tryptophan is rising.

Tryptophan oxygenase activity and the *v* mutants of *D. melanogaster* have been studied intensively, due to the presence in the genome of this organism of a suppressor gene. Mutations at this suppressor locus partially restore tryptophan oxygenase activity in some *v* mutants. This topic is outside the scope of this review but is discussed in detail in reviews by Linzen (1974), MacIntyre and O'Brien (1976), and O'Brien and MacIntyre (1978).

3.2 KYNURENINE FORMAMIDASE

The next enzyme of the xanthommatin pathway, kynurenine formamidase, catalyses the conversion of formylkynurenine to kynurenine. In all dipteran species studied so far, no mutant has been found which is blocked solely at this step (Glassman, 1956; Linzen, 1974; Summers and Howells, 1978), although Grigolo (1969) found that the tryptophan oxygenase-deficient mutant *green* of *Musca* also lacked kynurenine formamidase and that the mutant *carnation* had a much reduced activity of this enzyme. There are a number of possible explanations for the apparent rarity of kynurenine formamidase mutants. Since the spontaneous conversion of formylkynurenine to kynurenine occurs relatively rapidly (Linzen, 1974), it is possible that the enzyme activity may not be necessary *in vivo* for ommochrome formation. However, the fact that kynurenine formamidase has been found in all species tested tends to suggest that this enzyme is needed. Alternatively, the enzyme could catalyse some other essential metabolic reaction so that mutants lacking this activity may not be viable. However, there is no evidence to support this proposal. A more attractive explanation emerged from the recent work of Moore and Sullivan (1978). They found that two distinct genetic loci code for different functional types of kynurenine formamidase in *D. melanogaster*, so that the occurrence of a mutation at both loci (which is unlikely) would be necessary before a mutant phenotype would be expressed. It is interesting to note that in *Musca* there is apparently only one type of kynurenine formamidase (Moore and Sullivan, 1975), and mutants with reduced kynurenine formamidase activity have been found, as mentioned above.

Kynurenine formamidase, like tryptophan oxygenase, is found in the soluble fraction of the cell. Its activity in crude cell extracts is generally 50–100 times greater than that of tryptophan oxygenase (Linzen and Schartau, 1974; Sullivan and Kitos, 1976; Summers and Howells, 1978). Detailed developmental profiles of activity are available for *Protophormia* (Linzen and Schartau, 1974) and for *D. melanogaster* (Sullivan and Kitos, 1976). The essential features of the two profiles are similar. Activity is present throughout larval life, reaches a peak at around pupariation, declines somewhat during the first half of the period of development in the puparium but then rises again so that a second peak of activity is attained in early adult life. Activity then declines as the adults age. Sullivan and Kitos (1976) have noted that this profile (like that for tryptophan oxygenase) is superimposable on those for a number of enzymes of general metabolic function in *D. melanogaster.* In comparison with the profile for tryptophan oxygenase in this species, the profile for kynurenine formamidase peaks earlier both in late larval and in early adult life.

3.3 KYNURENINE HYDROXYLASE

In the third step of the pathway, the conversion of kynurenine to 3-hydroxy-kynurenine is catalysed by kynurenine hydroxylase. The structural gene for this enzyme in *D. melanogaster* is apparently located at the *cinnabar* (*cn*) locus. *Cinnabar* mutants accumulate kynurenine (Green, 1949) and extracts from these mutants fail to convert kynurenine to 3-hydroxykynurenine *in vitro* (Ghosh and Forrest, 1967; Sullivan *et al.*, 1963). The *cn* phenotype is nonautonomous when 3-hydroxykynurenine is provided in the diet (Beadle and Law, 1938; Schwabl and Linzen, 1972), by injection (Ephrussi, 1942) or by transplantation of the eye disc into a wild type host (Beadle and Ephrussi, 1936). Dose dependence studies have shown that the activity of kynurenine hydroxylase is proportional to the number of cn^+ genes present in the genotype (Sullivan *et al.*, 1973). The equivalent gene in *Musca* is *ocra* (Milani, 1975), in *Lucilia* is *yellow* (Summers and Howells, 1978) and in *Apis* is *ivory* (Dustmann, 1968, 1975).

Kynurenine hydroxylase has been studied recently in some detail in three insect species, the flies *D. melanogaster* (Sullivan *et al.*, 1973, 1974) and *Protophormia* (Linzen and Schartau, 1974) and the silkworm *Bombyx mori* (Ogawa and Hasegawa, 1975). In *D. melanogaster* and in *Bombyx* the enzyme was found to be located in particulate fractions which sedimented between 10 000 and 100 000 g (Sullivan *et al.*, 1974; Ogawa and Hasegawa, 1975). It is interesting to note that in yeast (Bandlow, 1972), in fungi (Cassady and Wagner, 1971) and in rat liver (Okamoto *et al.*, 1967), kynurenine hydroxylase has been found to be associated with the outer mitochondrial membrane. It seems probable that this enzyme has a similar subcellular location in insects.

Developmental profiles have been obtained for kynurenine hydroxylase activity in both *D. melanogaster* and *Protophormia*. In *D. melanogaster* during larval life, activity reaches a peak in early third instar but then falls to low levels at pupariation (Sullivan *et al.*, 1973). About 24 hours after pupariation in this species the level of activity begins to rise sharply to reach a peak at about 60 hours (Sullivan *et al.*, 1973; Ryall and Howells, 1974). This peak of kynurenine hydroxylase activity coincides with the rapid phase of xanthommatin deposition in the eyes (Ryall and Howells, 1974). After the peak, activity declines and is quite low again by the time of adult emergence. As discussed earlier, the profiles in *D. melanogaster* for the first two enzymes of the pathway (tryptophan oxygenase and kynurenine formamidase) are quite similar to those of general metabolic enzymes, suggesting that their major role concerns the overall degradation of excess tryptophan. By contrast, the developmental variations in kynurenine hydroxylase activity can be related readily to features of xanthommatin biosynthesis, suggesting that this enzyme is synthesised specifically for the production of this pigment.

The profile obtained for *Protophormia* (Linzen and Schartau, 1974) is quite unlike that for *D. melanogaster*. Activity appears during the second half of larval life, reaches a maximum at the end of the feeding period, but then falls rapidly so that its activity at pupariation is quite low. Activity remains low throughout adult development and disappears after emergence. These variations do not correlate well either with the levels of 3-hydroxykynurenine during development, or with the onset of xanthommatin synthesis. Further studies on the kynurenine hydroxylases of blowflies are needed, particularly in view of the unsuccessful attempts by Summers and Howells (1978) to detect the activity of this enzyme in extracts of *Lucilia*.

3.4 XANTHOMMATIN FORMATION

The final enzyme of the pathway, phenoxazinone synthase, catalyses the conversion of 3-hydroxykynurenine to xanthommatin. (Phillips *et al.*, 1973; Yamamoto *et al.*, 1976). This has proved to be a difficult enzyme to study *in vitro* because reactions it catalyses can occur both spontaneously in the presence of oxygen and be stimulated non-specifically by other enzymes (Ryall *et al.*, 1976). In *D. melanogaster*, the phenoxazinone synthase involved in xanthommatin biosynthesis is particulate, sedimenting at 30 000 g, and in adults is located almost entirely in the head (Phillips *et al.*, 1973; Yamamoto *et al.*,

1976). This activity seems likely to be associated with the pigment granules on which xanthommatin is deposited (see section 2). A similar particulate enzyme has also been found in *Lucilia* (Summers and Howells, 1978).

A developmental profile for the particulate phenoxazinone synthase activity has been obtained for *D. melanogaster* (Yamamoto *et al.*, 1976). Activity first appears midway through the period of development in the puparium and then rises in two distinct phases to reach a maximum about one day before adult emergence. There is then little change in activity over the period of emergence and early adult life. The time at which activity of phenoxazinone synthase is first detected 46–48 h after pupariation) coincides precisely with the appearance of xanthommatin in this species (Ryall and Howells, 1974), suggesting that the onset of xanthommatin biosynthesis is brought about by the synthesis or activation of this final enzyme. A similar developmental profile has been obtained for this enzyme in *Lucilia* (Summers, 1979).

Mutants blocked in the final step of the pathway would be expected to accumulate excessive amounts of 3-hydroxykynurenine. In *D. melanogaster* there are several mutants which have elevated levels of this metabolite (Howells *et al.*, 1977). Among these, the mutant *cardinal* appears to be the most likely to be defective primarily in this final step, since it only begins to accumulate 3-hydroxykynurenine late in adult development, after the normal time of onset of xanthommatin synthesis. Particulate fractions from *cardinal* do show much reduced levels of phenoxazinone synthase activity compared to wild type. However, extracts from other xanthommatin-deficient mutants such as *vermilion*, *cinnabar*, *scarlet* and *white* also lack phenoxazinone synthase activity *in vitro* (Phillips *et al.*, 1973; Yamamoto *et al.*, 1976). The mechanisms by which these mutations affect phenoxazinone synthase activity have yet to be established, but it seems likely that regulatory mechanisms operate in the pigment cells to ensure that the final enzyme is only activated when all other steps leading to xanthommatin formation are operative. Reduced levels of phenoxazinone synthase activity have also been reported for several xanthommatin-deficient mutants of *Lucilia* (Summers and Howells, 1978) and only in the mutant *tangerine* of this species does it seem that the blockage of the pathway occurs specifically at the final step. It is possible that other mutants of *D. melanogaster* which accumulate excess 3-hydroxykynurenine (such as *karmoisin*) are also blocked in the final step. It should be noted that the conversion of 3-hydroxykynurenine to xanthommatin is a complex, multistep reaction (Bolognese and Scherillo, 1974; Ryall *et al.*, 1976) and therefore may involve more than one protein.

With regard to nondipteran species, several mutants of *Apis* fail to convert 3-hydroxykynurenine to xanthommatin (*pearl*, *cream* and *chartreuse*).

However, Dustmann (1968, 1975) has suggested that only *chartreuse* specifically lacks the final enzyme. He proposes that the other two mutants fail at this step because they lack the capacity to bind 3-hydroxykynurenine to particles within the pigment cells. Cölln and Klett (1978) have found that mutants at the *alb* locus of *Ephestia* fail to convert 3-hydroxykynurenine to xanthommatin and may therefore be deficient in the final enzyme.

3.5 OTHER OMMOCHROME PIGMENTS

In most insect species, xanthommatin is the only ommochrome pigment found in the eyes. Notable exceptions are the species of Orthoptera, where ommins and ommidins have been reported to be attached to pigment granules in the eyes. These species rely largely on ommochrome pigments for body colouration and the eyes tend to have the same colour as the body. In the honey-bee *Apis*, ommins as well as xanthommatin have been reported to be attached to granules in the eyes (Dustmann, 1968, 1975). The bee has no coloured pteridine pigments in its eyes and hence additional ommochrome screening pigments may have evolved to replace these. In other groups, such as the Lepidoptera, ommochromes are used for the distinctive markings of the body, as well as being found in the eyes. The nature and distribution of these other ommochrome pigments is discussed in detail in the review by Linzen (1974).

3.6 EVOLUTIONARY ASPECTS OF XANTHOMMATIN SYNTHESIS IN DIPTERA

In all dipteran species which have been examined, the accumulation of xanthommatin in the eyes commences virtually midway between pupariation and adult emergence (Table 1). Similarly, the level of xanthommatin at

TABLE 1 Onset of xanthommatin biosynthesis in dipteran species

Species	Onset of xanthommatin synthesis[a]	Length of life in the puparium[b]	References
D. melanogaster	48–50 h	96 h	Schultz (1935); Ephrussi (1942); Shoup (1966); Phillips & Forrest (1970); Ryall & Howells (1974)
D. virilis	72 h	136 h	Felton & Howells (unpublished)
Calliphora	4–5 days	9 days	Linzen (1963)
Protophormia	2–3 days	5 days	Linzen & Schartau (1974)
Lucilia	3–4 days	7 days	Summers & Howells (1978)

[a]Time after pupariation [b]Time from pupariation to adult eclosion

emergence in these species (when expressed on a live weight basis) is remarkably constant, in spite of differences between species in size and in natural habitat (Table 2). This conservation during dipteran evolution both of the time of onset of xanthommatin synthesis and also of the relative level per insect emphasises the crucial role played by this screening pigment in vision in this group of insects.

TABLE 2 Xanthommatin level at emergence in dipteran species

Species	Weight/insect[a]	Xanthommatin at emergence[b]	Reference
D. melanogaster	1	1.4	Ryall & Howells (1974)
D. virilis	2	1.2	Felton & Howells (unpublished)
Lucilia	25	1.5	Summers & Howells (1978)
Protophormia	40	1.7	Linzen & Schartau (1974)
Calliphora	120	1.5	Linzen (1963)

[a]Approximate fresh weight of an adult insect, expressed in mg
[b]Expressed as $\mu mol \cdot g^{-1}$ fresh weight; calculated from figures given in references where necessary

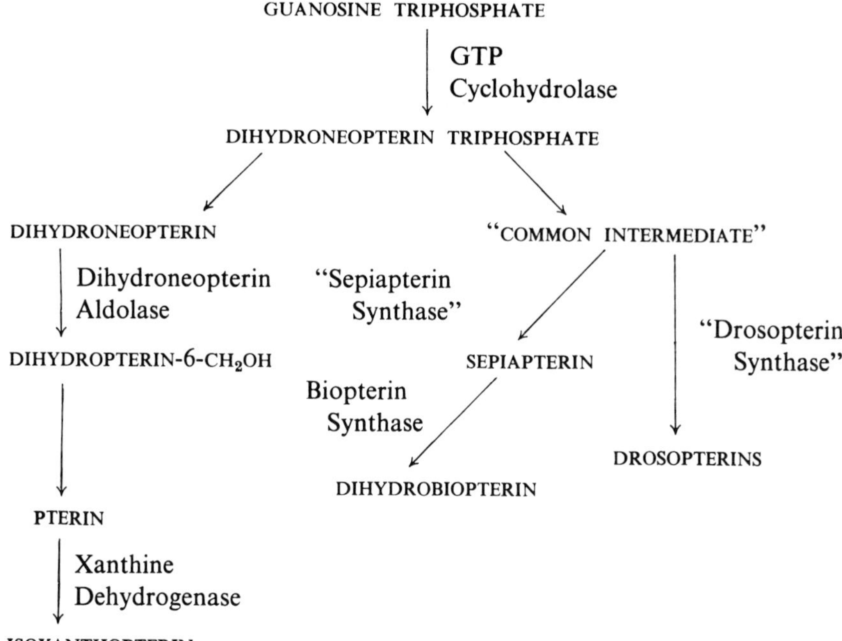

Fig. 18 Proposed branched pteridine biosynthetic pathways in *D. melanogaster*

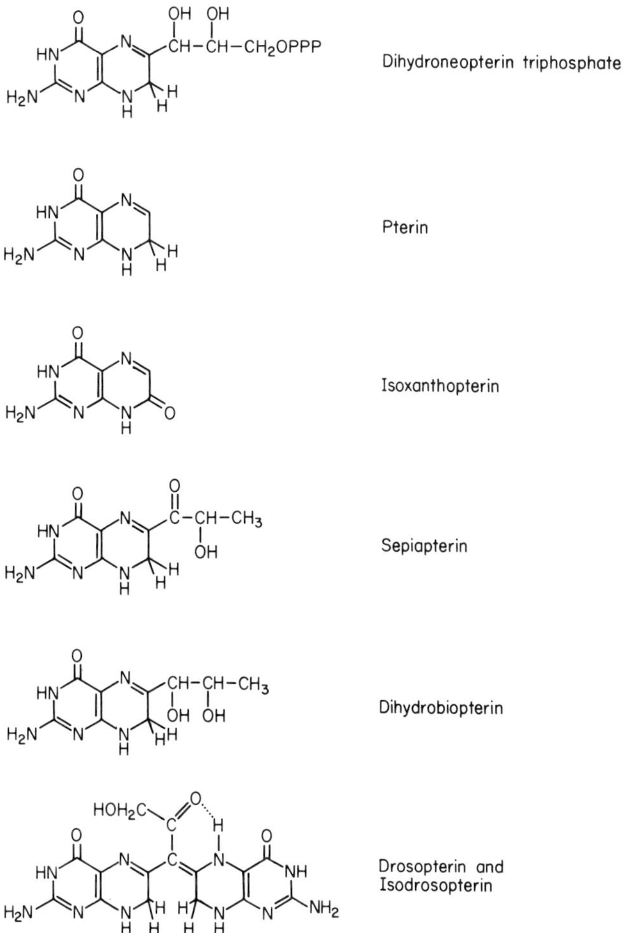

Fig. 19 Chemical structures of key pteridines involved in the branched pathways given in Fig. 18

4 The pteridine biosynthetic pathways

Pteridine biosynthesis in insects is less well understood than ommochrome biosynthesis. Many of the enzymes have not been studied in detail. In *D. melanogaster* the following major pteridines are found: biopterin, sepiapterin, pterin, isoxanthopterin and the drosopterin eye pigments. The drosopterins are probably a mixture of six closely related compounds, as will be discussed later. A branched pathway interrelating these various pteridines is presented in Fig. 18 and structural formulae for the major compounds given in these pathways are shown in Fig. 19. In relation to these pathways, it seems certain that isoxanthopterin and the drosopterins are two end-products of a branched biosynthetic pathway and that pterin is the immediate precursor of isoxanthopterin. Biopterin is probably another end-product but the position of sepiapterin is less clear. It is almost certainly the precursor of biopterin, but whether it is also an intermediate in the synthesis of the drosopterins remains to be clarified.

4.1 GTP CYCLOHYDROLASE

Pteridines are formed from guanosine triphosphate (GTP). The enzyme GTP cyclohydrolase catalyses the first committed step of the pathway, and has been studied in *D. melanogaster* recently by Fan and Brown (1976), Fan *et al.* (1976) and Evans and Howells (1978). Developmental profiles show that there are peaks of GTP cyclohydrolase activity at pupariation and at adult emergence (Fan *et al.*, 1976; Evans and Howells, 1978). These peaks coincide with peaks of rapid pteridine synthesis during development. Thus, isoxanthopterin accumulates at about the time of pupariation, while drosopterins are deposited rapidly in the eyes over the period of adult emergence (Fan *et al.*, 1976). The latter (adult) peak of GTP cyclohydrolase activity probably results mainly from the production of this enzyme in the eyes, since the specific activity of extracts made from heads of young adults is more than 5-fold greater than that of extracts from thorax-abdomens (Evans and Howells, 1978). At both of the developmental peaks, the enzyme is located in the soluble fraction of the cell.

Two eye colour mutant strains of *D. melanogaster* which have abnormal levels of GTP cyclohydrolase activity are *raspberry* (*ras*) and *prune* (*pn*) (Evans and Howells, 1978). Neither of these loci seems likely to contain the structural gene for the enzyme, since activity is higher than wild type at the early pupal peak but lower than wild type at emergence for both mutants. Evans and Howells (1978) also found that *rosy* mutants had a reduced activity of GTP cyclohydrolase, an unexpected result in view of the evidence that the gene for the enzyme xanthine dehydrogenase is located at the *rosy* locus. The implications of this result are discussed below.

4.2 THE PATHWAY LEADING TO ISOXANTHOPTERIN

The best studied step in the segment of the pathway from dihydroneopterin triphosphate to isoxanthopterin is that catalysed by xanthine dehydrogenase (XDH). This enzyme catalyses two reactions: the conversion of pterin to isoxanthopterin in pteridine biosynthesis (Fig. 18) and the conversion of xanthine to hypoxanthine and then to uric acid in purine catabolism. The evidence that places the structural gene for XDH at the *rosy* (*ry*) locus seems quite conclusive. Not only does the substrate of this enzyme (pterin) accumulate after pupariation in *ry* mutants (Mitchell *et al.*, 1959) but also the activity of XDH is proportional to the number of ry^+ alleles present in the genome (Glassman *et al.*, 1962; Grell, 1962). In addition, electrophoretic variants of XDH all map at the *ry* locus (Yen and Glassman, 1965) and *ry* mutant have only negligible amounts of material which is immunologically cross-reactive with antibody prepared against purified XDH (Gelbart *et al.*, 1974). In view of this evidence, the effects of mutations at the *ry* locus on the activity of GTP cyclohydrolase (see section 4.1) must be pleiotropic, suggesting an involvement of a product of one of the XDH-catalysed reactions in determining GTP cyclohydrolase activity. It has never been clear why *ry* mutants, which are blocked in the production of a colourless pteridine (isoxanthopterin), also have reduced levels of drosopterins and hence an abnormal eye colour. The indirect effect of *ry* mutations on GTP cyclohydrolase activity provides a possible explanation.

The *ry* locus of *D. melanogaster* has been studied in considerable detail in recent years by both genetic and biochemical techniques. Fine structure genetic maps of the gene have been prepared and a regulatory region, in addition to the region coding for XDH protein, has been positioned within the map (Chovnick *et al.*, 1976). Much is also known about the biochemical relationships between the product of the *ry* gene (XDH protein) and those of the *maroon-like*, *low-XDH* and *cinnamon* genes in determining XDH activity. However, these aspects are peripheral to eye pigmentation and are therefore outside the scope of this review. Detailed treatments of these topics are given in recent reviews by Finnerty (1976) and O'Brien and MacIntyre (1978).

XDH, like GTP cyclohydrolase, occurs in the soluble fraction of tissue extracts. A developmental profile of activity for *D. melanogaster* shows that activity is present throughout the life cycle, but that there are peaks just after pupariation and at about the time of adult emergence (Munz, 1964). The peak of activity at pupariation correlates with the accumulation of isoxanthopterin which occurs at this stage of development (Fan *et al.*, 1976). The rise in activity over the latter half of development in the puparium is probably related more to uric acid production than to pteridine metabolism, since uric acid accumulates during this period, reaching a peak at adult emergence (Altmann and Howells, unpublished).

The other steps in the branch of the pathway from dihydroneopterin triphosphate to isoxanthopterin have not been studied in insects. Those given in Fig. 18 are the ones which have been established for the corresponding pathway in *Escherichia coli* (Brown, 1971).

4.3 SEPIAPTERIN AND BIOPTERIN SYNTHESIS

The conversion of dihydroneopterin triphosphate to sepiapterin in extracts of *D. melanogaster* (the sepiapterin synthase activity) has been studied in some detail in recent years (Fan *et al.*, 1975; Krivi and Brown, 1979). The reaction requires Mg^{2+} and NADPH, and dihydroneopterin cannot replace dihydroneopterin triphosphate as substrate, suggesting that the phosphate groups are important in the conversion. As described by Krivi and Brown (1979), it is a complex multistep reaction which involves two separate enzymes, called "Enzyme A" and "Enzyme B" by these authors. "Enzyme A" appears to catalyse the conversion of dihydroneopterin triphosphate to an unstable intermediate by the nonhydrolytic elimination of the phosphates and "Enzyme B" is then thought to convert the intermediate to sepiapterin in a reaction involving NADPH.

The sepiapterin synthase activity of soluble extracts of *purple* (*pr*) mutants is reduced to between 8 and 30% of wild type (Yim *et al.*, 1977; Krivi and Brown, 1979). No other eye colour mutant tested showed reduced activity. Krivi and Brown (1979) showed that it is "Enzyme A" activity which is deficient in the *pr* mutants, "Enzyme B" activity being at normal levels. Data presented by Yim *et al.* (1977) show that sepiapterin synthase activity is proportional to the number of pr^{+} alleles in the genome, suggesting that the structural gene for Enzyme "A" is probably located at the *pr* locus.

Profiles for the variation of sepiapterin synthase activity during the developments of *D. melanogaster* have been obtained (Tobler *et al.*, 1979; Krivi and Brown, 1979). Activity is present during larval life but declines sharply just prior to pupariation, so that the level in early pupae is quite low. The activity then remains low until about one day before adult emergence when it increases sharply to reach a peak early in adult life. These changes in activity correlate well with developmental changes in the accumulation of sepiapterin, which occurs to some extent during late larval life and to a much greater extent over the period of adult emergence (Fan *et al.*, 1976).

Dihydrobiopterin is produced from sepiapterin. This reaction has been studied in extracts from *D. melanogaster* and the enzyme catalysing it has been called biopterin synthase (Fan and Brown, 1979). In addition to converting sepiapterin to dihydrobiopterin, this enzyme can also use oxidized (dehydro-) sepiapterin as substrate to produce biopterin. Activity of this enzyme is present throughout the life cycle in *D. melanogaster*, rising to a peak about two days after adult emergence. This correlates well with the accumulation of biopterin, which rises over the period of adult emergence (Fan and Brown, 1979).

4.4 DROSOPTERIN SYNTHESIS

The precise metabolic relationship between sepiapterin and the drosopterins is still unclear. Kaufman (1967) proposed that sepiapterin was a precursor of the drosopterins, since heads of the mutant *sepia* of *D. melanogaster* (which lack drosopterins) contain 10–20 times the normal level of sepiapterin. However, recent studies by Dorsett *et al.* (1979) have shown that the conversion of dihydroneopterin triphosphate to both sepiapterin and the drosopterins can be catalysed by extracts of heads of *D. melanogaster*, and suggest that sepiapterin and the drosopterins are alternative products of a common precursor (as shown in Fig. 18). This common intermediate might be the product of the reaction catalysed by "Enzyme A" of sepiapterin synthase, since *pr* mutants, which are deficient in "Enzyme A" activity, have reduced levels of both sepiapterin and the drosopterins (Wilson and Jacobson, 1977).

The remainder of the pathway leading to the drosopterins is still unknown. It is certainly a multistep reaction sequence and more than one enzyme may be involved. Since dihydroneopterin triphosphate (and sepiapterin) contain a single pteridine ring system, whereas the drosopterins are dipteridines (see Fig. 19), the reaction must involve a condensation of two molecules of the precursor with the elimination of one of the three-carbon side chains. Since the drosopterins seem to be deposited on pigment granules (see section 2.6), these reactions may occur on the granules themselves. Two different eye colour mutants of *D. melanogaster* seem to be defective somewhere in these terminal steps. The mutant *sepia* (*se*), which is totally blocked in drosopterin synthesis, accumulates large amounts of sepiapterin, presumably because the common precursor is directed into that alternative segment of the pathway. *Henna-recessive* (Hn^r) mutants are partially blocked in drosopterin synthesis and also accumulate excess sepiapterin (although not nearly to the extent seen in *se*). Since the double mutant *se*; Hn^{r3} accumulates sepiapterin in vast excess like *se* (Wilson and Jacobson, 1977), these authors propose that *se* mutations block drosopterin synthesis at an earlier step than Hn^r mutations.

4.5 CHEMICAL STRUCTURES OF THE DROSOPTERINS

The drosopterins appear to consist of six different compounds, which may be made up of two families of three compounds. The members of each family apparently have identical empirical formulae, being structural isomers of each other. Drosopterin, isodrosopterin and neodrosopterin make up one family. They are dipteridine compounds (Fig. 19) with a single bridging carbon atom linking the two ring system through position 6 in each case. A two-carbon side chain extends from the bridging carbon atom. Drosopterin and isodrosopterin are stereoisomers about the carbon-carbon single bond of the bridging carbon atom, whereas neodrosopterin is a *cis-trans* isomer of the other two about the carbon-carbon double bond of the bridging carbon atom (Rokos and Pfleiderer, 1975). These stereochemical rearrangements occur non-enzymically in aqueous solution. The aurodrosopterins (which have not been studied in the same detail) probably also comprise a similar three member family of stereoisomers (Dorsett *et al.*, 1979). The aurodrosopterins have very similar structures to the drosopterins, being dipteridines with a single bridging carbon atom between the two ring systems. They probably differ from the drosopterins in the structure of the side chain which extends from the bridging carbon atom (Rokos and Pfleiderer, 1975). It seems possible that drosopterins and aurodrosopterins are interconverted enzymically but no enzyme has yet been identified.

4.6 PTERIDINE-DEFICIENT MUTANTS

A large number of the eye colour mutants of *D. melanogaster* are deficient in the production of pteridine pigments. For a few of these mutants (*ry*, *pr*, *se*, *Hn^r*, *pn* and *ras*) some relevant biochemical data is available and this has already been discussed. However, for the other pteridine mutants, little specific information is known. Many are deficient in both pteridine and ommochrome production and may, therefore, be defective in some process involved in coordinating the production of the two classes of pigment. This feature of eye pigmentation will be considered in detail in the later sections of this chapter.

Except for the white-eye mutants (which have been found in many insect species and which lack both pteridine and ommochrome eye pigments), few pteridine-deficient eye colour mutants are known in insect species other than *D. melanogaster*. Recently the *grape* eye colour mutant of the blowfly *Lucilia* was found to have a much lower than normal level of pteridines in its eyes, but the nature of the biochemical defect in *grape* has yet to be identified (Summers and Howells, 1980a).

4.7 EVOLUTIONARY ASPECTS OF PTERIDINE EYE PIGMENTS IN DIPTERA

An interesting evolutionary aspect of pteridine eye pigment production in Diptera is that only species belonging to the genus *Drosophila* appear to have substantial quantities of the red drosopterin pigments. In other species of Diptera, sepiapterin seems to be the major eye pteridine pigment (see Ziegler and Harmsen, 1969).

In Table 3, the levels and times during development of onset of synthesis of the major eye pteridines for two species of *Drosophila* (*D. melanogaster* and *D. virilis*) and for the blowfly *Lucilia* are compared. The following two features are of interest. Firstly, the levels of drosopterins (per g fresh weight) in the two species of *Drosophila* are 6–8-fold greater than the level of sepiapterin in *Lucilia*. Moreover, the level of sepiapterin in *D. melanogaster* (0.4 μmol·g^{-1}—Evans and Howells, 1978) is higher than that in *Lucilia*. Thus drosopterin production in the *Drosophila* species in no way replaces sepiapterin production, rather, it appears to be superimposed upon it.

TABLE 3 Onset and levels of major eye pteridines in dipteran species

Species	Onset of synthesis of eye pteridine[a]	Length of life in the puparium[b]	Level of eye pteridine[c]	Reference
D. melanogaster (drosopterins)	68–70 h	96 h	1.5	Evans & Howells (1978)
D. virilis (drosopterins)	72–76 h	136 h	1.2	Felton & Howells (unpublished)
Lucilia (sepiapterin)	5–6 days	7 days	0.2	Summers & Howells (1980a)

[a]Time after pupariation [b]Time from pupariation to adult eclosion
[c]Expressed as μmol·g^{-1} fresh weight

It is interesting, in this light, that whereas sepiapterin probably occurs in a soluble form in the cytoplasm of pigment cells, drosopterins are deposited on Type II pigment granules (see section 2.6). Secondly, the times during development at which the synthesis of the major eye pteridine is initiated is different in the different species. In *D. virilis* drosopterin synthesis begins roughly midway between pupariation and adult emergence, in *D. melanogaster* about two-thirds of the way through, while in *Lucilia* sepiapterin production commences only just before adult emergence. This developmental variability in the time of onset of synthesis of pteridine eye pigments can be contrasted with the relatively constant time at which xanthommatin synthesis is initiated in these species (see Table 1). On the basis of this evolutionary variability, it seems reasonable to suggest that pteridines play a less crucial role than xanthommatin as screening pigments in the eyes of dipteran species.

5 Interactions between the xanthommatin and pteridine biosynthetic pathways

5.1 EVIDENCE FOR AN INTERACTION

In many eye colour mutant strains of *D. melanogaster*, a single genetic defect alters, to a greater or lesser extent, the production of both classes of pigment

(Hadorn and Kühn, 1953; Kühn and Egelhaaf, 1955; Ghosh and Forrest, 1967; Ziegler and Harmsen, 1969; Parisi, 1971; Parisi *et al.*, 1976a, b). The white eye phenotype, which is common to many species of insect, is the extreme example of this phenomenon. This phenotype results from the virtual absence of ommochromes and pteridines in adult eyes although body pigmentation is normal. White eye mutants in diptera include *white* of *D. melanogaster* (Lindsley and Grell, 1968; Judd, 1976), *white* of *Musca* (Hiraga, 1964; Milani, 1975), *chalky* of *Calliphora* (Langer, 1967) and *white* of *Lucilia* (Whitten *et al.*, 1975).

TABLE 4 Levels of red and brown pigment in eye colour mutant strains of *D. melanogaster*. (From data in Nolte, 1959a, b)

Allele	Red pigment (% of wild type)	Brown pigment (% of wild type)
(*i*) The *garnet* (*g*) allelic series		
g^1	38	56
g^2	16	33
g^3	21	44
g^4	22	22
(*ii*) The *white* (*w*) allelic series		
w^1	0.3	2
w^{col}	17	22
w^{e2}	7	55
w^{a3}	6	11
w^{co}	6	33
w^{sat}	3	122
w^{e3}	2	32
w^e	1	22
w^a	1	11
w^t	0.3	3

The first quantitative information on levels of eye pigments in eye colour mutants of *D. melanogaster* was provided by Nolte (1954a, 1955, 1959a, b). Of the nineteen mutants examined in these studies, fourteen were different from wild type in the levels of both major pigments. Where different alleles at one locus were examined, for example, the four alleles of *garnet* (Nolte, 1959b), the production of each pigment was found to be affected to a different extent in each allele (Table 4*i*). This is particularly evident in considering the *white* allelic series (Table 4*ii*), where both pigments can be almost completely eliminated (as in *w* and w^t), or reduced to different extents in the various other *white* alleles. In w^{sat}, for example, only drosopterin production is reduced, the xanthommatin level being as high as (or higher than) in wild type, while in w^{col} the levels of both are reduced to about 20%. Nolte (1954a) also notes that cardinal (*cd*) mutants show the reverse trend to that in w^{sat}. In this case, the xanthommatin level is only 14% of the wild type whereas the drosopterin level is raised. Parisi *et al.* (1976a) also report that *cd* has raised drosopterin levels. Similarly, Schwink and Mancini (1973) found that the pattern of pteridines in *cd* was different from that of wild type. Thus among the eye colour mutants of *D. melanogaster*, levels of ommochromes and pteridines are frequently both altered, and they can be changed, relative to each other, in a variety of different ways.

Among the eye colour mutants of *Lucilia*, the dark-eyed mutant *grape* has a reduced production of both xanthommatin and sepiapterin, the levels at emergence being respectively, 45% and 12% of wild type (Summers and Howells, 1978, 1980a). It is, therefore, similar to *D. melanogaster* mutants like *garnet*, *claret* and *lightoid* which have reduced levels of both major pigments. One interesting feature of the *grape* mutant is that the reduced production of sepiapterin is suppressed, to some extent, in double mutants additionally blocked in the xanthommatin pathway. Thus in *yellowish*; *grape*, xanthommatin is as low as in *yellowish* (about 1% of wild type) but sepiapterin production is partially restored to about 30% of the wild type level (Summers and Howells, 1980a). This further illustrates the complex interdependence of the two biosynthetic pathways in dipteran species.

The interrelationship of production of the two pigment types is also evident in mutant strains of the moth *Ephestia*. Here the nonautonomous mutant, *a*, is blocked in the first step of the xanthommatin biosynthetic pathway and the pteridine content is also abnormal. Injection of kynurenine into adults restores xanthommatin biosynthesis and also normalises the pteridine levels (Hadorn and Kühn, 1953; Kühn and Egelhaaf, 1955). A white eye colour mutant of *Ephestia* has also been described (Caspari and Gottlieb, 1975) in which the synthesis of both types of eye pigment is disrupted through mutation at a single locus.

5.2 THEORIES CONCERNING THE INTERACTION

A number of different proposals have been put forward to explain the interaction between pteridine and ommochrome production, and the *white* mutants have often been seen as the key to answering this question. Firstly, Nolte (1952) suggested that there might be a common substrate for the two pathways, so that abnormalities in its production would affect both pathways simultaneously. However, as the steps involved in pigment synthesis are now better known (Figs 16 and 18), it is clear that this proposal is incorrect. Another suggestion, put forward by Kühn (1956), was that both pigments might be deposited on the same granules, so that interdependence of synthesis would be due to a competition for sites on the granules. He further proposed that the absence of pigments in *white* mutants might result from a defect in the production of granules. The results of the ultrastructural studies on pigment cells carried out by Shoup (1966) do not support these proposals. For *D. melanogaster*, two classes of pigment granules can be seen in the electron micrographs of sections of pigment cells, apparently one class for xanthommatin (Type I granules) and the other for drosopterins (Type II granules). For the blowflies, also, the ultrastructural evidence fails to support the proposals, since sepiapterin seems likely to exist free in the cytoplasm of pigment cells (Ziegler and Harmsen, 1969; Langer, 1975; Pyliotis and Summers, 1980), rather than being granule-bound.

There have also been several suggestions that the interaction between the two pathways might be explained in terms of enzyme cofactors. Thus Ghosh and Forrest (1967) noted that phenylalanine hydroxylase from rat liver has a pteridine cofactor, and suggested, by analogy, that kynurenine hydroxylase might also have a pteridine cofactor. Thus a failure of pteridine synthesis would reduce the availability of this cofactor, so disrupting the ommochrome pathway as well. However, Sullivan *et al.* (1973) found that the *in vitro* levels of kynurenine hydroxylase activity in stocks carrying different alleles of *white* were normal, in spite of varying levels of pteridines in these mutants. In addition, *white* mutants of *D. melanogaster* accumulate significant amounts of 3-hydroxykynurenine during development (Howells *et al.*, 1977), showing that kynurenine hydroxylase functions *in vivo* in these strains. Hence, for the *white* mutants, at least, this proposal does not appear to be correct.

Another proposal involving cofactors was made by Parisi *et al.* (1976a), who suggested that dihydrobiopterin might be a precursor of sepiapterin and drosopterins and that xanthommatin could act as a cofactor for sepiapterin reductase, the enzyme which catalyses the dihydrobiopterin-sepiapterin conversion. Recent advances in our understanding of pteridine biosynthesis suggest that it is unlikely that sepiapterin is produced from dihydrobiopterin in a sepiapterin reductase-catalysed reaction (see section 4.3), but it is interesting that Krivi and Brown (1979) have proposed that dihydrobiopterin might be an intermediate in the conversion of dihydroneopterin triphosphate to sepiapterin (catalysed by the enzyme sepiapterin synthase). However, with the wild type enzyme, no dihydrobiopterin accumulates, so they suggest that this proposed intermediate must be enzyme-bound. There is no evidence that xanthommatin participates in or regulates this reaction. Parisi *et al.* (1976b) have also suggested that the level of xanthine dehydrogenase activity in cells might control the availability of the reduced nicotinamide-adenine dinucleotide cofactor required for kynurenine hydroxylase activity, thus linking the two pathways. Again, this explanation seems unlikely to be correct since XDH activity is low in tissues like the eyes

(Ziegler and Harmsen, 1969) where the activity of kynurenine hydroxylase is high (Sullivan *et al.*, 1973; Sullivan and Sullivan, 1975).

Recently, it has become apparent that pigment precursors must be transported from tissue to tissue during development, and often stored pending the synthesis or activation of enzymes further down the pathway. Defects in the processes involved in the uptake of pigment precursors by specific tissues (Sullivan *et al.*, 1974), or their storage (Howells and Ryall, 1975), might explain a variety of eye colour phenotypes and the pathways could be interrelated through common use or control of such processes. In order to assess these proposals, in the next section of this chapter detailed consideration is given to the interrelationships between different tissues in the production of ommochrome and pteridine pigments.

6 Interrelationships between different tissues in eye pigment production

6.1 TISSUES INVOLVED IN EYE PIGMENT PRODUCTION

In Table 5 information is listed, for several dipteran species, on the different tissues in which pigment biosynthetic enzymes are active and in Table 6 the times during development are given at which activity for these enzymes can be detected. For xanthommatin production, the following summary can be made. The conversion of tryptophan to kynurenine appears to take place throughout much of development. In larvae the fat body seems to be the major site of kynurenine production, whereas in developing adults, both the fat body and the eyes seem to be important. However, the next enzyme of pathway (kynurenine hydroxylase), which catalyses the conversion of kynurenine to 3-hydroxykynurenine, appears to be found mainly in the malpighian tubules of larvae and in the eyes of developing adults. Thus, during larval life kynurenine formed in the fat body must be transported to the tubules for conversion to 3-hydroxykynurenine. In addition, during adult development, tryptophan oxygenase activity peaks well before kynurenine hydroxylase activity, so that kynurenine produced as a result of this early tryptophan oxygenase activity must be stored prior to its transport to the eyes.

3-Hydroxykynurenine accumulates during larval life as the result of larval kynurenine hydroxylase activity and seems to be synthesised and stored mainly in the cells of the larval malpighian tubules. The larval storage of this metabolite seems to be of relatively minor quantitative importance in the production of adult xanthommatin in *D. melanogaster*, since it can account for no more than 25% of the pigment later produced (Ryall and Howells, 1974). Nevertheless, larval 3-hydroxykynurenine storage does have

TABLE 5 Tissue localisation of pigment biosynthetic enzymes in Diptera

Enzyme	Tissue	Species	References
(*i*) Xanthommatin biosynthetic enzymes			
Tryptophan oxygenase	Larval fat bodies	*D. melanogaster*	Rizki (1961)
	Larval malpighian tubules	*D. melanogaster*	Kaufman (1962)
	Adult eye discs	*D. melanogaster*	Clancy (1940); Nissani (1975)
	Adult fat body, testes, larval and adult malpighian tubules	*Protophormia*	Linzen & Schartau (1974)
Kynurenine formamidase	Most tissues	All species examined	Linzen (1974)
Kynurenine hydroxylase	Larval malpighian tubules, adult eyes	*D. melanogaster*	Sullivan *et al.* (1973); Sullivan & Sullivan (1975)
	Eye discs	*D. melanogaster*	Danneel (1941); Horikawa (1958)
	Larval malpighian tubules	*Calliphora*	Hendrichs-Hertel & Linzen (1969)
Phenoxazinone synthase	Adult heads	*D. melanogaster*	Phillips *et al.* (1973); Yamamoto *et al.* (1976)
	Adult heads	*Lucilia*	Summers (1979)
(*ii*) Pteridine biosynthetic enzymes			
Xanthine dehydrogenase	Larval fat body; adult haemolymph	*D. melanogaster*	Ursprung & Hadorn (1961)
	Adult haemolymph, malpighian tubules	*D. melanogaster*	Munz (1964)
GTP cyclohydrolase	Adult heads	*D. melanogaster*	Fan *et al.* (1976);
	Adult heads and body tissues of early pupae	*D. melanogaster*	Evans & Howells (1978)
Sepiapterin synthase	Adult heads	*D. melanogaster*	Dorsett *et al.* (1979); Krivi & Brown (1979)
Drosopterin synthase	Adult heads	*D. melanogaster*	Dorsett *et al.* (1979)

TABLE 6 Developmental profiles of activity for pigment biosynthetic enzymes

Enzyme	Activity profile	Species	References
(*i*) Xanthommatin biosynthetic enzymes			
Tryptophan oxygenase	Active throughout development. Peaks in young larvae and mature adults	*Protophormia*	Linzen & Schartau (1974)
	Active from late larval life onwards. Peaks at pupariation and in mature adults	*D. melanogaster*	Sullivan & Kitos (1976)
	Active throughout life in the puparium. Peak soon after pupariation	*Lucilia*	Summers (1979)
Kynurenine formamidase	Active throughout larval life and life in the puparium. Low activity in adults	*Protophormia*	Linzen & Schartau (1974)
	Active throughout development. Peaks at pupariation and in young adults	*D. melanogaster*	Sullivan & Kitos (1976)
	Active from pupariation onwards	*Lucilia*	Summers (1979)
Kynurenine hydroxylase	Active in late larval life and throughout life in the puparium. Peak at pupariation	*Protophormia*	Linzen & Schartau (1974)
	Active in late larval life and throughout life in the puparium. Major peak in late pupae	*D. melanogaster*	Sullivan *et al.* (1973)
	Active throughout life in the puparium. Peak in late pupae	*D. melanogaster*	Ryall & Howells (1974)
Phenoxazinone synthase	Activity appears mid-way between pupariation and adult emergence. Peak in early adults	*D. melanogaster*	Yamamoto *et al.* (1976)
	Activity appears mid-way between pupariation and adult emergence	*Lucilia*	Summers (1979)

TABLE 6 (*continued*)

Enzyme	Activity profile	Species	References
(*ii*) Pteridine biosynthetic enzymes			
Xanthine dehydrogenase	Active throughout development. Peaks at puparation and in early adults	*D. melanogaster*	Munz (1964); Altmann & Howells (unpublished)
GTP cyclohydrolase	Active from late larval life onwards. Minor peak at pupariation, major peak at adult emergence	*D. melanogaster*	Fan *et al.* (1976); Evans & Howells (1978)
Sepiapterin synthase	Activity peaks in late larval life and in early adults; low activity in early pupae	*D. melanogaster*	Tobler *et al.* (1979); Krivi & Brown (1979)

qualitative importance and appears to be vital for pigmentation. It is presumably released from the tubules during the tissue rearrangement which occurs early in metamorphosis (Gilmour, 1961; Wigglesworth, 1972) and could serve as an activator of phenoxazinone synthase (Muth, 1969; Linzen, 1974; Yamamoto *et al.*, 1976), or be used in the early stages of pigment synthesis before kynurenine hydroxylase in the eyes becomes fully active.

Phenoxazinone synthase activity has been found mainly in the head region, probably localised in the pigment cells of the eyes, so that the transport of both kynurenine (substrate for kynurenine hydroxylase) and 3-hydroxykynurenine (substrate for phenoxazinone synthase) into these pigment cells must be important. Cell lineage analysis by Nissani (1975) suggests that the eyes of wild type *D. melanogaster* are capable of producing kynurenine from tryptophan but depend on external kynurenine and/or 3-hydroxykynurenine supply to produce normal quantities of xanthommatin. Thus, in the case of xanthommatin production, the storage of precursors in specific tissues and their movement from tissue to tissue via the haemolymph seem to be of vital importance.

The situation with regard to pteridine pigmentation is less clear due to a lack of information concerning the tissue-specificity of the enzymes involved. For xanthine dehydrogenase, activity is known to be low in the eyes and gonads of some insects, yet high levels of its product (isoxanthopterin) are present in these tissues (Ziegler and Harmsen, 1969), suggesting that they may take up this metabolite from the haemolymph. GTP cyclohydrolase (Evans and Howells, 1978), sepiapterin synthase (Krivi and Brown, 1979) and drosopterin synthase (Dorsett *et al.*, 1979) are located in the heads of adult *D. melanogaster*, which probably means that the conversion of GTP to sepiapterin and to the drosopterins can occur entirely within the pigment cells. The formation of GTP in the pigment cells may depend on the uptake of guanine from the haemolymph (Sullivan *et al.*, 1979) which can be converted to GTP via nucleotide salvage enzymes (Murray, 1971). The key enzyme required for the salvage of guanine (hypoxanthine-guanine phosphoribosyltransferase) was recently found in extracts both of *Drosophila* tissue culture cells and also of whole adults (Becker, 1978). However, it has yet to be demonstrated that this enzyme is present in the eyes.

TABLE 7 Pigment precursors and related compounds found in dipteran larval malpighian tubules

Compound	References
Tryptophan	Wessing & Bonse (1962); Bonse (1969)
Kynurenine	Wessing & Bonse (1962)
3-Hydroxykynurenine	Wessing & Danneel (1961); Wessing & Bonse (1962); Wessing & Eichelberg (1972)
Riboflavin	Gilmour (1961); Wessing & Eichelberg (1968); Nickla (1972); Wigglesworth (1972)
"Pteridines"	Gilmour (1961); Wessing & Eichelberg (1968); Bonse (1969)
Pteridines, including isoxanthopterin, xanthopterin, pterin, sepiapterin, biopterin, riboflavin, 1 unidentified fluorescent compound	Hadorn *et al.* (1958); Handschin (1961); Ziegler & Harmsen (1969)

Of the various tissues involved in eye pigmentation, the role played by the larval malpighian tubules appears to be particularly significant. The correlation of abnormal phenotypes in both the eyes and the larval tubules is well documented among the colour mutants of *D. melanogaster* (Lindsley and Grell, 1968) and, in the white-mottled alleles of *white* in *D. hydei*, the adults have blotchy pigmentation in the eyes and the larvae have mottled malpighian tubules (van Brugel, 1973). A range of compounds which are either eye pigment precursors or related to eye pigments in chemical structure have been found in larval malpighian tubules (Table 7). Many are coloured and/or brightly fluorescent under ultraviolet light. The pigmentation of two other tissues in *D. melanogaster* (the adult gonads and the ocelli) is also frequently affected by mutations which cause abnormal eye colour. For example, the mutant *white* has a colourless testis sheath whereas in wild type this is yellow. The ocelli of many eye colour mutant strains are colourless, although they are brown in wild type due to the presence of xanthommatin.

6.2 STUDIES ON THE UPTAKE AND STORAGE OF PIGMENT PRECURSORS

It was clear from the early studies on tissue transplantation of *D. melanogaster* that at least some of the larval and adult tissues involved in eye pigment production were able to take up pigment precursors. Thus, developing eye discs from three nonautonomous mutant strains (*vermilion*, *cinnabar* and *rosy*) formed pigment normally in wild type hosts, indicating that the products of the enzymes missing in these mutants had been taken up from the haemolymph by the implanted eye discs. Since *vermilion* and *cinnabar* are deficient in xanthommatin synthesis, while *rosy* has a reduced level of drosopterins, eye discs were apparently able to take up both ommochrome and pteridine precursors.

The uptake of kynurenine and 3-hydroxykynurenine by larval malpighian tubules and by adult eyes of *D. melanogaster*, and the storage of these compounds in these tissues, have been studied in several different ways. Eichelberg and Wessing (1971) and Wessing and Eichelberg (1972) observed the storage of 3-hydroxykynurenine in the cells of the larval malpighian tubules using electron microscopy and proposed that this pigment precursor is deposited in granular form. Sullivan and Sullivan (1975) carried out *in vitro* uptake experiments and showed that isolated adult eyes and larval malpighian tubules could take up ^{3}H-kynurenine. Summers and Howells (1980b), working with the blowfly *Lucilia*, showed that following injection of either intact larvae or adults with ^{3}H-3-hydroxykynurenine, the larval malpighian tubules of the adult eyes take up this compound. Eyes were also shown to accumulate ^{3}H-kynurenine in these studies.

With respect to pteridine precursors, the situation regarding uptake and storage is much less clear. In wild type *D. melanogaster*, riboflavin (which is related in chemical structure to the pteridines) is stored in vacuoles in the cells of the larval malpighian tubules (Wigglesworth, 1972). Nickla (1972) found that the level of this compound in the larval tubules was reduced in two mutants (*light* and *clot*) which have reduced levels of drosopterins in their eyes. Since riboflavin plays no direct role in eye pteridine synthesis, these results seem to indicate that riboflavin and pteridines may share a common uptake/storage system, Recently, Sullivan *et al.* (1979) have shown that isolated larval malpighian tubules of *D. melanogaster* are able to take up several radioactively labelled purines (including guanine, which can be considered to be a pteridine precursor) as well as riboflavin. Thus cells of the larval tubules appear to have uptake mechanisms for pteridine precursors and for related compounds like riboflavin. Unfortunately, no data is yet available concerning the uptake of these types of compound by adult eyes.

6.3 MUTANTS AFFECTED IN THE UPTAKE AND STORAGE OF PIGMENT PRECURSORS

Several authors have discussed the possibility that some of the eye colour mutants of *D. melanogaster* might be defective in their ability to take up or store pigment precursors. Bonse (1969) noted the lack of fluorescent substances in the malpighian tubules of larvae of *white* and suggested that this was due to a failure by the cells of the tubules to accumulate tryptophan, kynurenine and pteridines. Sullivan *et al.* (1973, 1974) showed that although *scarlet* failed to produce normal levels of 3-hydroxykynurenine, nevertheless, extracts prepared from this mutant had wild type levels of activity of the three enzymes required for the synthesis of this precursor. They therefore suggested that *scarlet* mutants might have a defect in the mechanism required for transporting xanthommatin precursors into cells able to metabolise or store these compounds (principally cells of the larval malpighian tubules and of the adult eyes). This explanation is compatible with the results of tissue transplant experiments (Beadle and Ephrussi, 1936), in which *scarlet* eye discs behaved autonomously, failing to make pigment in wild type hosts.

In order to examine further their "transport defect" hypothesis concerning *scarlet*, Sullivan and Sullivan (1975) carried out *in vitro* uptake experiments and showed that adult eyes and larval malpighian tubules isolated from *scarlet* took up ^{3}H-kynurenine at reduced rates compared with the same tissues isolated from wild type. In addition, they found that tubules and eyes from *white*, and also from two other eye colour mutants of *D. melanogaster* (*lightoid*, *claret*), had similarly reduced rates of uptake. Consequently, Sullivan and Sullivan (1975) proposed that not only in *scarlet*, but also in *white*, and perhaps in a number of other eye colour mutants, the abnormal eye colour phenotypes could be attributed to defects in transport mechanisms for pigment precursors.

In other studies with these mutants, Howells and Ryall (1975) observed that *scarlet* does accumulate 3-hydroxykynurenine during development, but only after pupariation when excretion of solid materials is not possible. Thus they suggested that the absence of this metabolite in larvae and in adults of this mutant was probably due to its rapid excretion. Subsequently, larvae of *scarlet* were shown to excrete exogenously supplied ^{3}H-kynurenine and ^{3}H-3-hydroxykynurenine at rates 5–10-fold higher than those of wild type (Howells *et al.*, 1977). Similar elevated larval excretion rates were also obtained for *white*, and seem likely to be a characteristic feature of transport mutants in *D. melanogaster*.

The cells of the larval malpighian tubules are not only involved in the synthesis and storage of pigment precursors, but also play an important role in the excretion of such metabolites. Thus, for *scarlet* and *white*, we have the apparently conflicting findings that, on the one hand, their larval tubules are unable to take up pigment precursors *in vitro* but, on the other, in intact larvae they are apparently able to remove these compounds from the haemolymph for excretion. Recently, Sullivan *et al.* (1980) obtained evidence, by the use of nonmetabolisable compounds in uptake experiments with isolated tubules and also by examining the ultrastructure of the cells of the tubules using the electron microscope, that intercellular channels exist between the cells of the tubules. These appear to provide passages which would permit metabolites to pass from the haemolymph into the central lumen of the tubule, without having to enter the cells. Therefore, in terms of the transport defect hypothesis, the excretion data for ^{3}H-kynurenine and ^{3}H-3-hydroxykynurenine can be explained as follows: For wild type, in which the uptake mechanism for kynurenine and 3 hydroxykynurenine is functional, the labelled metabolites would be absorbed (to a large extent) by the cells of the tubules during intercellular passage, so that little would reach the lumen for excretion. For *white* and *scarlet*, in which there are defects in the uptake system of the tubules, there would be little absorption, so that the labelled metabolites would pass between the cells and into the lumen giving enhanced rates of excretion.

Eye colour mutants which appear to have defects in their ability to take up or store xanthommatin precursors have also been found in *Lucilia* (Summers and Howells, 1980b). The *white* mutant in this species seems to be similar in its properties to *white* of *D. melanogaster*. Thus, when *white* larvae of *Lucilia* were raised on a diet supplemented with ^{3}H-3-hydroxykynurenine, little labelled material was accumulated in the cells of the larval malpighian tubules (2% of the total radioactivity taken up, compared with 43% for wild type). Similary, following the injection of ^{3}H-3-hydroxykynurenine into young adults of *white*, the rate of uptake of label by the eyes was much reduced compared with that in wild type. Mutations at two other loci in *Lucilia* (*topaz* and *grape*) also appear to affect uptake or storage of pigment precursors. The *topaz* mutants were found to be defective in their ability to store 3-hydroxykynurenine in the larval tubules, but took up this metabolite into adult eyes at normal rates, whereas *grape* showed the reverse characteristics. Hence, for the *white* and *grape* mutants, the decreased levels of xanthommatin in adult eyes can be attributed to the reduced rates of uptake of 3-hydroxykynurenine by the pigment cells. However, this explanation does not apply to the *topaz* mutants. Not only do the rates of 3-hydroxykynurenine uptake by their eyes seem to be normal, but they also seem to have normal levels of 3-hydroxykynurenine at the time of onset of xanthommatin synthesis (Summers and Howells, 1978). Thus, for the *topaz* mutants, the only defects in xanthommatin metabolism discovered so far has been the reduced level of 3-hydroxykynurenine storage in the larval malpighian tubules. As discussed earlier (see section 6.2), larval 3-hydroxykynurenine storage seems to be critically important for adult xanthommatin production in *D. melanogaster*. The results presently available for the *topaz* mutants strongly suggest that a similar situation applies in *Lucilia*.

There have been no reports for nondipteran species, of mutants defective in the uptake of pigment precursors, although xanthommatin-deficient mutants defective in precursor storage have been described. The *wa* mutant of *Ephestia* (Caspari and Gottlieb, 1975) and the *cream* and *pearl* mutants of *Apis* (Dustmann, 1968, 1975), appear to lack the capacity to bind 3-hydroxykynurenine to granules in the eye pigment cells. The formation of granular 3-hydroxykynurenine in these cells is therefore suggested to be a necessary prerequisite for (but, nevertheless, distinct from) the enzymological conversion of this intermediate to xanthommatin, which probably also takes place in association with pigment granules. It is interesting, from a comparative viewpoint, that granular 3-hydroxykynurenine has not been observed in the pigment cells of any of the mutants of *D. melanogaster*.

The position regarding the biochemistry and genetics of uptake and storage for pteridine pigment precursors is less certain. Several pteridines (and also riboflavin) are reported to be stored in the larval malpighian tubules of *D. melanogaster* (see Table 7). *In vitro* uptake experiments by Sullivan *et al.* (1979) have shown that isolated larval tubules from *white*, *brown* and *pink-peach* of *D. melanogaster* are defective, compared with those from wild type, in their ability to accumulate labelled guanine. Tubules from *white* also showed a reduced uptake, compared with wild type, for labelled riboflavin (Sullivan *et al.*, 1979). Defective riboflavin uptake probably also applies to tubules from *brown*, since Lovelock and Howells (unpublished) have found that larvae of both *white* and *brown*, when supplied with ^{14}C-riboflavin in the diet, excrete this compound much more rapidly than do larvae of wild type. The possible relationship between uptake and storage mechanisms for riboflavin and for pteridines has been discussed earlier (see section 6.2). Thus the reduced production of drosopterins in the eyes of *white*, *brown*, *pink-peach* (and possibly other drosopterin-deficient eye colour mutants of *D. melanogaster*) may be a consequence of a defective uptake system for pteridine precursors in the pigment cells.

6.4 UPTAKE AND STORAGE OF PIGMENT PRECURSORS AND INTERACTIONS BETWEEN THE PIGMENT PATHWAYS

As noted earlier in this chapter, *white* mutants have often been seen as the key to gaining an understanding of the interaction between the ommochrome and pteridine pathways. The evidence discussed in the previous section shows that mutations at the *scarlet* and the *white* loci interfere with the accumulation of xanthommatin precursors by the cells of the larval malpighian tubules and by the pigment cells of the adult eyes, and that mutations at the *brown* and *white* loci affect accumulation of pteridines and their precursors by the same cells. Thus *scarlet* mutants seem to be defective only in the transport of xanthommatin precursors and *brown* mutants only in the transport of pteridine precursors, while the *white* mutants appear to be doubly defective in the transport of both types of pigment precursor. Consequently, a consideration of the likely modes of action of these three genes in relation to precursor transport, and of possible interrelationships between them, may shed light on the interaction between the two pigment pathways.

Information relevant to the action of these genes has come from studies on temperature-sensitive mutant alleles. The availability of temperature-sensitive mutants permits temperature-shift experiments to be performed, in which cultures of the mutant are shifted from the permissive to the restrictive temperature (and *vice versa*) at different stages of the development, in order to determine the temperature-sensitive periods (TSPs) during development for a particular gene (Suzuki, 1970). The *white* and *scarlet* genes have very similar TSPs during adult development, with respect to xanthommatin production, which begin somewhat before the onset of synthesis of this pigment and end before the completion of its deposition (Howells 1979). A likely interpretation of these findings is that the protein products of both these genes are needed just prior to and during the time of onset of xanthommatin synthesis but are not needed over the later stages of pigment deposition. The TSPs for the two genes seem to coincide with the period during which the pigment cells rapidly take up xanthommatin precursors from the haemolymph. The TSP for the brown gene (with respect to drosopterin production) begins later than that for the *white* and *scarlet* genes but precedes the time of onset of drosopterin synthesis. It probably coicides with the period during which the pigment cells rapidly take up pteridine precursors. The results obtained in these studies with the temperature-sensitive mutants therefore support the proposal that the *white*, *scarlet* and *brown* gene products participate in systems necessary for pigment precursor uptake by the eye pigment cells.

There are at least two general ways in which the products of the three genes could interact to provide such uptake systems. Interaction could occur at a regulatory level; for example, the product of the *white* gene could be involved in activating the expression of both the *scarlet* and *brown* genes. The products of these latter two genes might be the specific uptake permeases. Thus, in *white* mutants, which would be defective in this system of regulation, the uptake of both types of pigment precursor would be abnormal. The results obtained with the temperature-sensitive mutants, discussed above, indicate that the *scarlet* gene product becomes active at an earlier stage of adult development than the *brown* gene product. Consequently the proposed regulatory product of the *white* gene would have to act sequentially on these two genes rather than simultaneously. An alternative type of interaction suggested by Sullivan *et al.* (1979), could come about through a coordinated system of transport for both types of precursors. For example, the products of the *white*, *scarlet* and *brown* genes could associate within the plasma membrane of the pigment cells (and of the cells of the malpighian tubules) to provide a coordinated uptake system. In such a system, the *scarlet* and *brown* gene products might be the specific uptake permeases, while the *white* gene product might be required to correctly orient these permeases in the membrane. Again, in *white* mutants, in which the amount or the structure of the *white* product was altered, the production of both types of pigment would be affected. As yet, there is no direct experimental evidence to support one or the other of these proposals.

There are still a number of features concerning transport mutants of *D. melanogaster* which may be relevant to the interaction between the pathways but which still need clarification. Firstly, the extent of pigment production seems to be very sensitive to the level of activity of the products of the *white*, *scarlet* and *brown* genes. As noted earlier, there are many partially pigmented alleles of *white* (see Table 4), in which drosopterins and also, usually, xanthommatin are present at reduced levels. Partially pigmented alleles are also obtained relatively frequently at the *scarlet* (Howells, 1979) and *brown* loci (Lovelock and Howells, unpublished). This seems to imply that it is the rate of uptake of precursors by the pigment cells which determines the rate at which the pigments are deposited. An interesting, but as yet unexplained, observation concerning the partially pigmented *white* alleles is that the level of drosopterins always seems to be reduced to a much greater extent than the level of xanthommatin (Table 4). Also unexplained, at present, are the actions of genes such as *lightoid*, *light*, *claret*, *pink* and *garnet* which appear to influence the transport systems for pigment precursors. Hence, in spite of the progress made in recent years, it is clear that our understanding of uptake, storage and transport phenomena, in relation to eye pigmentation and to the interaction between the two pigment pathways, is still far from complete.

7 Future prospects

During recent years, substantial advances have occurred in our understanding of the way in which the eye screening pigments in insects are made.

The enzymology of the ommochrome pathway is now well studied and has been investigated in a number of different insect species. Much has been learned about the importance of tissue interactions in xanthommatin production, about the uptake and storage of pigment precursors and about mutants defective in these processes. Great advances have also taken place in the area of pteridine biosynthesis since the review of Zeigler and Harmsen (1969) was published. At that time, the pathways of pteridine biosynthesis were only poorly understood, xanthine dehydrogenase being the only enzyme of pteridine biosnythesis which had been studied in any detail. Today the biosynthetic relationships between most of the major pteridines found in insects are reasonably well understood, and some information is available about properties of many of the enzymes of pteridine biosynthesis.

Despite these advances, there are still many deficiencies in our understanding of eye pigmentation. In the xanthommatin pathway, the nature of the final enzyme has still to be rigorously defined. Since this enzyme is crucial to the onset of xanthommatin synthesis, renewed efforts to characterise it further must be made. With regard to the pteridine pathways, information on the properties of the enzymes is still quite limited, and, in particular, little is known about the sepiapterin and drosopterin synthases, which catalyse the terminal reactions of two branches of the pathways.

In the areas of tissue interactions and of the uptake and storage of pigment precursors, there is still much to be explained. It remains to be shown whether genes such as *scarlet* and *brown* of *D. melanogaster* do code for permeases, or serve some other function, such as influencing the structure of storage granules. The nature of other transport mutants and the role played by the *white* gene in influencing the transport of precursors for both pigments must still be clarified. An important problem, requiring further study, concerns the nature of the relationship between the cells of the larval malpighian tubules and the pigment cells of the adult eyes. Although it is possible that the cells of the tubules simply provide a store of pigment precursors for use during the early stages of adult pigment synthesis, some results suggest that this storage is essential, in some as yet undefined way, for the later terminal differentiation of the pigment cells. Another area in which our knowledge is still very limited, concerns the importance of tissue interactions and of uptake and storage during pteridine synthesis.

Further study of the ultrastructure of the pigment cells is also required, in order to resolve some important questions concerning the structure of the pigment granules. Although these granules appear to be membrane-bound and to have an organised substructure in electron micrographs, it has been suggested that the xanthommatin granules may be devoid of protein and consist merely of aggregates of pigment (Wiley and Forrest, 1979). The organisation and structure of xanthommatin granules in xanthommatin-deficient mutants also seems to be influenced, in some way, by the presence or absence of pteridines. Whether this has relevance to the interaction which occurs in the production of both types of pigment needs to be investigated. Similarly, it needs to be shown in what ways the drosopterin binding granules in *D. melanogaster* are different from the granules which bind xanthommatin.

Perhaps the most fundamental unanswered questions about the eye pigmentation system in insects concern its applicability to the study of gene regulation in higher eukaryotes. Several genes involved in eye pigmentation in *D. melanogaster* (for example, *white*, *rosy* and *vermilion*) are well characterised genetically, so that attempts to determine their nucleotide sequences, using recombinant DNA techniques, could provide valuable information about the ways in which genes are organised and perhaps regulated in higher eukaryotes. The *white* gene has now been cloned (Bingham *et al.*, 1981) and research on the *rosy* gene is well advanced. It should soon be possible to compare directly the genetic and the molecular maps of these genes. However, it is in the study of the molecular mechanisms involved in tissue differentiation and cell specialisation that the eye pigmentation system may have unique potential. In few other developmental systems is there such a range of mutants available in which the final phenotypes of the cells are altered, often in quantifiable ways. It is to be hoped that as our understanding of eye pigmentation advances, it will become possible to utilise more fully the advantageous features of this developmental system to provide insights into the mechanisms of tissue differentiation and cell specialisation.

Acknowledgements

We thank N. M. Summers for preparing the diagrams for Figs 1 and 2. Unpublished work described in this article was supported by a grant from the Australian Research Grants Committee.

References

Badaway, A. A.-B. (1977). The functions and regulation of tryptophan pyrrolase. *Life Sci.* **21,** 755–768

Baglioni, C. (1959). Genetic control of tryptophan peroxidase-oxidase in *Drosophila melanogaster. Nature* (*London*) **184,** 1084–1085

Baillie, D. L. and Chovnick, A. (1971). Studies on the genetic control of tryptophan pyrrolase in *Drosophila melanogaster. Molec. gen. Genet.* **112,** 341–353

Bandlow, W. (1972). Membrane separation and biogenesis of the outer membrane of yeast mitochondria. *Biochim. Biophys. Acta* **287,** 105–122

Beadle, G. W. and Ephrussi, B. (1936). The differentiation of eye pigments in *Drosophila* as studied by transplantation. *Genetics* **21,** 225–247

Beadle, G. W. and Law, L. W. (1938). Influence on eye color of feeding diffusible substances to *Drosophila melanogaster*. *Proc. Soc. exp. Biol. Med.* **37**, 621–623

Becker, J.-L. (1978). Regulation of purine biosynthesis in cultured *Drosophila melanogaster* cells: I. Conditional activity of hypoxanthine-guanine-phosphoribosyltransferase and 5-nucleotidase. *Biochimie* **60**, 619–625

Bingham, P. M., Levis, R. and Rubin, G. M. (1981) Cloning of DNA sequences from the *white* locus of *D. melanogaster* by a novel and general method. *Cell* **25**, 693–704

Bolognese, A. and Scherillo, G. (1974). Occurrence and characterisation of a labile xanthommatin precursor in some invertebrates. *Experientia* **30**, 225–226

Bonse, A. (1969). Über das Auftreten von Pterinen, Tryptophan und dessen Derivate in verschiedenen Organen der Mutante *white* von *Drosophila melanogaster*. *Z. Naturf.* **24b**, 128–131

Brown, G. M. (1971). The biosynthesis of pteridines. *Adv. Enzymol.* **35**, 35–77

Bullock, T. H. and Horridge, G. A. (1965). *Structure and Function of the Nervous Systems of Invertebrates*. Freeman and Company, San Francisco and London

Butenandt, A. (1952). The mode of action of hereditary factors. *Endeavour* **11**, 188–192

Brugel, F. M. A. van (1973). Cell clustering and pleiotropy in white variegated eyes and malpighian tubes of *Drosophila hydei*. *Genetics* **75**, 323–334

Carlson, S. D. and Chi, C. (1979). The functional morphology of the insect photoreceptor. *Ann. Rev. Ent.* **24**, 379–416

Caspari, E. (1933). Über die Wirkung eines pleitropen Gens bei der Mehlmotte *Ephestia kühniella* Zeller. *Wilhelm Roux' Arch.* **130**, 535–581

Caspari, E. W. and Gottlieb, F. J. (1975). The Mediterranean meal moth, *Ephestia kühniella*. In: *Handbook of Genetics* (R. C. King, ed.) Vol. 3, pp. 125–147. Plenum, New York

Cassady, W. E. and Wagner, R. P. (1971). Separation of mitochondrial membranes of *Neurospora crassa*. I. Localisation of L-kynurenine-3-hydroxylase. *J. Cell Biol.* **49**, 536–541

Chovnick, A., Gelbart, W., McCarron, M., Osmond, B., Candido, E. P. M. and Baillie, D. L. (1976). Organisation of the *rosy* locus in *Drosophila melanogaster*: Evidence for a control element adjacent to the xanthine dehydrogenase structural element. *Genetics* **24**, 233–255

Clancy, E. B. (1940). Production of eye color hormone by the eyes of *Drosophila melanogaster*. *Biol. Bull.* **78**, 217–225

Cölln, K. and Klett, G. (1978). Über die Funktion des *alb*-locus in der Ommochrom-Endsynthese bei *Ephestia kühniella* Z. *Wilhelm Roux' Arch.* **185**, 127–136

Danneel, R. (1941). Die Ausfärbung überlebender *v*- und *cn*-*Drosophila*-Augen mit Produkten des Tryptophanstoffwechsels. *Biol. Zbl.* **61**, 388–399

Danneel, R. and Zimmerman, B. (1954). Über das Vorkommen und Schicksal des Kynurenins bei verschiedenen *Drosophila* Rassen *Z. Naturf.* **9b**, 788-792

Dorsett, D. L., Yim, J. J. and Jacobson, K. B. (1979). Biosynthesis of drosopterins in the head of *Drosophila melanogaster*. In: *Chemistry and Biology of Pteridines* (R. L. Kisliuk and G. M. Brown, eds) pp. 99–104. Elsevier North Holland, New York

Dustmann, J. H. (1968). Pigment studies on several eye colour mutants of the honey bee, *Apis mellifera*. *Nature* **219**, 950–952

Dustmann, J. H. (1975). Die Pigmentgranula im Komplexauge der Honigbiene *Apis mellifera* bei Wildtyp und verschiedenen Augenfarbmutanten. *Cytobiologie* **11**, 133–152

Eakin, R. M. (1972). Structure of invertebrate photoreceptors. In: *Handbook of Sensory Physiology* (H. J. A. Dartnall, ed.) Vol. 7/1, pp. 625–684. Springer Verlag, Berlin

Eichelberg, D. and Wessing, A. (1971). Electron microscopic studies of the renal tubules (Malpighian tubules) in *Drosophila melanogaster*. II. Transcellular membrane-bound transport. *Z. Zellforsch. Mikrosk. Anat.* **121,** 127–152

Ephrussi, B. (1942). Analysis of eye color differentiation in *Drosophila. Cold Spring Harb. Symp. quant. Biol.* **10,** 40–48

Evans, B. A. and Howells, A. J. (1978). Control of drosopterin synthesis in *Drosophila melanogaster*: Mutants showing an altered pattern of GTP cyclohydrolase activity during development. *Biochem. Genet.* **16,** 13–26

Fan, C. L. and Brown, G. M. (1976). The partial purification and properties of guanosine triphosphate cyclohydrolase from *Drosophila melanogaster. Biochem. Genet.* **14,** 259–270

Fan, C. L. and Brown, G. M. (1979). Partial purification and some properties of biopterin synthase and dihydropterin oxidase from *Drosophila melanogaster. Biochem. Genet.* **17,** 351–369

Fan, C. L., Krivi, G. C. and Brown, G. M. (1975). The conversion of dihydroneopterin triphosphate to sepiapterin by an enzyme system from *Drosophila melanogaster. Biochem. biophys. Res. Commun.* **67,** 1047–1054

Fan, C. L., Hall, L. M., Skrinska, A. J. and Brown, G. M. (1976). Correlation of guanosine triphosphate cyclohydrolase activity and the synthesis of pterins in *Drosophila melanogaster. Biochem. Genet.* **14,** 271–280

Finnerty, V. (1976). The simple cistrons: *rosy* and *maroon-like*. In: *The Genetics and Biology of Drosophila* (E. Novitski and M. Ashburner, eds), Vol. 1, pp. 721–765. Academic Press, New York

Fuge, H. (1967). Die Pigmentbildung im Auge von *Drosophila melanogaster* und ihre Beeinflussung durch den *white*$^+$-Locus. *Z. Zellforsch.* **83,** 468–507

Gelbart, W., McCarron, M., Collins, P. and Chovnick, A. (1974). Characterisation of XDH structural mutants within the *ry* locus in *Drosophila melanogaster. Genetics* **77,** s24–s25

Ghosh, D. and Forrest, H. S. (1967). Enzymatic studies on the hydroxylation of kynurenine in *Drosophila melanogaster. Genetics* **55,** 423–431

Gilmour, D. (1961). *The Biochemistry of Insects.* Academic Press, New York

Glassman, E. (1956). Kynurenine formamidase in mutants of *Drosophila. Genetics* **41,** 566–574

Glassman, E., Karam, J. D. and Keller, E. C. (1962). Differential response to gene dosage experiments involving the two loci which control xanthine dehydrogenase of *Drosophila melanogaster. Z. Vererb.* **93,** 399–403

Green, M. M. (1949). A study of tryptophane in eye color mutants of *Drosophila. Genetics* **34,** 564–572

Green, M. M. (1952). Mutant isoalleles at the *vermilion* locus in *Drosophila melanogaster. Proc. Nat. Acad. Sci. USA* **38,** 300–305

Green, M. M. (1955). Homologous eye colour mutants in the honey bee and *Drosophila. Evolution* **9,** 215–216

Grell, E. H. (1962). The dose effect of *ma-l*$^+$ and *ry*$^+$ on xanthine dehydrogenase activity in *Drosophila melanogaster. Z. Vererb.* **93,** 374–377

Grigolo, A. (1969). Kynurenine formamidase (formylase) activity in normal strains of adult *Musca domestica* and in strains mutant for eye colour. *Redia* **51,** 169–178

Hadorn, E. and Kühn, A. (1953). Chromatographische und fluorometrische Untersuchungen zur biochemischen Polyphänie von Augenfarb-Genen bei *Ephestia kühniella. Z. Naturf.* **8b,** 582–589

Hadorn, E., Graf, G. E. and Ursprung, H. (1958). Der Isoxanthopterin-Gehalt transplantierer Hoden von *Drosophila melanogaster* als nicht-autonomes Merkmal. *Revue suisse Zool.* **65,** 335–342

Handschin, G. (1961). Entwicklungs- und organspezifisches Verteilungsmuster der Pterine bei einem Wildstamm und bei der Mutante *rosy* von *Drosophila melanogaster. Devel. Biol.* **3,** 115–139

Hendrichs-Hertel, U. and Linzen, B. (1969). Über die Kynurenin-3-Hydroxylase bei *Calliphora erythrocephala. Z. vergl. Physiol.* **64,** 411–431

Hiraga, S. (1964). Tryptophan metabolism in eye colour mutants of the housefly. *Jap. J. Genet.* **39,** 240–253

Horikawa, M. (1958). Developmental-genetic studies of tissue cultured eye discs of *D. melanogaster*. I. Growth, differentiation and tryptophan metabolism. *Cytologia* **23,** 468–477

Howells, A. J. (1979). The isolation and biochemical analysis of a temperature sensitive *scarlet* eye color mutant of *Drosophila melanogaster. Biochem. Genet.* **17,** 149–158

Howells, A. J. and Ryall, R. L. (1975). A biochemical study of the *Scarlet* eye-color mutant of *Drosophila melanogaster. Biochem. Genet.* **13,** 273–282

Howells, A. J., Summers, K. M. and Ryall, R. L. (1977). Developmental patterns of 3-hydroxykynurenine accumulation in *white* and various other eye color mutants of *Drosophila melanogaster. Biochem. Genet.* **15,** 1049–1059

Judd, B. H. (1976). Genetic units of *Drosophila*—Complex loci. In: *The Genetics and Biology of Drosophila* (E. Novitski and M. Ashburner, eds) Vol. 1, pp. 767–799. Academic Press, New York.

Kaufman, S. (1962). Studies on tryptophan pyrrolase in *Drosophila melanogaster. Genetics* **47,** 807–817.

Kaufman, S. (1967). Pteridine cofactors. *Ann. Rev. Biochem.* **36(I),** 171–184

Krivi, G. G. and Brown, G. M. (1979). Purification and properties of the enzymes from *Drosophila melanogaster* that catalyse the synthesis of sepiapterin from dihydroneopterin triphosphate. *Biochem. Genet.* **17,** 371–390

Kühn, A. (1956). Versuche zur Entwicklung eines Modells der Genwirkung. *Naturwissenschaften* **43,** 25–28

Kühn, A. and Egelhaaf, A. (1955). Zur chemischen Polyphänie bei *Ephestia kühniella. Naturwissenschaften* **42,** 634–635

Langer, H. (1967). Über die Pigmentgranula im Facettenauge von *Calliphora erythrocephala. Z. vergl. Physiol.* **55,** 354–377

Langer, H. (1975). Properties and functions of screening pigments. In: *Photoreceptor Optics* (A. W. Snyder and R. Mengel, eds) pp. 429–455. Springer Verlag, Berlin

Laughlin, S. B. (1975). The function of the lamina ganglionaris. In: *The Compound Eye and Vision of Insects* (G. A. Horridge, ed.) Ch. 15, pp. 341–358. Clarendon Press, Oxford

Leibenguth, F. (1970). Concerning non-darkening mutant *Habrobracon* (*Bracon hebetor*) eyes as a consequence of a new chromogen-reducing mechanism in insect larvae. *Experientia* **26,** 659–660

Leibenguth, F. (1971). Zur Pleiotropie des *wh*- and *el*-Locus bei *Habrobracon juglandis. Z. Naturf.* **26b,** 54–60

Lindsley, D. L. and Grell, E. H. (1968). *Genetic Variations of Drosophila melanogaster*, Publication No. 627. Carnegie Institution of Washington

Linzen, B. (1963). Eine spezifische quantitative Bestimmung des 3-Hydroxy-kynurenins. 3-Hydroxykynurenin und Xanthommatin in der Imaginalentwicklung von *Calliphora*. *Hoppe-Seyler's Z. physiol. Chem.* **333,** 145–148

Linzen, B. (1974). The tryptophan → ommochrome pathway in insects. *Adv. Insect Physiol.* **10,** 117–247

Linzen, B. and Schartau, W. (1974). A quantitative analysis of tryptophan metabolism during development of the blowfly *Protophormia terrae-novae*. *Insect Biochem.* **4,** 325–340

MacIntyre, J. R. and O'Brien, S. J. (1976). Interacting gene-enzyme systems in *Drosophila melanogaster*. *Ann. Rev. Genet.* **10,** 281–318

Maeno, H. and Feigelson, P. (1967). Spectral studies on the catalytic mechanism and activation of *Pseudomonas* tryptophan oxygenase (tryptophan pyrrolase). *J. Biol. Chem.* **242,** 596–601

Meyerowitz, E. M. and Kankel, D. R. (1978). A genetic analysis of visual system development in *Drosophila melanogaster*. *Devel. Biol.* **62,** 112–142

Milani, R. (1975). The house fly, *Musca domestica*. In: *Handbook of Genetics* (R. C. King, ed.) Vol. 3, pp. 337–399. Plenum, New York

Mitchell, H. K., Glassman, E. and Hadorn, E. (1959). Hypoxanthine in *rosy* and *maroon-like* mutants of *Drosophila melanogaster*. *Science* **129,** 268–269

Moore, G. P. and Sullivan, D. T. (1975). The characterisation of multiple forms of kynurenine formamidase in *Drosophila melanogaster*. *Biochim. biophys. Acta* **397,** 468–477

Moore, G. P. and Sullivan, D. T. (1978). Biochemical and genetic characterisation of kynurenine formamidase from *Drosophila melanogaster*. *Biochem. Genet.* **16,** 619–634

Munz, P. (1964). Untersuchungen über die Aktivität der Xanthinedehydrogenase in Organen und während der Ontogenese von *Drosophila melanogaster*. *Z. Vererb.* **95,** 195–210

Murray, A. W. (1971). The biological significance of purine salvage. *Ann. Rev. Biochem.* **40,** 811–826

Muth, F. W. (1969). Quantitative Untersuchungen zum Pigmentier ungsgeschehen in den Augen der Mehlmotte *Ephestia kühniella*. III. Die Pigmentbildung in *a*-Stamm nach Injektion hoher Dosen von Kynurenin und der Versuch einer Deutung. *Wilhelm Roux' Arch. Entw. Mech. Org.* **162,** 56–77

Nickla, H. (1972). Interaction between pteridine synthesis and riboflavin accumulation in *Drosophila melanogaster*. *Can. J. Genet. Cytol.* **14,** 105–111

Nissani, M. (1975). Cell lineage analysis of kynurenine producing organs in *Drosophila melanogaster*. *Genet. Res. Camb.* **26,** 63–72

Nolte, D. J. (1952). The eye pigmentary system of *Drosophila*. III. The action of eye color genes. *J. Genet.* **51,** 142–186

Nolte, D. J. (1954a). The eye pigmentary system of *Drosophila*. IV. The pigments of the vermilion group of mutants. *J. Genet.* **52,** 111–126

Nolte, D. J. (1954b). The eye pigmentary system of *Drosophila*. V. The pigments of the light and dark groups of mutants. *J. Genet.* **52,** 127–139

Nolte, D. J. (1955). The eye pigmentary system of *Drosophila*. VI. The pigments of the ruby and red groups of mutants. *J. Genet.* **53,** 1–10

Nolte, D. J. (1959a). The eye pigmentary system of *Drosophila*. VII. The *white* locus. *Heredity* **13,** 219–231

Nolte, D. J. (1959b). The eye pigmentary system of *Drosophila*. VIII. Series of multiple alleles. *Heredity* **13,** 233–241

O'Brien, S. J. and MacIntyre, R. J. (1978). Genetics and biochemistry of enzymes and specific proteins of *Drosophila*. In: *The Genetics and Biology of Drosophila* (T. R. F. Wright and M. Ashburner, eds) Vol. 2a, pp. 396–551. Academic Press, New York

Ogawa, H. and Hasegawa, K. (1975). Kynurenine 3-hydroxylase activity and follicle development in the silkworm, *Bombyx mori*. *Insect Biochem.* **5,** 119–134

Okamoto, H., Yamamoto, S., Nozaki, M. and Hayaishi, O. (1967). On the submitochondrial location of L-kynurenine-3-hydroxylase. *Biochem. biophys. Res. Commun.* **26,** 309–314

Parisi, G. (1971). Interaction between pterin pigments and ommochrome pigments in *Drosophila melanogaster*. *Boll. Soc. ital. Biol. sper.* **46,** 1015–1016

Parisi, G., Carfagna, M. and D'Amora, D. (1976a). A proposed biosynthesis pathway of drosopterins in *Drosophila melanogaster*. *J. Ins. Physiol.* **22,** 415–423

Parisi, G., Carfagna, M. and D'Amora, D. (1976b). Biosynthesis of dihydroxanthommatin in *Drosophila melanogaster*: Possible involvement of xanthine dehydrogenase. *Insect Biochem.* **6,** 567–570

Phillips, J. P. and Forrest, H. S. (1970). Terminal synthesis of xanthommatin in *Drosophila melanogaster*. II. Enzymic formation of the phenoxazinone nucleus. *Biochem. Genet.* **4,** 489–498

Phillips, J. P., Forrest, H. S. and Kulkarni, A. D. (1973). Terminal synthesis of xanthommatin in *Drosophila melanogaster*. III. Mutational pleiotropy and pigment granule association of phenoxazinone synthetase. *Genetics* **73,** 45–56

Pyliotis, N. A. and Summers, K. M. (1980). Ultrastructure of pigment granules in the eyes of adults of the ommochrome-deficient mutant *yellow* of the Australian sheep blowfly, *Lucilia cuprina*. *Eur. J. Cell Biol.* **20,** 297–300

Rizki, M. T. M. (1961). Intracellular localisation of kynurenine in the fat body of *Drosophila*. *J. Biochem. biophys. Cytol.* **9,** 567–572

Rokos, K. and Pfleiderer, W. (1975). Isolierung, physikalische Eigenschaften und alkalischer Abbau der Augenpigmente Neodrosopterin und Aurodrosopterin aus *Drosophila melanogaster*. *Chem. Ber.* **108,** 2728–2736

Ryall, R. L. and Howells, A. J. (1974). The ommochrome biosynthetic pathway in *Drosophila melanogaster*. Variations in the levels of enzyme activities and intermediates during adult development. *Insect Biochem.* **4,** 47–61

Ryall, R. L., Ryall, R. G. and Howells, A. J. (1976). The ommochrome biosynthetic pathway of *Drosophila melanogaster*. The Mn^{2+}-dependent soluble phenoxazinone synthase activity. *Insect Biochem.* **6,** 135–142

Schartau, W. and Linzen, B. (1976). The tryptophan 2,3-dioxygenase of the blowfly *Protophormia terrae-novae*: Partial purification and characterisation. *Hoppe-Seyler's Z. physiol. Chem.* **347,** 41–49

Schneider, L., Gogala, M., Draslar, K., Langer, H. and Schlecht, P. (1978). Feinstruktur und Schirmpigment-Eigenschaften der Ommatidien des Doppelauges von *Ascalaphus* (Insecta, Neuroptera). *Cytobiologie* **16,** 274–307

Schultz, J. (1935). Aspect of the reaction between genes and development in *Drosophila*. *Amer. Nat.* **69,** 30–54

Schwabl, G. and Linzen, B. (1972). Über die Bildung der Augenpigmentgranula bei *Drosophila v* und *cn* nach Verfütterung von Ommochromvorstufen. *Wilhelm Roux' Arch.* **171,** 223–227

Schwink, I. and Mancini, M. (1973). The drosopterin pattern in various eye color mutants of the fruitfly *Drosophila melanogaster*. *Arch. Genet.* **46,** 41–52

Shoup, J. R. (1966). The development of pigment granules in the eyes of wild type and mutant *Drosophila melanogaster*. *J. Cell Biol.* **29,** 223–249

Stavenga, D. G. (1975). Optical qualities of the fly eye—an approach from the side of geometrical, physical and wavelength optics. In: *Photoreceptor Optics* (A. W. Snyder and R. Mengel, eds) pp. 126–144. Springer Verlag, Berlin

Sturtevant, A. H. (1932). The use of mosaics in the study of the developmental effect of genes. *Proc. 6th Int. Congress of Genetics* **1.** 304–307

Sullivan, D. T. and Kitos, R. J. (1976). Developmental regulation of tryptophan catabolism in *Drosophila. Insect Biochem.* **6,** 649–655

Sullivan, D. T. and Sullivan, M. C. (1975). Transport defects as the physiological basis for eye color mutants of *Drosophila melanogaster. Biochem. Genet.* **13,** 603–613

Sullivan, D. T., Kitos, R. J. and Sullivan, M. C. (1973). Developmental and genetic studies on kynurenine hydroxylase from *Drosophila melanogaster. Genetics* **75,** 651–661

Sullivan, D. T., Grillo, S. L. and Kitos, R. J. (1974). Subcellular localisation of the first three enzymes of the ommochrome synthetic pathway in *Drosophila melanogaster. J. exp. Zool.* **188,** 225–233

Sullivan, D. T., Bell, L. A., Paton, D. R. and Sullivan, M. C. (1979). Purine transport by malpighian tubules of pteridine-deficient eye color mutants of *Drosophila melanogaster. Biochem. Genet.* **17,** 565–573

Sullivan, D. T., Bell, L. A., Paton, D. R. and Sullivan, M. C. (1980). Genetic and functional analysis of tryptophan transport in malpighian tubules of *Drosophila. Biochem. Genet.* **18,** 1109–1130

Summers, K. M. (1979). *Biochemical genetics of eye pigmentation in the Australian sheep blowfly, Lucilia cuprina.* Ph.D. Thesis, Australian National University, Canberra

Summers, K. M. and Howells, A. J. (1978). Xanthommatin biosynthesis in wild type and mutant strains of the Australian sheep blowfly, *Lucilia cuprina. Biochem. Genet.* **16,** 1153–1163

Summers, K. M. and Howells, A. J. (1980a). Pteridines in wild type and eye colour mutants of the Australian sheep blowfly, *Lucilia cuprina. Insect Biochem.* **10,** 151–154

Summers, K. M. and Howells, A. J. (1980b). Functions of the *white* and *topaz* loci of *Lucilia cuprina* in the production of the eye pigment xanthommatin. *Biochem. Genet.* **18,** 643–653

Suzuki, D. (1970). Temperature-sensitive mutations in *Drosophila melanogaster. Science* **170,** 695–706

Tobler, J. (1975). Dosage compensation and ontogenic expression of suppressed and transformed *vermilion* flies in *Drosophila. Biochem. Genet.* **13,** 29–43

Tobler, J., Bowman, J. T. and Simmons, J. R. (1971). Gene modulation in *Drosophila.* Dosage compensation and relocated v^{+} genes. *Biochem. Genet.* **5,** 111–117

Tobler, J. E., Yim, J. J., Grell, E. H. and Jacobson, K. B. (1979). Developmental changes of sepiapterin synthase activity associated with a variegated *purple* gene in *Drosophila melanogaster. Biochem. Genet.* **17,** 197–206

Trepte, H.-H. (1978). *Ivory*: A recessive white-eyed tryptophan metabolism mutant with intermediate F_2 and R_1 progenies in the flesh fly, *Sarcophaga barbata. Theor. appl. Genet.* **51,** 185–191

Trujillo-Cenóz, O. (1972). The structural organisation of the compound eye in insects. In: *Handbook of Sensory Physiology* (M. G. F. Fuortes, ed.) Vol. 7/2, pp. 5–62. Springer Verlag, Berlin

Ursprung, H. and Hadorn, E. (1961). Xanthindehydrogenase in Organen von *Drosophila melanogaster. Experientia* **17,** 230

Wessing, A. and Bonse, A. (1962). Untersuchungen über die Speicherung und Ausscheidung von freiem Tryptophan durch die malpighischen Gefässe von *Drosophila melanogaster*. *Z. Naturf.* **17b,** 620–622

Wessing, A. and Danneel, R. (1961). Die Speicherung von Oxykynurenine in den malpighischen Gefässen verschiedener Augenfarbmutanten von *Drosophila melanogaster*. *Z. Naturf.* **16b,** 388–390

Wessing, A. and Eichelberg, D. (1968). Die fluoreszierenden Stoffe aus den malpighischen Gefässen der Wildform und verschiedener Augenfarbmutanten von *Drosophila melanogaster*. *Z. Naturf.* **23b,** 376–386

Wessing, A. and Eichelberg, D. (1972). Electron microscopic studies of the renal tubules (Malpighian tubules) in *Drosophila melanogaster*. III. Intracellular storage of the amino acid 3-hydroxykynurenine. *Z. Zellforsch. Mikrosk. Anat.* **125,** 132–142

Whitten, M. J., Foster, G. G., Arnold, J. T. and Konovalov, C. (1975). The Australian sheep blowfly, *Lucilia cuprina*. In: *Handbook of Genetics* (R. C. King, ed.) Vol. 3, pp. 401–418. Plenum, New York

Wigglesworth, V. B. (1972). *The Principles of Insect Physiology*, 7th Edn. Methuen, London

Wiley, K. and Forrest, H. S. (1979). *Drosophila melanogaster* lacks eye pigment binding proteins. *Biochemistry* **18,** 473–476

Wilson, T. G. and Jacobson, K. B. (1977). Mechanism of suppression in *Drosophila*. V. Localization of the *purple* mutant of *Drosophila melanogaster* in the pteridine biosynthetic pathway. *Biochem. Genet.* **15,** 321–332

Yamamoto, M., Howells, A. J. and Ryall, R. L. (1976). The ommochrome biosynthetic pathway in *Drosophila melanogaster*: The head particulate phenoxazinone synthase and the developmental onset of xanthommatin synthesis. *Biochem. Genet.* **14,** 1077–1090

Yen, T. T. T. and Glassman, E. (1965). Electrophoretic variants of xanthine dehydrogenase in *Drosophila melanogaster*. *Genetics* **52,** 977–981

Yim, J. J., Grell, E. H. and Jacobson, K. B. (1977). Mechanism of suppression in *Drosophila*. Control of sepiapterin synthase at the *purple* locus. *Science* **198,** 1168–1170

Ziegler, I. (1961). Genetic aspects of ommochrome and pterin pigments. *Adv. Genet.* **10,** 349–403

Ziegler, I. and Harmsen, R. (1969). The biology of pteridines in insects. *Adv. Insect Physiol.* **6,** 139–203

The Physiology of Caste Development in Social Insects

J. de Wilde and J. Beetsma*

Department of Entomology, Agricultural University, The Netherlands

*Dedicated to the memory of our friend Martin Lüscher

1 Introduction

Castes in social insects are examples of environment controlled polymorphism. Within a colony, insects of the same genetic constitution show differences in gene expression, leading to a number of characteristic syndromes. The variation in gene expression is mostly discontinuous, resulting in well-defined functional (and often morphological) categories: Queens, soldiers, workers. In some cases, a more transient caste expression prevails. The members of the colony respond to the characteristics of caste individuals and so regulate their numbers. There often seems to be a "norm" for complete caste expression, beyond which the incomplete or defective caste individual is eliminated. The factors inducing caste syndromes are imposed partly from the outer world and partly from within the colony. External factors always play their role indirectly, influencing the members of the colony responsible for the care of offspring.

In this review we will discuss the development of castes from the point of view of developmental physiology: induction, reversibility and programming of caste differentiation will be the main topics. In view of the complexity of the inducing factors, the physiology of caste-inducing mechanisms, as far as they are known, will be included. In addition to caste induction, the physiological functioning of the castes and its social implications (e.g. dominance) will also be described.

It may help to consider the evolutionary origin of the two main groups of social insects: The termites (Isoptera), a lower insect order related to the Dictyoptera, and the social Hymenoptera: social bees, social wasps, and ants.

The Isoptera only comprise social species: solitary Isoptera are unknown. The colonies are bisexual, sex determination is of the XX–XY type, and both sexes participate in caste formation. According to Cleveland (1926) and Cleveland *et al.* (1934), caste formation in the Isoptera probably emerged from the particular feeding habits of this group. Feeding on cellulose and lacking the digestive enzyme cellulase, the lower termites have to depend on flagellates for symbiotic digestion. Since the flagellates are lost at each moult, the newly moulted larvae need reinfection with the microorganisms which are obtained from their nestmates by a system of proctodeal trophallaxis. In the higher termites, social production of cellulase by means of fungus combs has become a major feature. As the termites are hemimetabolic, their larvae are capable of participating in worker functions.

In the social Hymenoptera, colony formation probably emerged from the care of offspring, which is exclusively carried out by females. Next to a minority of functional females, a considerable and sometimes very large number of sterile females is produced, and is kept in this state as long as the queens are present. Their behaviour is directed by the "labour market" within the colony. Sex determination in the Hymenoptera is of the haplodiploid arrhenotokous type. Unfertilised eggs, from which the males emerge, are only deposited during part of the season. There is no period when male eggs only are laid. The presence of the male sex is, therefore, limited to part of the season. Male castes are unknown. In exceptional cases, workers and even queens can develop from unfertilised eggs. As the Hymenoptera are holometabolic, the larvae do not actively participate in the social functions of the colony, and care of the brood is exclusively carried out by adult workers. The "advantage" of social life is in the efficiency of mass specialisation (division of labour, e.g. in foraging) and cooperative defence.

Individuals of functioning colonies of social insects, with the exception of hibernating queens of species with unstable colonies, are only viable in isolation for a very short spell of time. A normal physiological life span is only reached within the framework of the colony. Colonies of social insects are, therefore, often considered as "superorganisms" in which functions such as nutrition, respiration, defence, and reproduction have both an individual and a social aspect. Castes are the expression of the "division of labour" attained in the social insects in the performance of these functions.

It is an interesting feature that taxonomically different groups can show similarity in caste systems in relation to the ecological conditions under which they have evolved. Termites and ants show many similarities in caste differentiation, in social reproduction and in behaviour. Bees and wasps show similarities in comb building, flower visiting and defensive behaviour. The inability of surviving in isolation and the tendency to eliminate aberrant individuals constitute key problems in the experimental study of caste differentiation. Progress in this field is therefore slow. Nevertheless, data recently obtained justify a survey of the present state of knowledge. Recent general reviews have been given by Weaver (1966), Schmidt (1974), Wilson (1974), Hermann (1979) and on termites by Krishna and Weesner (1969); on wasps by Jeanne (1980); on social bees by Röseler (1975) and on the honey-bee by Beetsma (1979). A recent symposium on the endocrine aspects was organised by Lüscher (1976).

2 Caste development as a deviation from solitary development

There is no example of solitary termites, and therefore comparison is usually made with the related Dictyoptera. Especially the feeding habits and symbiotic digestion with the aid of flagellates makes this comparison interesting.

The situation is different in the Apoidea and Vespoidea. Solitary species occur as well as species that form societies demonstrating a wide range of increasing sociality (Wilson, 1974; Michener, 1974). Eusocial bee colonies are restricted to the *Halictinae*, *Xylocopinae*, *Bombinae* and *Apinae;* in the wasps to the *Polistinae* and *Vespinae*. Generally, in primitively eusocial bees and in social wasps of the temperate zones, colony life is interrupted during winter. The queens are solitary during hibernation and during the subsequent founding of the colony in spring. During this period they display diapause, foraging, building, and care of offspring, just like their solitary sister species. The sterile worker caste is the novelty in their evolution.

In the honey bees and stingless bees this is different, because in none of their species is the queen able to perform the functions mentioned above, and she can only found a new colony with the aid of a number of workers; the colonies are perennial and reproduce by swarming. Some authors consider the queen as the novelty and the worker syndrome as most comparable with the situation in solitary bees. The fact that the honey-bee queen larva is provided with "royal jelly", a substance to which, without any scientific reason, several mysterious properties are ascribed, contributes to this opinion. But neither an isolated worker nor an isolated queen is able to survive for any length of time. While the queen larvae are fed abundantly, the honey bees have developed a system of piecemeal nourishment in worker rearing that enables up to 2000 workers to emerge each day and a total of 35–40 000 being reared at the same time, while the resulting workers are remarkably uniform in size and shape. The developmental time of these workers is $1\frac{1}{2}$ times as long as that of the queen, and they are endowed with a pattern of highly complicated behaviour including a "language" and capacities of learning not found in any solitary bee species.

In accordance with Lukoschus (1956a) we will therefore consider the worker as the most aberrant morph in the social bees, and concentrate on factors determining the worker syndrome. We will avoid a one-sided concentration on the formation of reproductives.

According to Brian's studies (see later), caste determination in the ant *Myrmica rubra* is essentially worker determination. Worker development is a diversion of normal female reproductive development, the dorsal set of imaginal discs failing to grow and differentiate (Brian, 1965). This state of affairs is in agreement with the view expressed above.

3 The caste syndromes and their functioning

It cannot be the aim of this review, nor would it be within our competence, to give a picture of the variety of caste systems, covering even the main features. Moreover, recent reviews mentioned in the Introduction give ample information on our present knowledge in this respect. We will limit ourselves to some illustrative examples.

3.1 TERMITES

It is a general rule that all castes apart from the queen and king are immature. An example of the developmental possibilities in lower and higher termites is given in Figs 1 and 2.

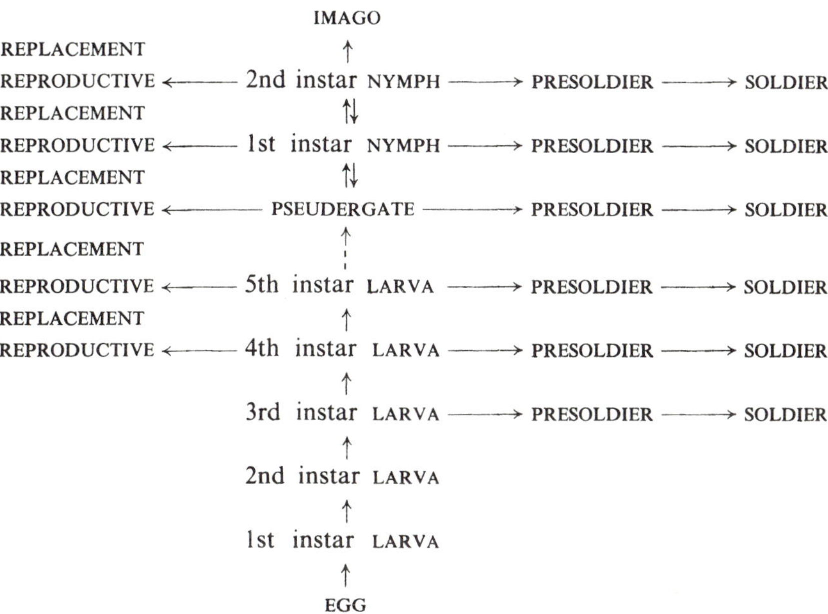

Fig. 1 Schematic representation of the developmental possibilities in the larvae of the termite *Kalotermes flavicollis*. Continuous arrows represent a moult, the broken arrow indicates a variable number of moults. (After Lüscher, 1976)

Among the lower termites, *Kalotermes flavicollis* is best studied. Figure 1 is based on the observations made by Grassi and Sandias (1896, 1897), Grassé and Noirot (1946a, b, 1947) and Lüscher (1952a, b). Next to larvae and nymphs (a term used to denote larvae with wingpads belonging to the line of reproductive development), the basic juvenile stages comprise pseudergates (a term introduced by Grassé and Noirot for large larvae which may develop progressively from 4th or 5th instar larvae or regressively from nymphs). Both sexes participate equally in all castes. Each basic juvenile stage and instar can develop in three directions: (*a*) the following stage in the line of reproduction leading towards adults, (*b*) replacement reproductives and (*c*) soldiers, via the obligatory intermediate phase of presoldier or "white soldier". In the larval and nymphal stages, regressive and stationary moults may occur. In the lower termites, the queen has a slightly inflated abdomen, but it does not attain the "physogastric" condition characteristic of the higher termites. The basic juvenile stages of the reproductive line are at the same time the workers.

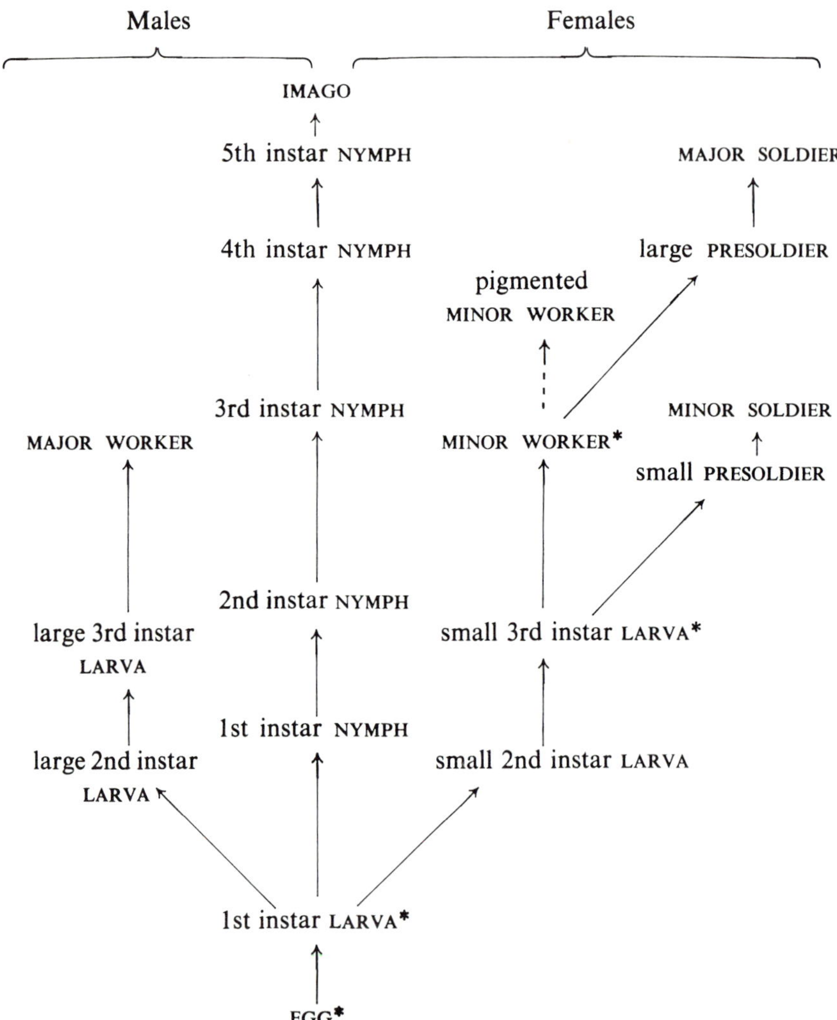

Fig. 2 Schematic representation of the developmental possibilities in the higher termite, *Macrotermes subhyalinus*. Continuous arrows represent a moult, the broken arrow represents pigmentation without a moult. The asterisks indicate the instar in which determination may occur. (After Lüscher, 1976, modified from Noirot, 1969)

The pattern of developmental possibilities in *Macrotermes bellicosus*, our example of the higher termites, is shown in Fig. 2, based on Noirot (1969). Next to the absence of pseudergates, the presence of two categories of workers and soldiers, minor and major ones, constitutes a main difference with the lower termites. Workers have three larval instars, and large workers are exclusively males while soldiers are exclusively females. Generalisation of this sexual specialisation is unwarranted, as there are other Termitidae where the reverse condition prevails. It should be noted that major and minor soldiers respectively emerge from minor workers and larvae. A very interesting early point of differentiation is in the first instar larvae, from where development may follow the reproductive or sterile line. In the lower termite, this occurs much later. In higher termites the reproductives (king and queen) are often encapsulated in a firm structure: the royal cell. The queen in most cases undergoes drastic physiological and morphological changes including body growth, the so-called physogastry. The abdomen of the queen may reach a volume of 500–1000% the original size. Two to several queens are observed in exceptional cases. The secretions and excretions of the queen are very attractive to the workers, who form a "court" around the queen not unlike the situation in the honey-bee.

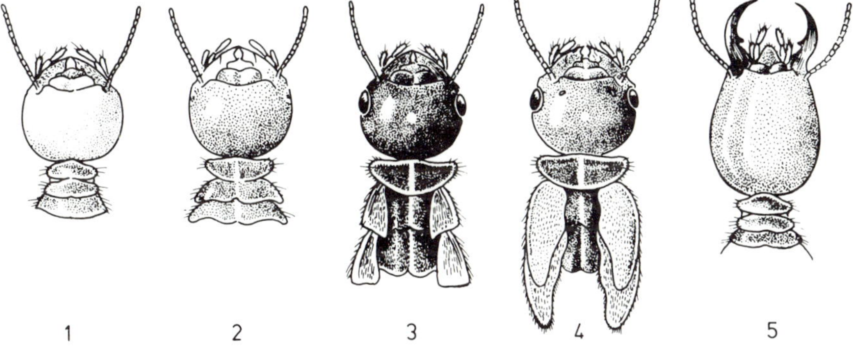

Fig. 3 Heads of the different castes of the termite, *Amitermes hastatus*. 1. worker; 2. tertiary queen; 3. primary queen; 4. secondary queen; 5. soldier. (After Skaife, 1954)

The soldiers are easily recognised by the enlarged head capsule, having characteristic large mandibles (Mandibulate soldiers), or have small mandibles but emit defensive secretions from a pore on the frontal projection of the head (Nasuti).

The lower termites comprise six families, among which are the *Kalotermitidae* and the Termopsidae. Only a few species have been thoroughly investigated: the American species *Zootermopsis angusticollis* and *Z. nevadensis* (Heath, 1927; Castle, 1934; Light, 1944; Miller, 1969), and the

European species *Kalotermes flavicollis* (Grassé and Noirot, 1946a, b, 1947, 1960; Lüscher, 1949, 1952a, 1956a, b; Lebrun 1967). Our knowledge of the social regulation of termite castes is mainly due to these studies. The caste system of *Zootermopsis* is practically identical to that of *Kalotermes*, to which we will mainly refer. In *Anacanthoterms* (Mednikova, 1977), caste development shows a tendency towards the higher termites.

The higher termites (family Termitidae) differ from the lower termites in two important respects. First, symbiotic flagellates do not play a role in digestion, but cellulose is either predigested by appropriate fungi (*Termitomyces* spp.) in the fungus combs or its intestinal digestion is carried out by anaerobe bacteria or by eating the white outgrowth of the fungal combs containing cellulase (Abo Khatwa, 1977). Second, their caste system is more elaborate but also more rigid. Here we lack the plasticity characterised by regressive moults and reversibility of lateral differentiation, prevailing in the lower termites. Caste differentiation is restricted to a few decisive steps in development. Immature forms do not participate in labour and are dependent on the worker castes (Fig. 3).

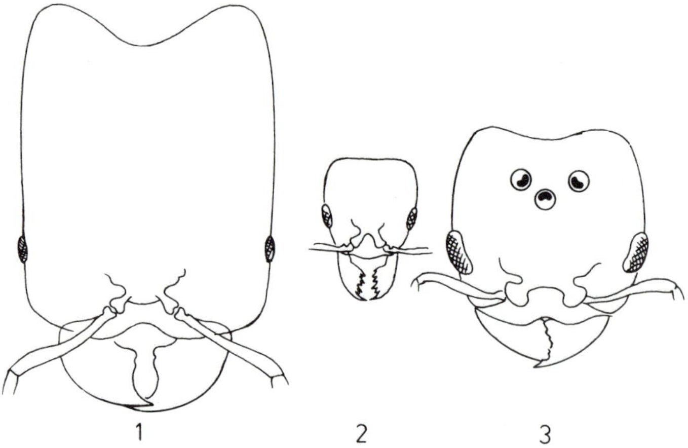

Fig. 4 Heads of the three castes of the ant, *Pheidole instabilis*. 1. soldier; 2. worker; 3. queen. (Redrawn after Wheeler, 1960)

As was mentioned above, the decision to develop along the reproductive line or in one of the two neuter (soldier, worker) lines occurs in the first larval instar (Kaiser, 1956). This has led to the supposition that this determination is blastogenic (Lüscher, 1976). However, since we know that caste determination in the honey-bee occurs in a covert way during early larval instars by the feeding condition of the larvae, the same could happen in the higher termites during the first larval instar. A point in favour of blastogenic determination is the extremely marked variation of the juvenile hormone concentration in the eggs of *Macrotermes subhyalinus* (Lüscher, 1976) to be discussed later. It was found by Greenberg and Stuart (1979) that in *Zootermopsis*, female alates can only have active ovaries when removed from the colony, while in neotenic imagines oogenesis will be inhibited under such conditions.

3.2 SOCIAL HYMENOPTERA

3.2.1 *General*

In the social Hymenoptera caste polymorphism is restricted to females. Queens and workers differ in social function and in physiology. The queens are the principal producers of reproductive eggs. Ant queens also produce trophic eggs, that are consumed by the queen or the larvae. Trophic eggs, that are laid by workers, are a part of the cell provisioning of stingless bees; the eggs are eaten by the queen. The workers are mainly involved in nest construction, collection and storage of food and water, care of all stages of brood and defence of the colony. Food is stored in storage cells (bees), in the larvae (wasps) or even in adult workers (ants). Repletes (living food reservoirs) can easily be distinguished in many ant genera. In *Solenopsis invicta*, the occurrence of repletes, in this case the major workers, can be demonstrated by feeding a dye (Glancey *et al.*, 1973).

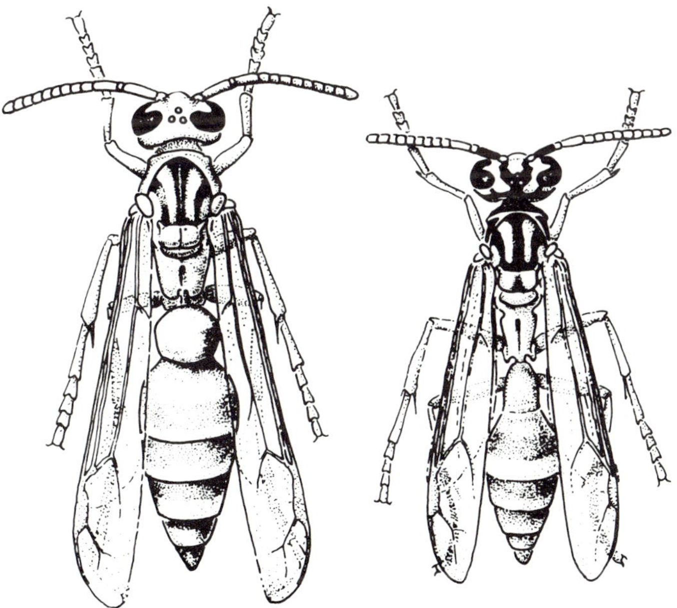

Fig. 5 Caste dimorphism in *Stelopolybia flavipennis*. Left, queen; right, worker. (After Evans and West-Eberhard, 1973)

3.2.2. *Differences between queens and workers*

Queens and workers appear as completely different morphs in ants, highly social bees and to a smaller degree also in vespine wasps (Blackith, 1958) (Figs 4–6). In polistine wasps, a wide range occurs between merely behavioural and physiological differences and complete queen-worker dimorphism (Richards and Richards, 1951).

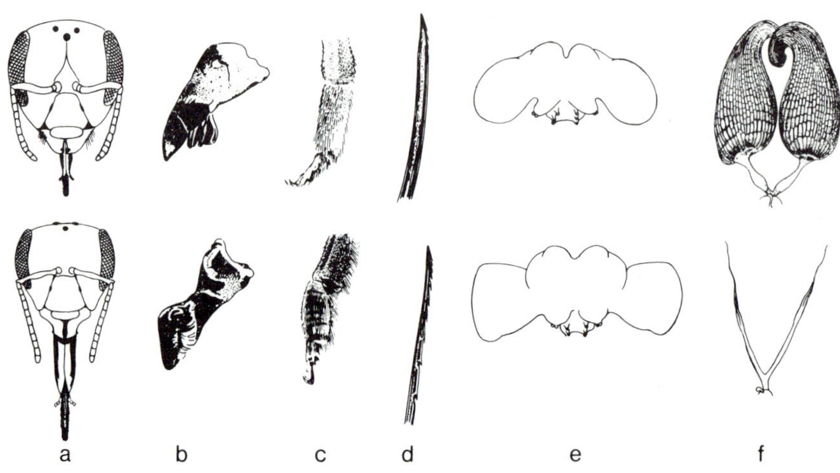

Fig. 6 Caste features of the honey-bee, *Apis mellifera*. (Top, queen; bottom, worker.) (*a*) Shape of head capsule and size of proboscis; (*b*) mandible; (*c*) tibia and basitarsus of hind leg; (*d*) sting; (*e*) brain; (*f*) ovary

It is characteristic for ants that workers never develop wings; queens usually develop wings but shed them after copulation. Within the same species two or three subcastes may occur: major workers (soldiers), media and minor workers. The ergatogynes are true intermediate morphs between the queen and worker castes. In primitively social bees, there is no or only a small difference in average size of queens and workers; e.g. in halictine bees (Knerer and Plateaux-Quénu, 1966; Plateaux-Quénu, 1967) and *Bombus* species (Cumber, 1949). In our opinion, it is a matter of debate whether individuals of intermediate size of halictine species should be classified as intercastes (Wilson, 1974). A large number of morphological and anatomical differences between honey-bee queens and workers have been summarised by Lukoschus (1956a). Some of these are shown in Fig. 6.

4 Caste inducing factors outside the colony

4.1 LOWER TERMITES

Seasonal effects on competence for progressive moults or for the formation of reproductives have been quantified by periodic sampling of *Kalotermes flavicollis* by Lüscher and Gast (Lüscher, 1974a) in southern France. Some results are shown in Fig. 7. First instar nymphs emerge in April or May and perform one or two stationary moults in the same year. In the next year they perform two progressive moults, becoming adults in August. Nuptial flights occur at this time in association with rainfall. Soldiers emerge from May–August. The number of progressive moults in nymphs and the number of replacement reproductives show parallel changes with a peak in early March. Which of the seasonal tokens provided by photoperiod, weather or food determine these activities, is as yet unknown.

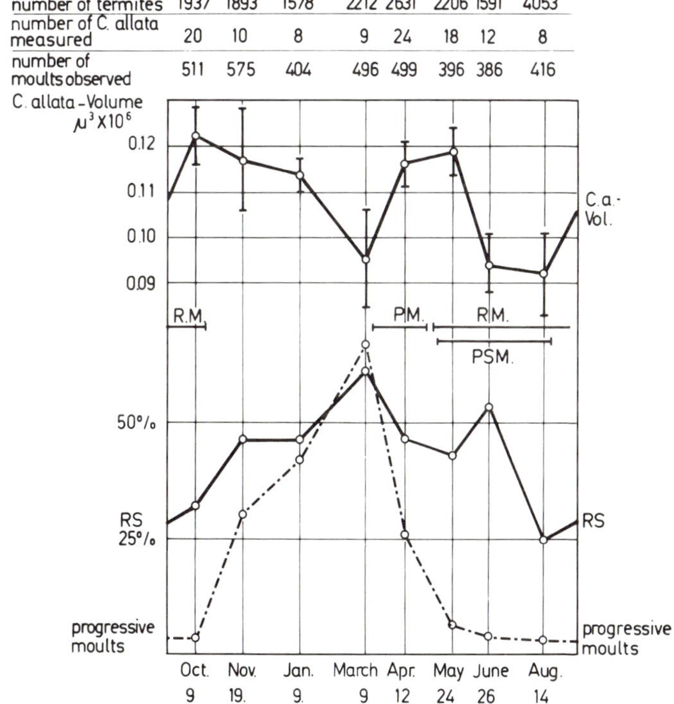

Fig. 7 Volume of CA, frequency of replacement reproductives and of progressive moults in first instar nymphs of the termite, *Kalotermes flavicollis*, under natural conditions in different seasons in southern France. C.a. Vol., volume of corpora allata; RS, percentage of individuals which developed into replacement reproductives after being isolated in groups of 50 without royal pairs; progressive moults, percentage of progressive moults (moults into 2nd instar nymphs in the remaining individuals of the group of 50 which did not develop into replacement reproductives). (After Lüscher, 1974a)

4.2 HIGHER TERMITES

In the higher termites so far studied, the formation of nymphs is determined by annual cycles. In the tropics nymphs differentiate during a limited period at the end of the rainy season. They subsequently develop into adults during the dry period and are ready for nuptial flights as soon as the rains have started. In exceptional cases, there are two reproductive cycles per year. In *Macrotermes subhyalinus*, the percentage of first instar larvae developing into nymphs is very low, and in some years no nymphs develop at all (Lüscher, 1976).

The other castes develop continuously, although their numbers may fluctuate considerably. This is especially true for the relative numbers of workers and larvae, but may also be valid for soldiers (Bodot, 1969, 1970, in *Cubitermes*).

4.3 SOCIAL HYMENOPTERA

It is generally accepted that the annual course of caste development is affected by seasonal fluctuations. Because social Hymenoptera evolved a rather stable microclimate in their nests and because food is stored, the care of the brood is not directly dependent on outside temperature. It appears that no universal and simple relation exists between the course of the favourable season for brood rearing and the size of the progeny produced or the period in which gynes emerge.

Gynes and workers of *Plagiolepis*, *Myrmica*, *Leptothorax*, and *Tetramorium* develop from hibernating 3rd instar larvae in spring or early summer, later only workers are produced (Brian, 1979).

In *Vespula germanica* and *V. vulgaris*, the worker/larva ratio increases when the colony grows (Archer, 1972; Spradbery, 1973), and colony size is significantly correlated with the size of the adults reared (Archer, 1972). As a rule, the first brood consists of workers, later brood nests also contain males and finally female reproductives (gynes) are produced. A similar situation is also found in primitively social bees. Whereas small workers are produced in spring, the summer nests yield large females that become either workers or gynes as in *Halictus ligatus* (Knerer and Plateaux-Quénu, 1966) or exclusively gynes as in *Evylaeus pauxillus* and *E. nigripes* (Knerer and Atwood, 1966). However, the average size of the progeny of *Lasioglossum zephyrum* decreases from early-summer to mid-summer, whereas the individuals produced in late-summer are the largest (Breed, 1975). In laboratory reared colonies of the same species, the size of the progeny appeared to be negatively related to day length and temperature, but positively related to the number of bees in the colony and the average size of the nurses (Kamm, 1974; Kumar, 1975). In bumble-bees, there also is no gradual increase in the average size of the workers.

At the end of the colony cycle a small number of large workers and gynes appear. Cumber (1949) observed that the queen reduces her oviposition rate prior to the production of gynes; the worker/larva ratio then rises to at least 1. A survey of the synchronisation of colony development and climatic cycles in social Hymenoptera is given by Brian (1977).

5 Caste inducing and regulating factors inside the colony

5.1 TERMITES

A primary factor regulating caste formation is the number of caste individuals already present. Regulation can take place by inhibition of caste differentiation of competent stages, by stimulation of competence, or by the elimination of caste individuals. Each caste appears to be regulated by the number of individuals of its own kind, and this even pertains to the two sexes. In addition, the strength of the colony as a whole is of importance. Methods for the investigation of caste determination in the higher termite *Macrotermes subhyalinus* were described by Bühlmann (1977a, b).

5.1.1 *Inhibitory effects*

It has been shown in *Kalotermes flavicollis* (Lüscher, 1974a) that alate adults do not develop in colonies with less than 250 individuals. A similar effect is observed in the formation of 2nd instar nymphs, which do not develop in colonies below 160 individuals. If first and second instar nymphs of larger colonies are introduced into colonies below the critical size, they undergo regressive moults. This system clearly prevents small colonies from being reduced still further. Furthermore, the presence of one pair of reproductive adults inhibits the formation of replacement reproductives. This has been shown in both *Zootermopsis* and *Kalotermes* (Light and Weesner, 1951; Grassé and Noirot, 1946a; Lüscher, 1951, 1952b, 1956a, b). In *Zootermopsis*, replacement reproductives are never present in intact colonies. They develop when the primary sexuals are eliminated.

The functioning female inhibits the differentiation of female replacement reproductives and this is potentiated by the presence of the male. In *Zootermopsis*, the functioning male preferentially suppresses the transformation of male larvae into replacement reproductives. In *Kalotermes*, at least two adult males are necessary to inhibit this transformation (Grassé and Noirot, 1960; Lüscher, 1964). The differentiation from nymphs to alates is not inhibited by the primary sexuals. In *Zootermopsis*, the above effects are only weakly exerted by the replacement reproductives themselves (Light, 1944).

Similar inhibitory effects regulate the formation of soldiers. In *Coptotermes formosanus*, Haverty (1979) observed that the percentage of soldier formation was strictly proportional to the size of the colony. In *Mastotermes*, on the contrary, Watson *et al.* (1977) found very constant soldier proportions in colonies widely differing in size.

Castle (1934) and Light (1943) showed that in *Zootermopsis angusticollis*, the first soldier emerging in a newly founded colony inhibits the differentiation of further soldiers. When the soldier is removed, a replacement individual is formed. In *Kalotermes flavicollis*, a significant increase of the percentage of larvae transforming into soldiers was shown in soldier-free colonies (Springhetti, 1969). Also Lenz (1976) mentions that in *Kalotermes* the differentiation of larvae or pseudergates into pre-soldiers was extensively reduced by the presence of soldiers. The inhibiting effect of the primary sexuals on the differentiation of larvae into replacement reproductives is much less persistent in *Kalotermes* (24 h) than in *Zootermopsis* (2–3 weeks) (Lüscher, 1952b, 1973).

Bordereau (1975) removed the royal pair from several colonies of *Macrotermes bellicosus* at a time when nuptial flights do not normally occur. Two months later, in the majority of colonies 2nd, 3rd and 4th instar nymphs were present. The same observations were made with *Macrotermes subhyalinus* (Lüscher, 1976). It thus appears that, as in the lower termites, the royal pair may have an inhibitory effect on nymph formation. This inhibitory effect is obviously less in large colonies; nymph formation occurs spontaneously when the colony size exceeds a critical level. Whether this effect is based on pheromones, is purely hypothetical (Noirot, 1974; Lüscher, 1976).

5.1.2 *Stimulating effects*

In *Kalotermes*, individual male sexuals stimulate the transformation of female larvae into replacement reproductives. This even happens with larvae or pseudergates which have lost their competence for this differentiation. Extracts of the heads of functional males selectively stimulate female larvae in the same way (Lüscher, 1964). This is apparently related to the need to regenerate queens in case they are lost. Results of experiments (Springhetti, 1969, 1970) with *Kalotermes flavicollis* indicate that the reproductives in a colony activate the formation of soldiers. Also pseudergates stimulate, to a lesser extent, other competent members of the group. This effect is increased with rising insect numbers (Springhetti, 1972).

5.1.3 *The nature of the regulating stimuli*

It has been presumed that the improved state of nutrition of the larvae, due to the removal of the primary sexuals, would be the crucial factor for regulating castes (Goetsch, 1940). It has also been hypothesised that the change in behaviour of the orphelined larvae would be responsible, via the "group effect" of the colony as a whole (Grassé, 1949). The most likely theory, which is best supported by experimental evidence, is that "token" substances are released by the differentiated castes. The substances have been referred to in the literature as ectohormones (Castle, 1934; Light, 1943), sociohormones, and, by the more generally accepted term of pheromones (Karlson and Lüscher, 1959).

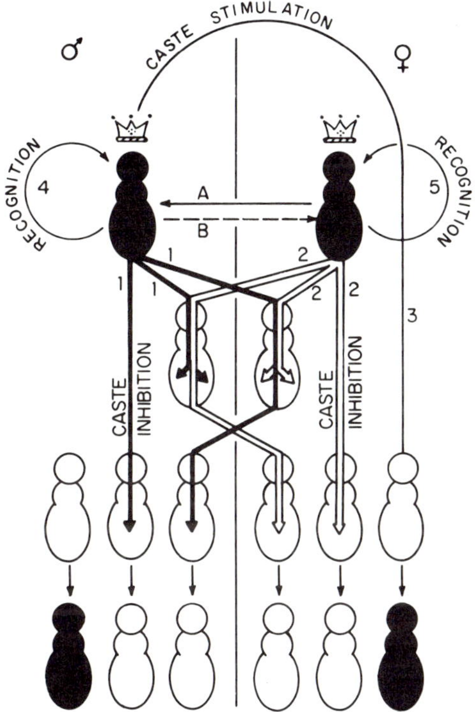

Fig. 8 Schematic representation of the known pathways in the control of reproductive caste formation in the termite, *Kalotermes flavicollis*. The crowned figures are the functional male (king) and female (queen) whereas the remaining figures are pseudergates. 1 and 2, queen and king pheromones inhibiting the formation of replacement reproductives; 3, male pheromone stimulating female pseudergates to change into the reproductive caste; 4, pheromone of supernumerary males, causing mutual aggression; 5, pheromone of supernumerary females with the same effect; A and B, stimuli of an unknown nature, inducing pheromones 1 and 2. (From Wilson, 1974, modified from Lüscher, 1961)

The inhibitory pheromone of the primary sexuals is distributed via the anus. This had already been shown in the experiments of Light (1944) with *Zootermopsis*, and confirmed by Lüscher (1955) with *Kalotermes*. The experiments consist of placing the sexuals in a diaphragm made in a wall which divides the two parts of a colony. At one side of the diaphragm, the abdomen was left free, at the other side the head. Only in the compartment containing the abdomen was the suppression of transformation of larvae into replacement reproductives observed. By covering different parts of the abdomen with varnish, Lüscher (1974a) showed that the pheromone is released by the anus. He established by similar experiments that the larvae pass on the pheromone by oral intake and anal release, probably in the frame of proctodaeal trophallaxis common in termites. Another approach was to make extracts of different parts of the body, already undertaken in the pioneer studies of Light (1944). Total extracts of female replacement reproductives significantly inhibited the formation of this caste in females. Extracts of heads gave a strong inhibition whereas extracts of the visceral organs were weaker. Extracts of the abdominal bodywall gave no inhibition at all. A scheme illustrating pheromonal involvement in the formation of reproductives in *Kalotermes* is given in Fig. 8. The nature of the suppressive effect of soldiers upon the formation of their own kind has not been further analysed, but a pheromonal effect is most likely.

None of the pheromones mentioned above in *Kalotermes*, *Zootermopsis* or any other lower termite species, has as yet been identified chemically.

5.1.4 *Mechanisms of caste elimination*

The elimination of replacement reproductives has been investigated in *Kalotermes flavicollis*. After removal of the primary sexuals, several replacement reproductives are formed, of which only one male and one female are retained (Grassi and Sandias, 1896, 1897). It has been shown that supernumerous replacement reproductives are eaten by the larvae and pseudergates. In the investigations by Ruppli and Lüscher (1964) and Ruppli (1969) it has been shown that, in the case of supernumerous replacement reproductives, fights take place between individuals of the same sex. Attacked individuals usually escape, but if one of the two partners is wounded during a fight, it is rapidly consumed by the larvae and pseudergates. Females may also eliminate surplus males. Elimination does not take place when the antennae of the replacement reproductives are removed, suggesting an olfactory recognition.

Lenz (1976) was able to increase, in *Kalotermes*, the number of presoldiers by the application of Juvenile Hormone analogues (see section 8). This did not affect the mortality rate of the existing old soldiers. But, as soon as the presoldiers moulted into soldiers, the mortality rate of the old soldiers increased from 11 to 50 percent. Whether this is due to malnutrition or to physiological stress, is unknown (see also Howard and Haverty, 1978).

5.2 SOCIAL HYMENOPTERA

Caste formation in the social Hymenoptera is regulated by the presence of queens. However, two aspects must be considered separately. Queens may, due to their dominant position in the colony, suppress the development of brood into gynes or the reproductive behaviour in adults. The latter aspect will be discussed in section 6.

5.2.1. *Queen pheromones and their effect on worker behaviour*

(*a*) *Ants* Although the role of the queens in the reduction or inhibition of queen rearing in ant colonies has been demonstrated, there is also evidence for the existence and function of queen pheromones.

When confined in cages containing blotting-paper, queens of *Solenopsis invicta* and *S. geminata* mark the substrate with substances that are highly attractive to workers. The attractiveness persists at least for 72 h and is reinforced by colony specific substances, because squares previously occupied by their own queen are twice as attractive for *S. invicta* workers than squares previously occupied by a foreign queen. The substances seem not to be species specific because *S. invicta* workers are strongly attracted by *S. geminata* queens; however in the reverse situation attractivity is poor. The queen substances, that can be extracted by hexane, are preferably called queen-tending pheromones (Jouvenaz *et al.*, 1974).

Concerning the inhibition of the production of sexuals in (polygynous) *Monomorium pharaonis*, two aspects have been studied. Whereas the presence of unmated queens does not suppress the production of new sexuals (Berndt and Nitschmann, 1979), mated queens give a complete inhibition except in the normal reproduction period. Berndt (1975) demonstrated the presence of an inhibitory substance on the body surface of mated queens. After a daily washing of the queens for 15 seconds with acetone, colonies started to produce sexuals as occurs in queenless colonies. Berndt (1975) suggests that the inhibitory substance is licked from the queen's body by the workers and distributed throughout the colony by trophallaxis. Although egg production of queens is continued in October, the author suggests that the pheromone producing glands become exhausted. The production of sexuals could not be delayed by mixing freshly killed queens with the food and only slightly by daily introducing freshly killed queens into the colony. On the other hand Buschinger and Kloft (1973) and Petersen-Braun (1977a, b) are not convinced by the abovementioned evidence and assume that production of sexuals depends on the quality of food the queens receive from the workers, (see section 5.2.2). Daily offering of dead *Myrmica rubra* queens to queenless colonies inhibited the development of queen pupae to a large extent. The average weight of all pupae was lowered when compared to queenless colonies. Only when the nurses could contact the queen bodies was the effect obtained. Carr (1962) concluded that the reduction of queen rearing was due to a relatively nonvolatile pheromone probably acting on feeding behaviour of the workers. Brian and Hibble (1963) also obtained inhibition of larval growth by applying ethanol extracts of queen heads to larvae and workers and by feeding it to them in sugar. The active compound occurred in the sterol fraction of the extracts (Brian and Blum, 1969). In 1975 Brian described the effects of the presence of dead or live queens on the feeding behaviour of workers. The presence of queens increased the amount of water and prey that older workers collected and fed to the larvae in autumn. Older workers were retained as nurses instead of becoming foragers. As a result, the growth rate of larvae was increased and workers developed. Larvae receiving less food developed slowly and entered into diapause. It seems that merely the quantity of food offered to the larvae is decisive. The proportion of water and prey collected is not changed, nor is the number of larvae that become workers affected by adding the contents of workers' pharyngeal glands (Brian, 1975b). In *Pheidole pygmaea*, Passera (1980b) found that, provided the workers have access to the queen or parts of the queen, no sexual larvae develop. The rate of inhibition is related to the size of the body-part offered. Taking away the queen's abdomen or antennae or fixing her to a substance with a wire, did not reduce her inhibitory efficacy. Sealing the buccal cavity of the queens with wax resulted in 6 out of 7 colonies with development of sexual larvae, production of reproductive eggs by the workers or both. Therefore, the author suggests that the queen's pheromones are produced by head glands.

(*b*) *Wasps* Pheromones have also been demonstrated in *Vespa orientalis* queens (Ishay *et al.*, 1965). Queens are very attractive, and workers demonstrate retinue behaviour; they lick her body, and especially her head and mouthparts. After queen removal, the workers start fighting, many even leave the nest and roam around. Brood care is disrupted, brood is eaten and after a few days some of the workers begin to lay eggs. Workers are attracted and calmed by an ethanol extract of queens. An important component of the queen's pheromones proved to be δ-*n*-hexadecalactone (Ikan *et al.*, 1969). This substance induces queenless workers to build large queen cells at the end of the season (September/October), a behavioural response that normally occurs only in the presence of an egg-laying queen. The opposite situation is found in the honey-bee.

(*c*) *The honey-bee* Pheromones of the honey-bee queen regulate worker behaviour, but one of them functions also as a sex attractant. Michener (1977) and Velthuis (1977) assume that the last mentioned function is the original one. When the queen is removed from the colony, the workers become restless and start fanning within 30 minutes; within a day they begin to transform a number of horizontal hexagonal worker cells containing young larvae into vertical round emergency queen cells (Butler, 1954). The young worker larvae in these cells previously fed on small amounts of worker jelly, receive more food of which the composition is changed (royal jelly).

It is assumed that swarm preparation, i.e. the construction of queen cups, which are provided with an egg by the queen, the subsequent care of the queen larvae and the completion of swarm queen cells, is due to a deficiency of "queen substance" (Butler, 1960). "Queen substance" is produced in the mandibular glands (Butler and Simpson, 1958) and is distributed over the queen's body when she grooms herself and is probably dispersed actively within the cuticle (Butler *et al.*, 1974). The main component of "queen substance" is *trans*-9-oxodec-2-enoic acid (9-ODA) (Butler *et al.*, 1961; Barbier and Lederer, 1960). This substance is perceived on the body of the queen by workers forming a retinue around her. These bees are contaminated with 9-ODA and function as "substitute queens". Sensory perception of 9-ODA (Beetsma and Schoonhoven, 1966; Kaissling and Renner, 1968) induces queenright behaviour of the workers (Verheyen-Voogd, 1959, Velthuis, 1972). This substance is not species specific; it is produced in the mandibular glands of different Apis species (Shearer *et al.*, 1970; Sannasi *et al.*, 1971). Other queen pheromones also play a role (Velthuis, 1970a; Vierling and Renner, 1977).

5.2.2. *Trophogenic factors*

(*a*) *Ants* In about 75% of *Pheidole pallidula* colonies, a part of the first brood produced in spring develops into sexuals. The initial rate of oviposition is high; large bipotential eggs are laid. Under laboratory conditions, a high rate of oviposition is maintained during 15 days, afterwards it decreases and small eggs yielding only workers are produced (Passera, 1980a). The production of bipotential eggs is independent of the number of nurses present and also from whether or not workers have nursed brood before. Even soldiers, which are nurses, are able to care for the sexual brood. Only in small colonies does the queen inhibit the development of sexuals; she causes the worker to neglect the sexual brood. According to Passera (1980a) in large colonies the ratio of workers to larvae has a greater impact on the development of sexuals (food condition) than the presence of the queen. The author concludes that caste in *P. pallidula* is determined by autogenic (blastogenic) determination followed by larval trophogenic determination. Soldiers can develop both from large and small eggs. Development of soldiers in the first spring brood is inhibited when the proportion of adult soldiers is more than 5 percent. When the number of workers increases, food conditions improve and soldiers are reared again (Passera, 1977).

In *Formica rufa rufo-pratensis minor*, Bier (1954) demonstrated that the blastogenic determination of eggs could experimentally be reversed by reducing the food conditions of larvae from winter eggs (normally yielding queens) and by improving the food conditions of larvae from summer eggs (normally yielding workers) when they were nursed by *F. rufa pratensis* workers. The queen has an important effect on these relations, an effect of the queen on feeding behaviour of the workers has also been demonstrated in *Plagiolepis pygmaea* (Bonavita and Passera, 1977). When workers rear 3rd instar post-hibernation larvae in queenright or queenless colonies, worker larvae result in the former, whereas queen larvae develop in the latter (Passera, 1969). Although the weight of both larval castes is the same during the first six days, it could be shown by ^{198}Au-labelling that prospective queen larvae receive more sugar containing crop contents per unit of body weight from the workers and grow faster than worker larvae reared in queenright colonies. It is suggested that the food of larvae of both castes differs only in quantity, though traces of head-gland products might be added (Bonavita and Passera 1977; Bonavita-Cougourdan and Passera, 1978). Caste induction seems to begin before differences in larval weight appear. At the 3rd day after hibernation, the number of protein fractions is different between larval castes (Passera, 1972). Introducing larvae of different ages from queenright colonies into queenless ones and the reverse (see Colombel, 1978) could yield additional information about the exact larval age at which caste differentiation is induced and determined.

When rearing Pharaoh's ant colonies under constant laboratory conditions Petersen-Braun (1977a) varied the composition of colonies and found that young and senile queens did not inhibit the rearing of sexuals, whereas complete inhibition is found in the presence of fully fertile queens. The author suggests that sexual development of larvae in the presence of the last mentioned queens is impossible because of the competition for worker glandular secretions between larvae and queens. Fully fertile queens are exclusively fed with these secretions, whereas young and senile queens receive honey and meat in addition. Caste development is affected by a blastogenic factor and food conditions. Young queens produce larger blastogenic worker-determined eggs, whereas senile queens lay smaller bipotent eggs. Sexuals can only be produced when young workers are present. Ledoux (1977) demonstrated in queenless *Aphaenogaster senilis* colonies that variations in temperature and feeding of vitamins considerably increased the number of alate females. In queenright colonies, the inhibitory action of the queen on the production of gynes is partially suppressed by these factors. The production of gynes (alate females) in *Odontomachus haematodes* has been studied by rearing larvae *in vitro* (Colombel, 1978).

In colonies containing an excessive number of queens, workers exclusively develop. In queenless colonies, some 29% of the larvae develop into gynes. After eggs are collected from normally functioning colonies, they are kept in liquid paraffin until the larvae hatch. First instar larvae are introduced into queenright and queenless colonies and collected at the beginning of the 2nd, 3rd and 4th larval instar to rear them *in vitro* up to the adult stage. Larvae previously reared under queenless conditions yield 6–15% of gynes; the percentage increases with the duration of their stay in the colony. However, the percentage of larvae developing into gynes decreases (5–0%) the longer they had previously been nursed in a queenright colony. It appears in combination with another experiment, in which larvae (early 3rd and 4th instar) were transferred from queenless to queenright colonies, that the programming of both queen or worker development cannot be changed during the last (4th) larval instar. Under these conditions, queen development can be changed nearly completely into worker development during the 3rd instar; only 1% of gynes resulted. Feeding behaviour is affected by the presence of queens.

Food quantity and quality may be adjusted according to larval size. By using ^{32}P-labelled food, Markin (1970) observed that the smaller worker larvae of *Iridomyrmex humilis* receive more head-gland products than larger ones. Only 20% of the smallest worker larvae had received sugar after 48 h, whereas the larger larvae all had consumed sugar or proteins.

In a queenless colony of *Odontomachus haematodes*, nurses immediately respond to the begging movements of the larvae. The larvae indicate their need of food by quickly extending and retracting the frontal part of their body and by moving their mouth parts. Some workers seem to be especially involved in nursing of large larvae; they lick and touch the larvae which in turn respond with begging movements. These interactions lead to an increased food intake of large larvae. Under queenright conditions, the workers more or less neglect the large larvae, which seem to receive about the same amount of food as smaller prospective worker larvae. Development of *O. haematodes* larvae into queens *in vitro* can probably be improved by selecting those larvae that respond quickly with begging movements after they are touched with the hairs of a brush. Slowly responding larvae can be fed less often (Colombel, 1978).

Le Masne (1953) found differences in the frequency and intensity of begging behaviour between queen and worker larvae. Larvae are regularly tested by the workers for their food requirements. Attractivity of larvae and pupae of worker army ants (*Neivamyrmex opacithorax*) is caused by secretions (Watkins and Cole, 1966). Cold hexane extracts of *Solenopsis saevissima* larvae applied to different substrate particles caused these to be transferred to the brood nest. The treated particles were not only collected but also groomed and rubbed with the antennae and palpi. After the pheromone has disappeared or because additional brood signals are lacking the particles are discarded. The effect of extracts of sexual larvae is stronger than that of worker larvae (partial reaction without dilution) and can be demonstrated with diluted extracts (1: 10 and some response when using 1: 100) (Glancey *et al.*, 1970). The pheromone of sexual brood (triolein) was identified by Bigley and Vinson (1975). In *Myrmica rubra*, the presence or absence of queens determines whether worker-biased or queen-biased larvae respectively are fed preferentially in spring. Queen-biased larvae are even attacked in the presence of queens (Brian, 1973). The two types of larvae can be distinguished by the workers because of the production of caste specific skin exudates (Brian, 1975a).

In considering the demonstrated and suggested factors that play a role in caste differentiation of ants we wholeheartedly agree with Delage-Darchen (1977): Caste differentiation is probably regulated by simple fundamental mechanisms. However, in every ant species studied various aspects of these mechanisms appear; this makes it impossible to come to a generalised conception of how castes in ants originate.

(*b*) *Wasps* When feeding ^{198}Au-labelled honey to colonies of *Paravespula vulgaris* and *P. germanica*, Montagner (1963) demonstrated that the quantity of food intake of the larvae increases with their length. However, when compared on a per unit of length basis, queen larvae receive more food than worker larvae. Larval food contains secretions from mandibular, labial and thoracic glands. Montagner (1966) found that 48 h after feeding the workers with labelled food, these glands become highly radioactive. Because this radioactivity is found afterwards in the crop, the author suggests that gland secretions are fed to the larvae by way of the crop. When treated workers simultaneously feed queen and worker larvae, queen larvae receive much more of the labelled gland secretion(s) than worker larvae. The author suggests that cell size might be a factor regulating the amount and the quality of the food offered to the larvae. The impact of cell size on adult body size has been demonstrated for *Vespa crabro* males reared in queen and worker cells. A distinct bimodality is obtained in the wing length of the males

(Spradbery, 1973). Preferential feeding of *P. germanica* queen larvae might also be due to the fact that they offer the nurses smaller amounts of concentrated saliva containing 3 times as much sugar. Worker larvae produce a lot of diluted saliva (Ishay, 1975). Brian (1979) suggests that the concentrated saliva could make queen larvae more attractive to the nurses. Both in *Paravespula* species (Montagner, 1967, see Spradbery, 1973) and in *Vespa crabro* (Ishay, 1975) it has been demonstrated that 3rd instar larvae from queen cells transferred into worker cells yielded workers. However, when 3rd instar larvae from worker cells are transferred to queen cells, the increased food provision causes these larvae to develop into queenlike intercastes and gynes. When *V. crabro* queen larvae are transferred at an older age, normal development seemed to be impossible; the larvae disappeared. In the reverse situation, only workers could be obtained. From these experiments it appears that the induction of queen development has to start earlier than the 3rd instar to obtain completely developed queens. After the 3rd instar, worker development seems to be determined. Ishay (1975) suggests that cell size indirectly determines the caste of the developing larvae. Workers of *Vespa orientalis* and *P. germanica*, when offered large cells from other species, induce the development of worker brood by reducing the size of these cells under queenright conditions. According to the author the workers can discriminate between larvae of both castes.

(*c*) *Bumble-bees* Both *Bombus hypnorum* and *B. terrestris* are pollen-storers. The queen starts a colony in spring, and intially only workers emerge from the eggs she deposits. The larvae are progressively provisioned by the queen and the workers. Queens of *B. terrestris* are distinguished from workers by their size and by differences in the size of their fat body, lipid and protein contents of haemolymph, and behaviour. In *B. hypnorum*, mainly the last mentioned physiological and behavioural parameters must be used to distinguish between both castes because of the occurrence of intermediate sizes (Table 1). Caste is not determined in the egg in *B. hypnorum*. When the number of larvae is experimentally reduced or when workers are added, queens develop even from the first batch of eggs deposited by the queen in spring. Queen rearing occurs when the worker/larva ratio is higher than 1: 2 (Röseler, 1967, 1970). The trophogenic aspect of caste differentiation was also demonstrated in *B. pratorum* (Free, 1955). However, the presence of the queen in *B. terrestris* inhibits queen rearing completely, even when the worker/larva ratio was increased to 3: 1. Older queens are no longer able to inhibit the development of queens. Röseler (1970) suggests that queen pheromones are involved in this process. By dividing a colony into a queenright and a queenless part and using different types of screens and by interchanging workers, the author demonstrated that the queen affects the feeding behaviour of the workers. In *B. hypnorum*, the older queen larvae are fed more often than older worker larvae. This increase in frequency of feeding in *B. terrestris* queen larvae occurs after the worker larvae start to pupate (Röseler and Röseler, 1974). In contrast to the honey bee, the duration of the larval and pupal stages in prospective queens of *B. terrestris* is extended in comparison to the situation in prospective workers. Röseler (1970) found that larvae taken from a colony when 3½ days old or more are determined to become workers; under queenless conditions they develop into workers only. The development of larvae that have been nursed in the colony for a shorter period can be changed into that of a queen when reared under queenless conditions. A scheme of the factors involved in caste determination in the above two *Bombus* species is given in Fig. 9.

TABLE 1 Caste characteristics in *Bombus hypnorum* and *Bombus terrestris*. (After Röseler, 1975b)

	B. hypnorum		*B. terrestris*	
Feature	Indifferent or size dependent	Caste-specific	Indifferent or size dependent	Caste-specific
External morphology	+		+	
Body size	+			+
Inner organs	+		+	
Larval duration	+			+
Pupal duration	+			+
Hibernation		+		+
Behaviour		+		+
Haemolymph		+		+
proteins, days 2–5				+
Oogenesis	+			

Although it is obvious that queen larvae receive more food than worker larvae, no data are available about differences in food quality e.g. contents of secretion from the hypopharyngeal or mandibular glands. Röseler (1974, 1976) suggests the possible role of a qualitative factor such as farnesol on caste differentiation.

(*d*) *Stingless bees*

(*i*) Trigonini. As far as the mechanism of caste differentiation in the genus *Trigona* has been studied, it seems to be regulated by larval food quantity. Workers develop in small cells whereas queens develop in large marginal cells of the horizontal combs. However in *T. julianii* normal queens grow up in queen cells, small queens in worker cells (Juliani, as described by Michener, 1974). Queen cells contain two times or more the quantity of food found in worker cells (Velthuis, 1976). Generally, the larvae are mass-provisioned. Experimentally, queen differentiation could be obtained in four African *Trigona* species (Darchen and Delage, 1970, Darchen and Delage-Darchen, 1971), by transferring full-grown worker larvae into newly provisioned worker cells. It seems unlikely that caste or species specific food components play a role. The same effect could be obtained by offering larval food taken from worker cells of another species. Queens developed even after adding a honey-pollen mixture to the food of worker larvae (Darchen, 1973). Restricting the food consumption of *Axestotrigona eburnensis* larvae in large queen cells resulted in adult workers (Darchen and Delage-Darchen, 1971). Miniature queens and workers are also obtained by undernourishment. The queen-determining effect of offering additional food to worker larvae reared *in vitro* has also been demonstrated for the Brazilian species *Scaptotrigona postica*. By offering different food quantities, queens, intercastes or workers could be obtained (Camargo, 1972). As in several *Meliponini* and other *Trigonini*, cells of *Leurotrigona muelleri* are arranged in irregular clusters, that are generally connected by wax pillars. In case a pillar is lacking and cells are constructed side by side, older larvae are able to enter the neighbouring cell and to consume another food portion. These larvae develop into queens (Terrada, as described by Velthuis, 1976).

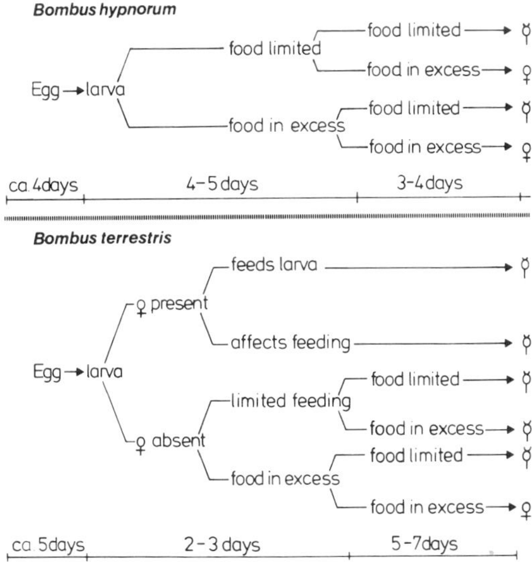

Fig. 9 Schematic representation of the factors involved in caste determination of *Bombus hypnorum* and *Bombus terrestris*. (From Röseler, 1975)

From these experiments it could be concluded that the programming of queen differentiation begins in the last larval instar. When additional food is provided, queen determination can be initiated. Questions still have to be answered about deviations in worker behaviour resulting in the construction and provisioning of large queen cells and in the case of *Leurotrigona muelleri* mentioned above where wax pillars between cells have been omitted.

(*ii*) Meliponini. In the genus *Melipona*, queens and workers are reared within the same type of cells. Queen brood is distributed at random between worker brood. Adult queens can be distinguished from workers by their smaller heads and thoraces. The hypotheses about the mechanism of caste differentiation include factors such as genetic constitution, food quantity and food quality.

Genetic factors. When determining the queen-worker production ratio in colonies of eight different *Melipona* species, Kerr (1969) found that the percentage of queens has a maximum of about 25. The only exception was *M. flavipennis* which had 7.3% of queens. This and other observations led to the well-known hypothesis that caste in *Melipona* bees must be determined by genetic factors. Assuming that two pairs of caste determining alleles are involved, that queens mate but once, and provided that a sufficient amount of larval food is present, only the double heterozygous females develop into queens. All individuals that are homozygous for one or both pairs of caste determining alleles result in workers. When a suboptimal food quantity is present, the queen-determined larvae develop into workers too. A similar queen-worker ratio was found by Camargo *et al.* (1976) when rearing larvae of four *Melipona* species *in vitro*.

Food quantity. Several environmental factors had no effect on the queen-worker ratio in *M. anthidioides* except the quantity of food deposited and consumed (Kerr *et al.*, as described in Schmidt, 1974). These authors found that the total amount of food deposited in a cell is inversely correlated with the number of workers involved. Cells provisioned by 10–12 workers contained less food and yielded only workers. However both queens and workers develop from cells provisioned by 4–8 workers; these cells contain the largest amount of food. The effect of the amount of food consumed could also be demonstrated by measuring the pupal weights of queens and workers of *M. quadrifasciata* (Kerr *et al.*, as described in Schmidt, 1974). Again the same queen-worker ratio (1:3) was found in pupae weighing between 78 and 108 mg. Pupae weighing less than 72 mg developed into workers only.

Food quality. The mechanism of caste differentiation of the Mexican species *M. beecheii* seems to be different. When Darchen and Delage-Darchen (1974, 1975) diminished or increased larval food quantity they obtained in separate combs of three colonies 0–20% and 36–71% of queens respectively. According to Darchen and Delage-Darchen (1975), differences in the quantity of glandular secretion present in the food could also play a role. Since several workers provision a cell, different food compositions could be expected. According to Velthuis (1976), this is not in agreement with the observation that under natural conditions no intercastes occur, unless caste differentiation depends on a very sensitive mechanism in the larvae which responds to small quantitative and qualitative differences in the food. In our opinion, it could as well be suggested that some workers, responding differently to the behaviour of the queen, deposit food of a different composition. If this food could be recognised by subsequent workers, they could add food of similar quality. This suggested mechanism results in two types of larval food and could explain why no intercastes develop.

(*e*) *The honey-bee* The primary factors affecting queen or worker production emanate from the queen. Under queenright conditions, the females produced are only workers. When the presence of queen pheromones has diminished or completely ceased, queens are reared while the rearing of worker brood is continued. Differences in shape, size and orientation of queen and worker cells may be secondary releasers of specific feeding behaviour of the nurses. By transferring worker larvae of different ages into queen cells, Weiss (1971) demonstrated that in these individuals queen characteristics only develop completely when these larvae are less than 3 days of age. Size and weight of the adult queens are affected by the food conditions beginning directly after hatching of the larvae (Woyke, 1971; Weiss, 1974). By transferring queen larvae of different ages into worker cells and the reverse, Weiss (1978) concludes that the duration of the bipotent phase during the development of both castes is different.

Differential feeding behaviour of nurse bees towards both castes was observed by Jung-Hoffman (1967). Through small windows in the wall or bottom of the cell, she could distinguish between three components of food deposited by the bees. Worker larvae were fed with a white, a clear and a yellow component in the ratio 2: 9: 3. The white component was thought to be a mixture of the secretions of the hypopharyngeal and the mandibular glands. The clear one originated from the hypopharyngeal glands and the honey stomach. The pollen-containing yellow component appeared in cells with worker larvae aged 3½ days or more. Queen larvae received only the white component during the first three days, and thereafter a combination of the clear and white components in the ratio 1: 1. Queen larvae, therefore, received proportionally more mandibular-gland secretion than worker larvae did. According to the observations by Lindauer (1952), queen larvae are provisioned about ten times as often as worker larvae.

In normal colonies, royal jelly (RJ) always contains more pantothenic acid and biopterin than worker jelly (WJ). According to Lingens and Rembold (1959), RJ of queen larvae (3 days old) contains 4 times as much pantothenic acid than WJ of larvae of the same age; the amount of biopterin in RJ was 3.5–17.5 times as high as in WJ of young larvae. According to Hanser and Rembold (1964), the mandibular gland secretion of queenless workers contains, on average, 8 times as much pantothenic acid and 115 times as much biopterin as is found with queenright workers. No difference was found in the secretion of the hypopharygeal glands. Rembold (1969a) concludes that worker larvae are fed with the product of the hypopharyngeal glands only, whereas queen larvae receive secretion of the mandibular glands as well.

Rembold (1969a) suggests that in a queenless colony, nurse bees specialise in feeding either queen or worker larvae, and that this specialisation would be irreversible. However, when comparing the abovementioned amounts of pantothenic acid and biopterin in RJ, WJ and the mandibular glands of queenright and queenless nurse bees the conclusion of Rembold (1965, 1969a) would, therefore, not seem to be justified. In our opinion, the results indicate rather that a number of nurse bees behave as "generalists", feeding larvae of both castes. Free (1960) suggests that each nurse bee deliberately varies the composition of her secretions to conform to the status of each larva she visits.

When Weaver (1955, 1958) reared worker larvae *in vitro* on stored RJ, he obtained queens, intercastes and workers. He concluded that RJ contains a very labile principle which induces queen differentiation. Rembold and Hanser (1964) extracted a dialysable factor from RJ which they named queen-determining principle. This partly purified factor has hydrophilic properties and a low molecular weight, but its structure has never been elucidated. According to Rembold (1969a) the factor originates from the mandibular glands, but direct proof is still lacking. However, the significance of substances from the mandibular glands on caste differentiation must be doubted. Hanser and Rembold (1964) have already demonstrated that pantothenic acid, biopterin and neopterin are not involved in caste differentiation.

Although the weight of queen and worker larvae seems to increase at nearly the same rate in the initial phase, the rate of respiration from the 2nd–3rd day onwards is higher in queen larvae (Melampy and Willis, 1939). In three-day-old queen larvae, 6 times as many mitochondria are found as in worker larvae (Osanai and Rembold, 1968). Therefore, the rate of food intake of young queen larvae must be considerably higher than that of young worker larvae. When rearing young worker larvae *in vitro* on ^{32}P-labelled RJ, Dietz and Lambremont (1970) found that larvae consuming a high amount of food developed into queens but when smaller amounts were consumed, workers resulted. In this respect, the difference in sugar content between RJ and WJ is remarkable. Shuel and Dixon (1959) found about 34% of sugars in the food of queen larvae 1–4 days old, whereas the food of worker larvae 0–1¼ days old contained only 12%. The sugar content increases to 47% in the food of worker larvae 3–4 days old.

Asencot and Lensky (1976), who obtained only a few adult workers after adding small amounts of glucose and fructose to WJ, obtained a majority of queens and intercastes when adding 200 mg of glucose and 200 mg of fructose per gram of WJ. The authors suggest that the sugars stimulate food intake in the honey-bee larva, as has been reported from many insect species. Adding 80 mg of glucose and fructose each per gram of WJ increased food intake from 212 till 366 mg. As a result, larvae weighing over 220 mg when spinning develop into queens, while larvae with a spinning weight below 180 mg become workers.

The results of Asencot and Lensky (1976) are in agreement with the observation that the sense organs of larval mouth parts are sensitive to sugars and salts, but not to other food components, including extracts obtained from Rembold, presumably containing the "queen-determining principle" (Goewie, 1978). Shuel *et al.* (1978) include in their studies the effects of RJ, WJ, additional sugar and 10-hydroxy-2-decenoic acid and topical application of juvenile hormone, when rearing larvae *in vitro*. The authors conclude that the contents of 10-hydroxy-2-decenoic acid, sugar and other nutrients play a role in caste differentiation.

When the experiments carried out with different social Hymenoptera are considered, the food condition of the larva (in addition to blastogenetic factors) seem to be an essential factor in queen development. Generally, female larvae have to consume a certain amount of food and surpass a certain weight to develop into queens. The larger amount of food consumed by larvae of prospective queens is not merely expressed in a larger size of the larvae or adults.

5.2.3 *Mechanisms of caste elimination*

Polistine wasp colonies are started by a single foundress, which may be joined by others subsequently, or by swarms consisting of several queens and many workers. In *Polistes gallicus*, one of the foundresses becomes the dominant principal egg-layer or queen; some of the other females initially may produce some eggs, but soon their ovaries regress and they function as workers. These females are eliminated as soon as the worker brood emerges. The colony becomes monogynous again (Pardi, 1948). In some swarm-founding *Polistinae*, 1–15% of queens are present at the beginning of nest construction. However, the vigour of the queens is constantly tested by the workers. The weaker queens are either driven from the nest or function as workers. Before the emergence of the first brood only one or a few queens are left (see Jeanne, 1980).

In the *Trigonini*, virgin queens are kept in reserve throughout the year. These queens are separated from the physogastric queen by the workers often by imprisonment in empty storage pots. The virgin queens attempt to replace the old queen by placing oral and anal secretions (probably containing pheromones) on to her body. Increased licking of the physogastric queen by the workers seldom leads to supersedure in *Paratrigona subnuda* (Imperatriz-Fonseca, 1977a, b). In the event that the functional queen is lost, she is replaced by a virgin queen (Kerr, 1969). According to Kerr (1974), *Melipona* colonies may lose up to 25% of their newly emerged individuals, when reared under optimal food conditions, this is because the surplus of young queens are killed by the workers. During winter, *M. quadrifasciata* and *M. marginata* produce respectively 3.45 and 6.50% of queens (Kerr, 1946). The constant production of queens (as in *Trigona* bees) is considered as a "life-insurance" for colony maintenance in case the functioning queen dies.

In the honey-bee, the queen receives less food from the workers when swarm queen cells containing an egg or a larva are present. These queen cells seem to affect worker behaviour (Morse *et al.*, 1966). When the queen cells are sealed, the queen's weight is at its minimum; she is no longer cared for and finally driven out of the nest. The first queens to emerge from their cells locate each other, and also queens still within their cells, by acoustic communication. They kill queens within their cells and fight amongst each other. The remaining young queens are chased off by the workers. Young or old queens leaving the nest are accompanied by a number of workers to found a new colony.

6 Dominance in the functioning of castes

In the regulation of caste development, dominance assumes an important place. The maintenance of colony structure is menaced by the tendency of immature individuals to reach the reproductive state. The dominance of the reproductive caste prevents this tendency to disintegration of the colony, and the attractiveness of the female reproductives to the worker caste helps the latter to stay within or return to the nest, or helps a swarm to settle.

In social Hymenoptera, female reproductive dominance is most complicated, and this is to be understood from the point of view of care of the offspring, from which the social structure of the Hymenoptera is most probably derived. The prolonged adolescence of workers by the influence of the queen is part of this situation. In termites, the situation is different, as care of offspring often hardly prevails and the young are frequently ready to undertake worker functions. Here, reproductive dominance is a property of the royal pair, and the soldier caste has its own type of dominance in suppressing the development of more soldiers.

6.1 TERMITES

As we have seen in section 5, it was found by Grassé and Noirot (1946a) and Lüscher (1951, 1952b, 1956a, b) in *Kalotermes flavicollis*, and by Light and Weesner (1951) in *Zootermopsis nevadensis* and *Z. angusticollis*, that reproductives have an inhibitory effect on the formation of replacement reproductives. These effects are due to pheromones which are given off with the faeces and circulated by trophallaxis (Lüscher, 1974a). As juvenile hormone (JH) and JH analogues have similar effects as reproductive pheromones (Wanyony, 1974), it was first conceived that JH itself is a pheromone, but in several respects the effects of both substances are different. It is now assumed that the reproductive pheromones stimulate the corpora allata (CA), resulting in a decrease of replacement reproductive competence, the latter depending on low CA-activity (Lüscher, 1977).

In *Kalotermes flavicollis*, when there is an excess of reproductives, they fight until only one pair survives. During this process, one individual remains aggressive and shows a dominance over the others, which is maintained for the whole fight. As a consequence, the other reproductives are wounded and eliminated by the workers. Males wound mostly males and to a much lesser extent females (Ruppli, 1969). In *Kalotermes flavicollis*, a complete inhibition of the formation of male and female reproductives is only possible if one male and one female are together in a colony. Females only prevent the formation of females. Because of this state of affairs, it is presumed that male and female pheromones are different (Lüscher, 1964). Although the degree of sex specificity in dominance differs from species to species (Nagin, 1972), the existence of two different pheromones seems assured (Lüscher, 1977).

Springhetti (1969, 1970) demonstrated a marked stimulation of presoldier and soldier development by reproductives in the absence of soldiers. It would seem that, in this case, the larval titre of JH is increased to a level inducing presoldier differentiation. But this only occurs in later stages of a moulting interval when soldier competence is increasing (Springhetti, 1972; Lüscher, 1974b).

6.2 SOCIAL HYMENOPTERA

Queen control over the activity of workers' ovaries has been demonstrated in all superfamilies. The queen's position can be maintained by physical dominance, by the production of inhibitory pheromones or by lack of "pro-fertile substances" produced by the worker (see Wilson, 1974).

6.2.1 *Ants*

After the functioning queen is removed from colonies of *Solenopsis invicta*, in 61% of these colonies physogastric replacement queens appear that are attractive to the workers (Tschinkel and Howard, 1978). Dissection of these queens established that 21% of the colonies (the smaller ones) had an inseminated queen producing workers. A part of the unmated dealates only produced some workers, probably by thelytokous parthenogenesis. The authors propose that the inseminated replacement queens are foundresses that remain within the colony and that *S. invicta* is often polygynous but functionally monogynous. These foundresses, that are not physogastric, do not produce eggs but have shed their wings, and are kept as a reserve. In contrast to other observations, Tschinkel and Howard (1978) found in laboratory reared colonies founded by 5 queens, that they remained polygynous well beyond the production of the first workers.

Reproductive dominance of the queen over other inseminated females, in functionally monogynous colonies, inhibiting their ovarian activity and behaviour has also been demonstrated or suggested for *Leptothorax* and other *Solenopsis* species. Passera (1978) demonstrated that queen pheromones regulate the type of eggs produced by the workers of *Plagiolepis pygmaea*. In queenright colonies, only trophic eggs are produced, whereas in queenless colonies both reproductive and trophic eggs appear. The production of reproductive eggs could not be inhibited by adding dead queens or queen extracts to queenless colonies. It appears that the queen's pheromones can be removed partially by daily dipping her into acetone for 15 seconds. In 5 out of 11 colonies with treated queens, the workers produced reproductive eggs, though far less than in queenless colonies. Treatment of queens also resulted in a considerable increase in the number of trophic eggs. Otherwise the queens seemed hardly to be affected by the treatment; they continued to produce small numbers of eggs.

6.2.2 *Vespine wasps*

The ovarian activity of workers of Vespine wasps depends on the presence of the queen and on previous social contact. Marchal (see Spradbery, 1973) found that young workers of *Vespula* species initiate oviposition within 10 days after queen removal. The workers' ovaries also became active when the queen was prevented from laying eggs for several days. Therefore, Spradbery (1973) suggests that an ovary-suppressing pheromone may be produced or emitted only during the act of oviposition. The effect of dominant workers on ovarian activity of subordinate workers has been found in *Vespa orientalis*. Ovaries are only activated in workers of high hierarchical rank, when isolated from the nest in groups, whereas isolated single workers all demonstrate this phenomenon. According to Motro *et al.* (1977, 1979) a previous social contact acts as a releasing stimulus for ovarian activity, because when workers are collected directly after emergence and isolated singly only 50% of the individuals respond.

6.2.3 *Polistine wasps*

Nests of *Polistes gallicus* are generally founded by a number of foundresses. Aggressive interactions between these mated females result in a dominance hierarchy headed by the queen. In *Metapolybia* species, West-Eberhard (1977) obtained evidence that queens are recognised by the workers because of the secretion of a pheromone from the head. The dominance system regulates the distribution of liquid food (originating directly from plants or obtained from larvae during trophallaxis). Subordinate auxiliaries more frequently offer food to dominant ones (and finally the highest ranking auxiliaries to the queen) than in the reverse situation.

According to Pardi (1948) the social status is related to the division of labour. The lowest ranking auxiliaries mainly collect food and wood pulp and construct cells. Higher ranking auxiliaries stay in the nest, inspect cells and produce eggs. The queen's main task is oviposition. The dominant individuals eat the eggs produced by subordinate ones in *P. fuscatus* and *P. canadensis* (West-Eberhard, 1969); also the queen of *P. metricus* eats the eggs laid by auxiliaries (Gamboa *et al.*, 1978).

In the association of foundresses, Pardi (1948) found that individuals with more active ovaries generally are more dominating. Ovarian activity is affected by trophic advantage of higher ranking individuals. The behaviour of *P. bimaculatus* workers did not change after ovariectomy and therefore Deleurane (1948) concludes that there is no causal relation between dominance behaviour and the degree of ovarian activity. After emergence of the first workers, the auxiliaries are sooner or later eliminated. Workers are always subordinate to the queen and to some auxiliaries; they perform all colony tasks except egg laying. Among the first emerging workers a dominance order according to age is established. In case the queen dies or becomes socially regressive the next ranking auxiliary takes over her position, the auxiliaries function both as workers and as reserve queens.

Later in the season an older dominant worker may take over the position of the queen and become fertile, producing only unfertilised eggs. Since the social status of individuals can be completely reversed when the most dominant one is removed or dies, West-Eberhard (1969) prefers to denote these phenomena as imaginal caste differentiation instead of caste determination. However, because merely behavioural and physiological changes occur in the adult insect, none of these qualifications apply. The observed dominance hierarchy merely is a result of differences in caste functioning.

6.2.4 *Halictine bees*

In *Lasioglossum zephyrum*, the queen physically dominates the worker; e.g. the queen leads pollen-carrying workers to the cells in the nest to be provisioned (Breed, 1977; Breed and Gamboa, 1977). Mating has no effect on caste functioning (Michener *et al.*, 1971). The queen eats the eggs laid by the workers (Brothers and Michener, 1974). When the queen is removed from the colony, in several cases she is soon replaced. The replacement queen may exhibit queenlike behaviour within one hour after queen removal. Reintroduction of the original queen results in dominance interactions with the replacement queen (Breed, 1977). The average size of the ovaries is largest in queens, smaller in guards and smallest in foragers. A relationship between ovarian development and dominance behaviour is suggested by Breed *et al.* (1978).

6.2.5 *Bumble-bees*

The ovaries of queenright brood-rearing *Bombus hypnorum* workers contain undeveloped oocytes up to the age of 10 days. At older ages, the percentage of individuals having completely developed eggs increases. Workers isolated from the queen develop mature eggs and oviposit at the age of 5 days. Röseler (1974) suggests two alternatives to explain the results. Either egg development does not take place because the food the young workers would need is used to feed the larvae, or the queen inhibits development of eggs. Dominance not only exists between the queen and the workers, but in the queenless colony also between workers. In *B. derhamellus* and *B. pratorum* (Free, 1955), the degree of ovarian activity in workers is related to dominance. Queenless workers not receiving pollen do not develop eggs and demonstrate no aggressive behaviour towards each other. It seems unlikely that ovarian activity induces dominant behaviour. Queens that lose dominance are still fertile (Röseler and Röseler, 1977). Workers of *B. hypnorum* and *B. terrestris* with activated ovarioles inhibit ovarian development in nest mates. About 50% of the individuals become laying workers. Röseler (1974) suggests that the inhibitory mechanism of the queen should be the same as that of laying workers.

6.2.6. *The honey-bee*

Ovarian activity in workers is suppressed by the presence of the queen. When the queen is removed from the colony, a restricted number of workers start laying eggs. However, when worker larvae or pupae are present, ovarian activity is considerably reduced (Jay, 1970). In queenless groups of workers, activation of the ovaries can be suppressed to some extent by the presence of a dead queen, a queen extract (Verheijen-Voogd, 1959; Butler *et al.*, 1961; Pain 1961), the odour of queen extract (van Erp, 1960), or the odour of 9-oxo-decenoic acid (Butler and Fairey, 1963).

7 The determination of patterns during development

In Holometabola such as Diptera and Hymenoptera, the body of the adult is almost completely reconstructed during metamorphosis from imaginal discs and histoblasts present in the larval body. These undifferentiated discs are derived from cleavage cells during blastoderm formation, and are set aside early in development (Geigy, 1931). They multiply their cells by proliferation, but do not differentiate until metamorphosis when they replace most of the larval organs and integument.

Imaginal discs, which form the head, the thorax and its appendages, and genital organs, are determined during their development. According to the definition by Hadorn (1967), determination is the programming of the developmental potential by activation of specific groups of genes. This programming is inheritable, and when it changes, it does so for the imaginal disc as a whole when another cluster of genes is activated. This phenomenon, named transdetermination (Hadorn, 1966) is only observed when discs are cultured *in vivo* for some length of time.

Within the disc, the cells receive information about their position in the field (specification). They will subsequently differentiate into a specific bristle, sensillum, etc. This specification is not inheritable, but is subject to extensive pattern regulation involving cell division (Wolpert, 1969; Schneidermann, 1979). The development of cuticular patterns by the epidermal cells, so important in differentiating between castes, is thought to be governed by gradient effects in which communication between the epidermal cells is extensive (Stumpf, 1967; Lawrence, 1970).

We do not know how determination becomes established initially in the embryo, but instructions are transferred by the cortical plasm of the egg. The determination of cephalo-caudal polarity and left-right symmetry takes place during oogenesis (Raven, 1961). Wigglesworth (1961) has drawn attention to the fact that in insect development, the daughter cells of the germinal vesicle split into clones of cells in which different elements of the gene system become derepressed. Morphogenesis implies the differentiation of two types of characters.

1 Characters determining the topographic differentiation of the body (e.g. head-thorax-abdomen):

2 Characters specific to the developmental stage or instar (larva-pupa-adult).

The polymorphism demonstrated by stages and instars of the same individual is complicated by characters related to sex (sexual dimorphism), which may be overt throughout development.

Caste differentiation implies a fourth set of characters specific to (in honey-bees) queen or worker, or (in ants and termites) worker – soldier – reproductive. This type of polymorphism belongs to the category of extrinsically controlled morphogenesis and the different forms ensuing have been termed oecomorphs (de Wilde, 1976*). They are formed in response to specific stimuli arising from the outer world, and serve specific functions. In this category belong the phases of locusts, morphs in aphids, seasonal dimorphism in Lepidoptera. Morphological features may be coupled to physiological syndromes such as diapause (*Araschnia levana*, Müller, 1955; *Papilio xuthus*, Ishizaki, 1958). Among the environmental factors involved are photoperiod, temperature, food and population density. The messages provided by these factors are being led through several steps of translation, and are mostly acting via the neuroendocrine system.

Caste differentiation depends on specific "trigger" stimuli acting during sensitive periods. Their relation to the determination of patterns is not a direct one. Although ontogenetic determination sets limits to the differentiation of characters during metamorphosis, reprogramming can occur, allowing for pupal or adult development. A similar reprogramming seems to occur in the development of caste features. Here again, certain groups of genes are apparently made operative, modifying the differentiation of larval or adult characters and the expression of the innate behaviour, to form larval or adult castes. In the ant *Myrmica ruginodis*, the imaginal discs of the larva can be divided into a dorsal and a ventral set; the dorsal group containing wing buds, genital discs and ocellar buds. In the case of queen development, both the dorsal and the ventral set start to grow and differentiate at pupal formation. In the case of worker development, only the ventral set continues, while the development of the dorsal set is arrested (Brian, 1957, 1965). This occurs abruptly, like ovarium degeneration in worker bee larvae. As we shall see, juvenile hormone (JH) is at the basis of the control of this type of polymorphism. Extrinsic control of caste differentiation is mediated, therefore, through an environmental impact on JH levels.

* The term "morph" was introduced by Hille Ris Lambers (1966) in a review on polymorphism in Aphids. For reasons of clarity, I have proposed the term *oecomorph* (de Wilde, 1975) to denote polymorphism in genetically identical individuals, as induced by extrinsic conditions.

8 Endocrine impact on caste development

8.1 HORMONAL IMPACT IN POSTEMBRYONIC PROGRAMMING

Cellular programming of the epidermis and the imaginal discs can be modified by hormones. Reprogramming of the epidermal cells during metamorphosis is initiated by 20-hydroxy ecdysone in the absence of JH. During metamorphosis of *Manduca sexta*, when the JH-titre is lowered, at least one or two new *m*-RNA's and proteins appear, and six or seven *m*-RNA's and proteins present in the larva are no longer utilised (Riddiford, 1980). In *Drosophila melanogaster*, the complete sequence of development of the imaginal discs from larval to pupal to adult can be induced *in vitro* by two days exposure to 20-hydroxy ecdysone (Mandaron *et al.*, 1977). Juvenile hormone changes the ecdysterone-induced synthesis of DNA, RNA and protein. The effect of JH is not simply inhibitory; it directs cellular processes in discs as it does in other epidermal structures (Riddiford, 1980). It appears to determine where in the genome ecdysone will exert its action (Riddiford, 1981).

8.2 HORMONAL IMPACT IN ECOLOGICAL ADAPTATION

The course of the JH-titre during the period preceding metamorphosis in some caterpillars (Varjas *et al.*, 1969) suggests a very intricate interplay, coinciding with behaviour phases in preparing pupation sites. An increase in JH-titre may delay a moult, and may even lead to larval diapause (Fukaya and Kobayashi, 1966; Chippendale, 1977). The green form of grasshoppers (Rowell and Frazer, 1967) and locusts (Joly, 1951, Staal, 1961) reflects high JH-activity, the brown or grey form a low activity. Another type of environmental polymorphism are apterous and alate morphs in aphids. In the cabbage aphid, *Brevicoryne brassicae*, a high JH-titre seems to suppress the formation of wingbuds (White, 1976). Diapause in adult insects, such as the Colorado potato beetle, is induced by a very low JH-titre, and this includes morphogenetic, behavioural and metabolic elements (de Wilde, 1971). In this beetle the "physiological states" of reproduction and diapause seem to be controlled by specific "state levels" of JH-activity (de Wilde, 1978). Seasonal dimorphism in Lepidoptera also has an endocrine basis, as exemplified by the studies of Endo (1980) on several butterfly species, where the cerebral neurosecretory cells (NSC) appear to be involved. Even types of genetic polymorphism such as sexual dimorphism show cases where the genes exert their effects via the endocrine pathway. In the very interesting experiments by Naisse (1966) it was shown that in the glowworm *Lampyris noctiluca*, sex determination involves the action of both corpora allata (CA) and corpora cardiaca (CC) and can be modified by interfering with these glands.

As a consequence of all these discoveries, the term "juvenile hormone" has become more and more questionable. We are dealing with a much more versatile "morphogenetic hormone". It is used by insects in the control of a large variety of functions. The CA, which secrete JH, may be under the control of environmental stimuli such as photoperiod, nutrition, crowding, and many others, depending on the strategy of survival of the species. This control is generally exerted via the cerebral neurosecretory cells.

8.3 ENDOCRINE INVOLVEMENT IN CASTE FORMATION IN TERMITES

The neuroendocrine regulation of moults in Hemimetabola differs from the Holometabola in the following respect, that hormonal activities completely determine the differentiation at each moult. This is not only true for the moulting stages and instars, but also for the caste features.

8.3.1 *Lower termites*

The cerebral NSC of *Kalotermes flavicollis* were described by Noirot (1957), the prothoracic glands (ventral glands, tentorial glands) by Lüscher (1960). The prothoracic glands (PG) degenerate in primary adult sexuals and in replacement reproductives (Herlant-Meeuwis and Pasteels, 1961; Lüscher, 1961) but are retained in soldiers, although no moults take place in this caste. Among the endocrine glands, the CA vary most in size and histological appearance. Their shape is roughly globular with a radius of 10 μm in newly moulted pseudergates and 25 μm in newly moulted replacement reproductives. The volumes of these glands therefore compare in the proportion 1 : 15, suggesting important fluctuations in activity. In the higher termite, *Macrotermes subhyalinus*, the naturally occurring juvenile hormone is JH-III (Meyer *et al.*, 1976).

We will first discuss the role of the PG and their moulting hormone (MH), ecdysone. As in other insects, injection or oral administration of MH during an intermoult stage may sometimes accelerate a moult and sometimes retard it. Lüscher and Karlson (1958) obtained these results in competent larvae of *Kalotermes*. The retarded moults resulted in a very low percentage of replacement sexuals. Lebrun (1967) implanted PG and CA obtained from cockroaches (*Periplaneta*, *Leucophaea*) in pseudergates of *Kalotermes*, and found them to be active. Implantation of PG prevented the pseudergates from differentiation into replacement sexuals. Injection of MH in nymphs resulted in intermediate stages between nymph and adult. It is thought that these effects result from an interplay between CA and PG. Corpora allata inhibit a moult, and thereby leave more time for the JH to induce its specific effect.

We will subsequently discuss the role of the CA in the formation of replacement reproductives. In the nymphs of colonies in the field, CA-volume shows a minimum in early spring and during summer. Corpus allatum volume is negatively correlated with the frequency of replacement reproductives (Fig. 6). Application of JH to pseudergates reduced the formation of reproductives. Subsequent experiments have proved that competence for replacement reproductive formation requires a low CA-activity (Lüscher, 1957). The "queen substance" of honey-bees, 9-oxo-decenoic acid which, in worker bees, gives rise to a reduced CA-activity, has the same effect in nymphs of *Kalotermes*, and this results in an increased formation of replacement reproductives (Lüscher and Walker, 1963; Sannasi and George, 1972). The CA, therefore, appear to increase their volume after the competence for reproductive formation is obtained. Their activity is apparently related to the gonadotropic effect of JH in the adult (Lüscher, 1974a). Lebrun (1969) has shown that implantation of CA in young adults accelerates sexual development.

We will conclude by discussing the endocrine relations of the formation of soldiers. Implantation of CA obtained from different castes (Lüscher, 1958a, b; Lüscher and Springhetti, 1960) or treatment with synthetic JH (Lüscher, 1969) induces pseudergates to differentiate into presoldiers. Pseudergates which already had started developing into replacement reproductive may be induced to form intercastes between presoldier and replacement reproductive (Lebrun, 1967). After the implantations and JH-treatments mentioned above, the CA of the treated individuals are reduced in size.

The above results of Lüscher (1969) and Lebrun (1967) were confirmed for *Reticulitermes lucifugus* by Hrdý (1973) and for *Zootermopsis nevadensis* by Wanyonyi (1974). The last author additionally found that the dosage of JH influences the size of the PG

The lower termites are characterised by a large degree of plasticity in caste development, including the occurrence of regressive moults. It has been shown in several cases that such moults are accompanied by an increased activity of the CA. Interesting quantitative effects of administered JH-analogues (JHA) are reported for *Zootermopsis* by Wanyonyi (1974; Wanyonyi and Lüscher, 1973). When the Zoecon juvenoid ZR-512 was used, low doses inhibited moulting, slightly higher doses prevented the development of replacement reproductives and alates. An even higher dose was required for regressive development, and the highest dose led to presoldiers and worker-like forms. After similar experimental evidence, Lenz (1976) stated: "For every developmental stage, a certain JH-titre is characteristic". The establishment of JH-titres, however, has still to be performed in the lower termites.

Howard and Haverty (1978) found in *Reticulitermes flavipes* an increased rate of presoldier formation after keeping undifferentiated larvae on methoprene (JHA) treated filterpads (see also Haverty and Howard, 1979). Hrdý and Křeček (1972) found the same in *Reticulitermes lucifugus*.

8.3.2 *Higher termites*

In *Neocapritermes* and *Nasutitermes* (Kaiser, 1956) and *Odontotermes* (Lüscher, 1976) the CA of larvae increased in volume when they developed into soldiers. In *Odontotermes* this increase in volume is as much as 15 fold. CA-volume in different castes of *Macrotermes subhyalinus* has been measured by Okot-Kotber (1977) and are given in Table 2 and shown in Fig. 10.

TABLE 2 Corpora allata volume and head width of presoldiers and adults of *Macrotermes subhyalinus*. (After Okot-Kotber, 1977)

Caste or stage	Sample size (n)	CA volume μm^3 ± S.D.	Head width mm ± S.D.
Queens (19.50 ± 1.59 g)	6	582 ± 297	—
Kings (0.213 ± 0.01 g)	6	19.0 ± 14.8	—
Alates W	5	3.41 ± 0.85	—
Alates W	5	2.78 ± 0.52	—
Major presoldiers	57	0.283 ± 0.148	3.39 ± 0.65
Minor presoldiers	36	0.147 ± 0.074	1.91 ± 0.31
Major soldiers	34	0.065 ± 0.023	4.49 ± 0.33
Minor soldiers	4	0.018 ± 0.007	—
Major workers	30	0.040 ± 0.029	2.48 ± 0.23

Activation of the CA in female 3rd instar larvae leads to minor soldiers and a similar activation in young female workers causes development of major soldiers. The measurement of CA-volume of first instar larvae, which may be crucial in relation to their future differentiation, has as yet not been carried out (Lüscher, 1976). As mentioned before, there is some reason to suppose that blastogenic determination of some developmental lines may occur; this is because of the variable contents of JH in the eggs of *Macrotermes*. From 22 samples of 10 000 eggs collected during different seasons, JH was extracted and its quantity determined by the *Galleria* bioassay. In samples collected in January and February, the JH-contents was much lower than in the other parts of the year. As the nymphs appear in April and May, they emerge from eggs laid in January and February (Lüscher, 1976).

As the JH in eggs is obtained from the queen, it is of interest to consider the JH-titre in the haemolymph of this reproductive caste. In *M. subhyalinus* the CA of the physogastric queen may reach enormous proportions, becoming more than 100 times as large as in the young dealate female. The JH-titre in the haemolymph of dealate females rises steeply until after two weeks, 300 *Galleria* Units (G.U.)/μl are present, assuring a high rate of incorporation into the oocytes. It is not surprising, therefore, that in young colonies, nymphs are never found. The old physogastric queen contains extremely high doses of JH-III in the haemolymph and values of 140 G.U./μl are measured in the majority of samples taken. Lanzrein *et al.* (1978) even found JH-III titres of 3–25 μg/ml. Observed throughout the seasons, it appears that the lowest values occur in August–November which is three months earlier than the period of low JH-activity in the eggs. It appears, therefore, that incorporation of JH into the eggs occurs very early during oogenesis.

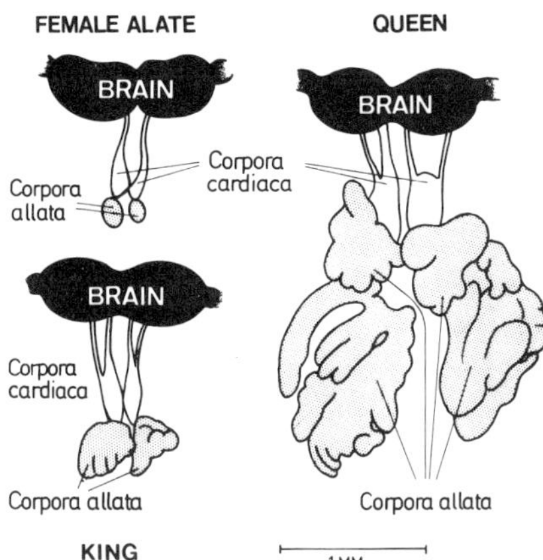

Fig. 10 The appearance of dissected and fixed neuroendocrine systems of adult *Macrotermes subhyalinus*. All systems are drawn to the same scale. (Redrawn from Lüscher, 1976)

Although statistically reliable data are still lacking, it is possible that neuters are programmed by a high JH-content of the egg. Even so, this programming is still reversible as shown in the experiments of Bordereau (1975) mentioned above. In the above context, this can be explained by a pheromone stimulating CA-activity and given off by the royal pair, being subsequently taken up by the predetermined first instar worker larvae. Another way in which larvae could increase their JH-content, is by taking the queen's anal fluid. This fluid, which is produced in a regular flow, is very attractive to the workers. It contains JH in varying, sometimes large quantities. Yet another way of providing JH seems to be the fungus combs in some termite species. It was observed by Sannasi *et al.* (1972) that the fungus combs and conidia of several species of *Odontotermes* contain rather high JH-activities but this has not been confirmed in *Macrotermes* (Lüscher, 1976).

It is of interest to mention some physiological peculiarities of the physogastric queens of the higher termites. In *M. subhyalinus*, vitellogenic protein is synthesised in the ovaries rather than in the fat body which apparently functions as an exocrine gland releasing a building pheromone (Wyss-Huber and Lüscher, 1975; Bruinsma and Leuthold, 1977). The respiratory rate of ovary homogenates of termite queens is greatly stimulated by ecdysone, and this effect is inhibited by JH. Lüscher (1976), therefore, supposed that ecdysone rather than JH promotes vitellogenesis and egg maturation but in a subsequent paper (Lanzrein *et al.*, 1977) the authors still consider the very high JH-titre in the haemolymph to be the regulating factor. Ecdysone was in fact detected in large amounts in the ovaries of the queens of *Macrotermes bellicosus*, *M. subhyalinus*, *Cubitermes fungifaber*, and *Cephalotermes rectangularis*, and is even thought to be synthesised by the ovaries (Bordereau *et al.*, 1976, 1977).

In whole queens, the following values were found: *M. bellicosus* (500 ng/g fresh weight); *Cu. fungifaber* (1500 ng/g fresh weight); *Ce. rectangularis* (1500 ng/g fresh weight). In *M. bellicosus*, 95% of the total ecdysteroids were found in the ovaries (1500 ng/g). In *M. subhyalinus* the ovaries contained 600 ng/g (Delbecque *et al.*, 1978). Interestingly, high levels were observed in the mature eggs (*Macrotermes* 2700 ng/g, *Cephalotermes* 3000 ng/g, *Cubitermes* 4000 ng/g). The most prevailing ecdysteroid was α-ecdysone. It may play a role in the terminal phase of oogenesis, but may also intervene in embryonic development (Bordereau *et al.*, 1977). The prevalence of α-ecdysone was also found by Lanzrein *et al.* (1977) as indicated in Table 3.

TABLE 3 Ecdysone titre in different organs of queens of *Macrotermes subhyalinus*. The contents in ovary or "royal fat body" are expressed as RIA activity equivalents per mg fresh weight or per μl for haemolymph. (After Lanzrein *et al.*, 1977)

	Ovary	"Royal fat body"	Haemolymph
Queen B	490 pg α-ecd 90 pg β-ecd		
Queen C	680 pg α-ecd 200 pg β-ecd	110 pg α-ecd < 5 pg β-ecd	28 pg α-ecd 8 pg β-ecd
Queen D	280 pg α-ecd 100 pg β-ecd	55 pg α-ecd < 5 pg β-ecd	10 pg α-ecd 12 pg β-ecd

The endocrine situation in the physogastric queens of the higher termites, as follows from the above, is very complex. The differential role of ecdysteroids and JH in oogenesis, embryogenesis and caste programming is far from being clarified. JH and ecdysteroids also occur in the eggs of solitary insect species such as cockroaches (Imboden *et al.*, 1978) and Lepidoptera (Bollenbacher *et al.*, 1978; Hsiao and Hsiao, 1979; Legay *et al.*, 1976), where they apparently induce the embryonic moult (Lagueux *et al.*, 1979). This function may be of a more general nature, and their role in an eventual blastogenic caste formation is speculative.

In fact, the existence of blastogenic caste formation in termites is not assured. It may still be, that the first larval instar is subject to pheromone, endocrine or other messages regulating the activity of its CA and, thereby, its developmental program as appears to occur in early larval stages of the honey bee, where blastogenic predetermination is excluded.

8.4 ENDOCRINE INVOLVEMENT IN CASTE FORMATION IN SOCIAL HYMENOPTERA

8.4.1 *Ants*

(*i*) Myrmica rubra. In *Myrmica*, α-ecdysone and JHA appear to have opposite effects on caste differentiation. Topical application and injection of farnesenic acid or farnesyl-methyl-ether delays metamorphosis in female sexual larvae, and as a result more and larger queens arise. On the other hand, injection of α-ecdysone during the phase of leg-segmentation in queen larvae accelerates the development. Metamorphosis is speeded up with the formation of small queens, intercastes and workers. Earlier injections of α-ecdysone are ineffective.

From experiments with postcephalic ligatures combined with JHA-application, Brian (1974b, 1976) concluded that JH inhibits larval growth and the differentiation of leg discs. The development of wing discs may be stimulated. The hormonal balance seems to depend on food conditions. After hibernation, large female larvae can develop into queens, provided a sufficient amount of food is offered. Intercastes and workers develop when larvae are deprived of protein for some days. The queen caste is determined as soon as the legs have two segments; after this, starvation has no effect (Brian, 1976). Differences in the accumulation of neurosecretory material in the CC between presumptive queen and worker larvae during the phase of leg segmentation in spring may be related to the production of MH. However, the PG, which are conspicuous in second instar larvae, cannot be found in older ones. Brian (1976, 1979) suggests that MH may be produced elsewhere in the body. Although the size of CA of larvae of both castes is the same in autumn, during hibernation the CA of prospective queen larvae are enlarged as compared with prospective worker larvae. The increase in size of the CA is in proportion to body size. The author assumes that worker development occurs after the CA have ceased to be active. In queen larvae, the activity of the CA is maintained for a longer period, and then decreases gradually until the phase of leg segmentation. In our opinion, further support of the above effects by the measurement of JH- and MH-titres would be desirable.

(*ii*) Pheidole pallidula. Feeding mealworms injected with JHA to *Pheidole* colonies that normally only produced workers in the summer months resulted in the appearance of presumptive reproductives (Passera and Suzzoni, 1978). This effect is also obtained by repeated topical application of 1.0, 0.1 and 0.01 μg of JH (mixture of isomers) to temporarily isolated queens, irrespective as to whether or not reproductives had already been reared. After the last application of JH, production of larvae of reproductives ends immediately or continues up to 4 weeks after treatment. As in untreated small queenright colonies the larvae of reproductives never attain the pupal stage; they disappear (Passera and Suzzoni, 1979). Eggs produced both by JH-treated queens during the isolation periods and JH-treated eggs given to untreated queenright and queenless colonies resulted in the development of all or a high percentage of the larvae into larvae of reproductives. Topical application of JH to larvae of the first three instars or feeding JH-injected mealworms to workers had no effect. Therefore, Passera and Suzzoni (1978) conclude that JH is not transferred when the workers lick the eggs and that JH induces the development of larvae of reproductives during oogenesis or embryogenesis. In our opinion, it would be interesting to measure the JH-titre both in eggs and in the queens after hibernation and during the summermonths, to determine whether or not the queen's JH is accumulated in the eggs. It seems that Passera and Suzzoni are very close to a possible explanation (maybe in addition to the nutrimentary condition within the egg) of the phenomenon of blastogenetic caste control.

(*iii*) Monomorium pharaonis. Successful eradication of *Monomorium pharaonis* nests were obtained by offering honey-liver baits containing methoprene (Edwards, 1975a, 1977; Hrdý *et al.*, 1977) and hydroprene (Hrdý *et al.*, 1977). The following effects were observed in laboratory experiments. All brood died in 4–8 weeks, pupae and large larvae disappeared before smaller larvae and eggs. The queens were irreversibly sterilised and the ovaries atrophied. The workers died within 20 weeks after treatment, queens remained alive for more than 20 weeks (Edwards, 1975b). Effects on brood development and oviposition were also observed by Hrdý *et al.* (1977). According to Rupes *et al.* (1978), the development of young larvae (L_1 and L_2) is inhibited by methoprene, but in contrast to the results of Edwards (1975b) the development of L_3 and pupae is unaffected. Senescent workers die, the longevity of workers is shortened. Also, in contrast to Edwards (1975b), Rupes *et al.* (1978) conclude that the oviposition of queens seems to be affected reversibly. Degradation of Pharaoh's ant colonies by methoprene are the result of interference with reproductive functions and social interactions (Rupes *et al.*, 1978).

(*iv*) Atta sexdens. High amounts (50 μg) of some JHA topically applied to the last larval instar of *Atta sexdens* larvae prevented pupation. Topical application to pupae resulted in abnormal adults. Considerable adult worker mortality was obtained after contact with JHA at lower concentrations. It is suggested that the queen of an *Atta cephalotes* colony is sterilised after mixing the food with hydroprene.

(*v*) Solenopsis invicta. Application of JHA to *Solenopsis* brood or entire colonies interferes with brood development and worker behaviour. Wendel and Vinson (1978) demonstrated that a ^{14}C-labelled JHA is distributed within the colony; the workers contain the largest amount of JHA and within 36 h the amount of JHA in the larvae increases up to 7% of the recovered dose. Only traces of JHA are found in pupae, the queen and the eggs. The half-life of the JHA(R-20458) in the colony is 2 days or less.

Methoprene (ZR515) or hydroprene (ZR512) offered to colonies prevented metamorphosis of larvae (Cupp and O'Neal, 1973; Troisi and Riddiford, 1974). Vinson and Robeau (1974) could not confirm this result. However, they observed that the number of pupae fell immediately, partly due to cannibalistic behaviour of the workers. Individually JHA-treated larvae and pupae are also removed or destroyed by the workers (Cupp and O'Neal, 1973; Vinson *et al.*, 1974). Feeding or topical application of methoprene and hydroprene to individual larvae result in abnormalities in immature stages and occasionally in adult workers (Cupp and O'Neal, 1973). The same effects are obtained when JHA is distributed in the colony by trophallaxis, whereas hydroprene is less effective. The first three larval instars are affected, whereas 4th instar larvae develop into normal workers. Individually JHA-treated female reproductive pupae reduced the number of resulting adults and the number of adults acquiring adult pigmentation (Vinson *et al.*, 1974). Abandonment of the larvae by the workers within 12 h after contact with higher doses of hydroprene applied to filter paper may be due to a masking of the brood-rearing pheromone (Glancey *et al.*, 1970) of the larvae by the odour of JHA (Troisi and Riddiford, 1974) or to a direct effect on pheromone production by the larvae. Also, aggressive behaviour between workers, resulting in a high worker mortality, is thought to be caused by interference with pheromone communication (Troisi and Riddiford, 1974).

By application of JHA to colonies, information has been obtained about the endocrine mechanism of caste differentiation. When screening the effect of several JHA on caste differentiation and colony development, Banks *et al.* (1978) found that some JHA caused a shift in the ratio between worker and sexual larvae. The production of worker brood stopped and the number of alates increased. The process of caste differentiation can be changed during the 1st larval instar and possibly also in the 2nd instar, older worker larvae remain unaffected and develop into normal workers. The resulting sexual larvae died or were killed by the workers. Vinson and Robeau (1974) found that only female reproductive pupae appeared, from which many were destroyed and therefore only few reached the adult alate stage. Banks *et al.* (1978) found that most colonies produced female sexuals after JHA-treatment, however ten colonies produced only males and few colonies produce both sexes. Troisi and Riddiford (1974) found that queens from JHA-treated colonies mainly produced alate males, suggesting that the process of fertilisation had been affected but that this was a reversible effect. Banks *et al.* (1978) also found that when JHA is diluted sufficiently by trophallaxis, major workers were produced instead of alates. Robeau and Vinson (1976) observed that application of JH stimulated the production of major workers, intercastes and alate females.

From the above mentioned experiments, it appears that JHA-treatment of colonies or separate brood stages interferes with brood development and worker behaviour. On the one hand, brood may be killed after contact with JHA to be subsequently removed by the workers. On the other hand, removal of brood by the workers could be due to their recognition of an aberrant (queen) development of the brood. Worker behaviour itself might also be changed by contact with JHA.

8.4.2 *Bumble-bees*

The function of JH in the process of caste differentiation could be demonstrated in *Bombus hypnorum*. Topical application of 5 or 16 μg of a mixture of JH isomers given 3 times within a day during the last larval instar resulted both in a delay of pupation and in the development of large queens in queenright colonies (Röseler and Röseler, 1974; Röseler, 1976). Larvae that had received the highest amounts resulted in individuals with malformations. Neighbouring untreated larvae, which probably received smaller amounts of JH through the cocoons, also developed into queens and did not show such malformations. When amounts of 0.1 – 1.0 μg were topically applied to prepupae of presumptive workers, Röseler (1976) obtained queens of worker size. These queens could only be distinguished by differences in haemolymph composition and in behaviour. The author concludes from these experiments that body size and the degree of expression of caste characteristics must not be considered as interdependent effects of the amount of food consumed by the larvae. In our opinion this statement is premature because JH-production might be regulated by the amount of food consumed. In addition, Copijn *et al.* (1979) found that after topical application of JH to young honey-bee worker larvae the degree of expression of queen characteristics could be improved by offering an extra amount of food during late larval development. The result that the prepupal stage is the most sensitive period to obtain queens by topical application of JH is in agreement with the finding that the JH-titre of prospective queens becomes higher than that of prospective workers after the first day of the prepupal stage (Röseler, 1977b).

Topical application of JH to the last larval instar of *B. terrestris* delayed the pupation up to 6 days, still no queens were obtained. JH-treatment of presumptive worker prepupae also resulted only in workers. Also, malformations were observed (Röseler and Röseler, 1974). After Röseler (1977b) found that in prospective worker prepupae the JH-titre is higher than in prospective queens of the same stage, it is obvious that topical application of JH to worker prepupae could not result in the development of queens. Whereas prospective queens have a higher JH-titre than prospective workers during the last larval instar, a positive effect should have been expected after JH application. The author concludes, that the presence of the queen in *B. terrestris* colonies does not allow the development of new queens. Even when presumptive queen larvae are introduced into a colony with a dominant queen, they are "redetermined" into workers of queen size (Röseler, 1976).

8.4.3 *Stingless bees*

Not until the beginning of the pupal stage of *Melipona quadrifasciata*, when castes can be distinguished by degree of eye development, do the sizes and the numbers of cells of CA appear as two separate groups. (Kerr *et al.*, 1975). However, larval instars were not considered. It is likely that the observed differences in the CA indicate differences in JH production and release. Campos *et al.* (1975), Velthuis and Velthuis-Kluppell (1975) and Engel (1979) studied the effect of topical application of JHA and JH to larvae 4 days old or spinning larvae of *M. quadrifasciata* reared *in vitro*. All larvae received 170 mg of food to secure at least a queen-worker segregation of 1 : 3. Surprisingly, large amounts of ZR 512 (18 μg/larva) and small amounts of an isomer mixture of Cecropia JH (0.035 and 2×10^{-3} μg/larva) applied to spinning larvae induced queen development in all female individuals. Very small amounts of JH (1 and 2×10^{-7} μg/larva) yielded about equal numbers of queens and workers. In compari on with similar experiments with *Apis mellifera* (see Beetsma, 1979), induction of queen differentiation in *M. quadrifasciata* is obtained with very low amounts of JH, and very high amounts of ZR 512. In addition, with *Apis mellifera* only queenlike intercastes could be obtained, whereas in *M. quadrifasciata* very few individuals developed into intercastes, treated larvae mainly became either queens or workers. On the basis of their experimental observations, Velthuis and Velthuis-Kluppell (1975) propose two main alternatives for the mechanism of caste differentiation. 1. The production of JH is regulated by the amount of food consumed and only the queen determining gene combination can be activated by JH or express their activity in combination with a high JH-titre. 2. One or both of the queen-determining genes is involved in JH-synthesis. But finally they suggest that another mechanism, not directly under genetic control, could play a role. Both explanations depart from a hypothetical genetic background of caste formation which they did not test themselves (see p. 192).

From the above experiments it appears that, whatever combinations of caste determining genes occur, all individuals can be induced to develop into queens after JH-application to the larvae. This means that gene complexes, directly involved in the development of queen characteristics, are activated. In the light of the genetic hypothesis, it would be necessary to assume that the amount of available food triggers JH production in only a fraction of the female larvae.

The experiments of Camargo (1977a, b) are of interest because an equal segregation of males and females (queens and workers) was obtained in *M. quadrifasciata* by inbreeding. Individuals homozygous for the sex allele are diploid males, hemizygocity yields haploid males, whereas heterozygocity results in females. It is assumed that 25% of the diploid male larvae are double heterozygotic for the (female) caste determining genes. Giving additional food to these larvae did not result in the appearance of female characteristics and therefore Camargo (1977a) suggests that the homozygocity for the sex allele has an epistatic effect on the caste determining genes. Topical application of 10^{-6} μg of JH to diploid male prepupae resulted in the development of queen characteristics. The author concludes from these experiments that diploid males have a higher threshold than the females in their response to JH. Queen characteristics in diploid male individuals could only develop when, in addition to their own JH-production, additional JH was applied.

According to Kerr *et al.* (1975) the amount of food affects the number of cells developing in the CA and hence the production of JH, provided the glands can be activated. The authors suggest that the coding for synthesis of JH-specific RNA is present in the two caste determining genes. This code is only complete in the double heterozygotes. "Juvenile hormone synthetase" production depends on the action of this specific RNA. In this hypothesis, the amount of food consumed and the production of JH-synthetase regulates the production of JH. A high or low JH-titre activates the other genes for femaleness, resulting in a queen, or the genes for maleness, resulting in a worker, respectively. Including a "juvenile hormone synthetase" in this hypothesis is certainly an oversimplification when considering the complete mechanism of JH-synthesis (de Kort and Granger, 1981).

At present, only indirect evidence exists for genetic caste determination. Therefore, genetic experiments should be undertaken concerning the mendelian inheritance of the caste determining genes and their effects (Velthuis, 1976). Several results underlying the hypothesis of Kerr and colleagues concerning the role of genetic factors in caste differentiation in *Melipona* bees are discussed by Darchen and Delage-Darchen (1977).

8.4.4 *The honey-bee*

(*a*) *Ontogenesis in the castes* The duration of embryonic development and the intermoult periods of the first four larval instars are the same in queen and worker. When the larva is 3½ days old, it moults and enters the fifth instar. At day 6, both castes enter the prepupal stage, and after this, the rate of development becomes very different. In the worker larva, the pupal moult is at day 7, in the queen larva at day 8. Subsequently, the adult queen emerges at day 12 and the adult worker at day 17. The difference is mainly in the duration of the pupal stage, which is 5 days for the queen and 9 days for the worker. A time table of the development of the two castes, according to Bertholf (1925), is given in Fig. 11. After the workers have closed the cell, the larva spins the cocoon. Before pupation, the worker larva and queen assume horizontal and vertical positions respectively. Emergency queen cells are constructed by enlarging a worker cell and by extending it in a downward direction.

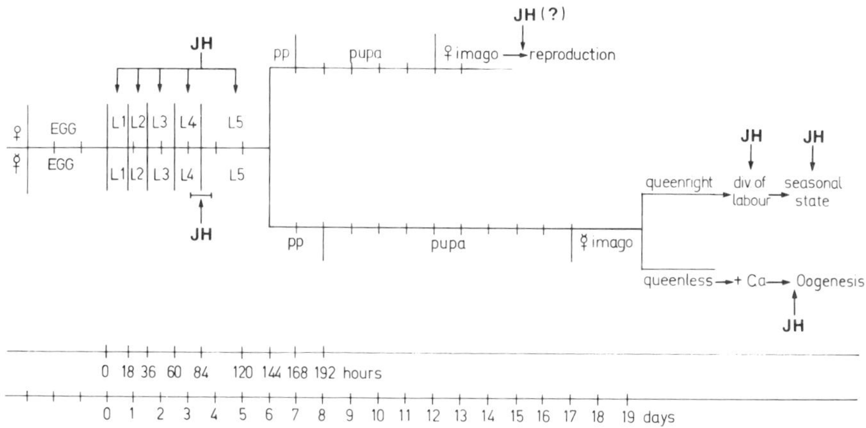

Fig. 11 Timetable of duration of stages and instars in queen and worker development in the honey-bee. At the top is indicated the involvement of JH in determining stage morphogenesis and oogenesis. At the botton are shown the effects of JH on caste morphogenesis and worker functions

Regression of the rate of development in the worker larva, which apparently is started before cocoon spinning, is most accentuated in the ovaries. At 72 h, the full set of ovariole primordia is present equally in both castes, and the formation of ovariole "cysts" has started. Ovarian regression begins in the worker larva, reducing the number of ovariole primordia. At 96 h, only 4–6 primordia are left, which subsequently develop into germaria. The queen retains her full number of 150–180 germarial primordia in each ovary (which subsequently develop into full ovarioles). This regression in worker ovary development is one of the first signs of a program which must have been defined earlier in larval development. The external appearance of queen and worker larva does not differ at this time, and attempts by Goewie (1976) to find differences in the face masks by scanning electron microscopy, have not succeeded. Likewise we failed to confirm the difference in the number of conical papillae claimed to have been found by this author. At 72 h the CA of the queen larva is already much larger than that of the worker larva (Canetti *et al.*, 1964).

The nutritive state of the larva during the early phase of its development shows considerable caste-specific differences. Fat body cells of worker larvae during the first three days, when investigated histologically, show little stores and large watery vacuoles. Only after 3½ days, during the "gorging phase" (Bertholf, 1925) when worker jelly is supplemented by nectar and pollen, some accumulation of glycogen and fat occurs. Queen larvae, on the contrary, right from the beginning accumulate considerable stores of glycogen in their fat body. Wirtz (1973) concludes that worker larvae are "left short of food" during the first three days. The fact that the body weight of a worker larva during the first three days does not differ from that of queen larva, must be explained by a higher water content of the former and a higher rate of respiratory metabolism of the latter (Melampy and Willis, 1939; Osanai and Rembold, 1968; Rembold and Hanser, 1964). In any event, nutrition in both castes must be sufficient during the first 3½ days of larval life, to pass the "periode de nutrition indispensable" (Bounhiol, 1938) which is essential for a moult. The synchrony between the first four moults in both castes does not leave room for any other conclusion.

(*b*) *The role of the juvenile hormone in caste programming* Until 1973 our ideas on endocrine involvement in caste programming were subject to much contradictory information. It was known that the CA in presumptive queen larvae are considerably larger than those of presumptive worker larvae (Canetti *et al.*, 1964), but in relation to body weight the difference appeared to be less conspicuous (Lukoschus, 1956b) or even absent (Pflugfelder, 1948). Lukoschus (1956b) observed a marked increase in volume of the CA of queen larvae older than 3½ days as compared with worker larvae. He related this difference to the fact that worker larvae after 3½ days receive modified worker jelly as a food. Haydak (1943), Lukoschus (1955), and Shuel and Dixon (1960) suggested that a difference in hormone balance in early stages of larval development is a causal factor in caste differentiation. Ber-Lin Chai and Shuel (1970) implanted CA of queen larvae (3–4 days old) into worker larvae (4 days old), and observed a reduced extent of ovary regression, resulting in adults with more ovarioles than sham-operated and control workers.

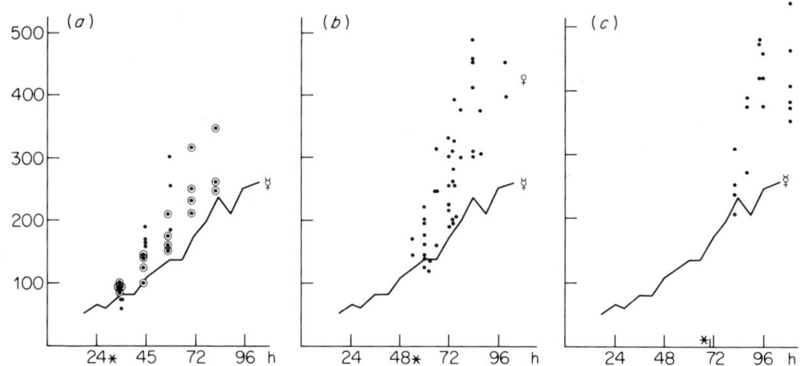

Fig. 12 Surface area (arbitrary planimiter values) of CA of honey-bee larvae after being transferred to queen cells on the first (*a*) 2nd (*b*) and 3rd day (*c*) of larval development. The very steep rise of CA growth after transfer in the third day indicates the reactivity of the larva during the critical period of caste induction. The solid line represents the surface area of the CA of worker larvae; ⊙, worker larvae transferred into 3-day-old queen cells containing royal jelly. (After Wirtz, 1973)

Wirtz (1973), in a series of conclusive experiments, proved beyond doubt that caste programming is under the control of the CA. The evidence in favour of this hormonal control is the following:

1 Upon grafting worker larvae into queen cells, it was shown that the optimum age at which the volume of the CA of these larvae increases fastest, in response to the new conditions, is at the end of the third day (Fig. 12).

2 It was subsequently shown that the ultrastructural activity pattern in CA of queen larvae shows characteristic differences from that in worker larvae, pointing towards a much higher specific rate of JH-synthesis in the former. The difference diminishes at 123 h and has disappeared when the larvae reach the prepupal phase.

3 By means of the *Galleria* bioassay, it was shown that the JH-titre in the haemolymph of queen larvae is much increased as compared with worker larvae. When worker larvae 44 h old were grafted into queen cells, at the age of 72 h, the titre was more than ten times that in comparable worker larvae. It remained at about the same high level until the beginning of the prepupal phase, when it dropped to a level equal to that in the worker prepupa (Fig. 13).

4. Wirtz, (1973; see also Wirtz and Beetsma, 1972) was able to induce queen characteristics in the resulting adults by external application of JH-I to worker larvae. The optimal degree of expression of queen characteristics which could be obtained under these conditions, was reached when the larvae were treated with 1 μg of JH-I shortly before the end of the third day. The

worker bees continued to feed these larvae in the same way as worker brood, and did not discriminate between treated and untreated larvae until the time of cocoon spinning.

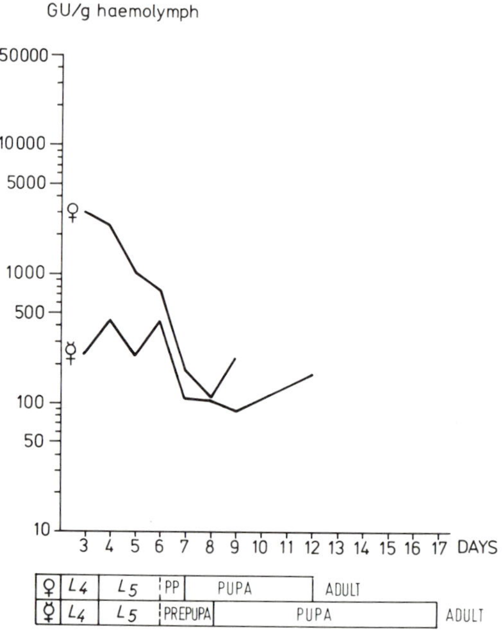

Fig. 13 A comparison of the JH-titre of queen (upper curve) and worker larvae (lower curve), determined by means of the quantitative *Galleria* wax test. Note the considerable difference between the titres during the period of caste induction. The boxes at the bottom represent the timetable for the development of queens (top) and workers (bottom); PP, prepupa. (After Copijn *et al.*, 1979)

It follows from the above data that a high titre of JH switches the developmental program in such a way that the expression of caste characteristics is directed towards queen formation. (Figs 14–16). These results were confirmed by Asencot and Lensky (1976) by applying 1 μl of JH-I (mixture of isomers) to worker larvae 2–3 days old reared *in vitro* on worker jelly. Earlier *in vitro* experiments of Rembold (1969b), concerning the effect on caste differentiation of extracts of silkworm (*Bombyx mori*) pupae, possibly also belong in this category. We consider it probable that the JH content of these extracts promoted queen induction in larvae that otherwise would have developed into workers. However, after topical application of JH-I to young worker larvae, when reared on royal jelly, Rembold *et al.* (1974) obtained fewer queens and more intercastes than from untreated larvae. Under conditions where a high percentage of larvae have been induced to become queens, it cannot be expected that application of JH to young worker larvae reared on royal jelly will improve the degree of expression of queen characteristics. We suggest that increasing the high level of the JH-titre of prospective queens inhibits the queens' own CA and thereby interferes with caste development. After Trautmann *et al.* (1974) found that JH-III is the only naturally occurring JH in the adult honey-bee (a result that is discussed now because of the low sensitivity of the method of detection),

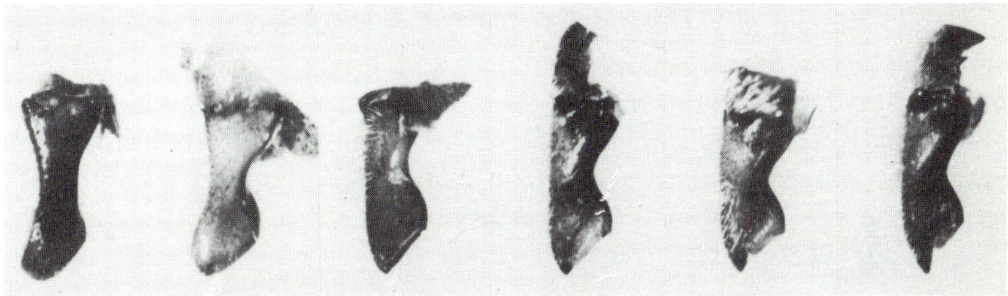

Fig. 14 Effect of JH on queen expression in mandibular development of the honey-bee. The mandible of a normal worker is shown on the left and that of a normal queen on the right. The 4 examples in the middle represent different degrees of queen expression in mandibles of workers treated with JH on the 3rd day of larval development. (After Ebert, 1978)

Goewie and Beetsma (1976) demonstrated that topical application of 1 or 2 μg of JH-III to young worker larvae had hardly any effect on the development of queen characteristics. In contrast to the results of topical application of JH-I, larger doses (3–14 μg) of JH-III caused only some features of an individual to become queenlike, others being unaffected and completely workerlike. These different results cannot be attributed to a differential rate of breakdown of the two types of JH. While Mane and Rembold (1977) are in favour of a direct control of JH titre by tissue esterases, de Kort *et al.* (1977) found that JH-I is not degraded by JH-specific esterases in the haemolymph of larvae of the three last instars; a very low JH-III-esterase activity is found only in the 5th instar. As it is very unlikely that enzyme activities differ for JH-I and JH-III (Kramer and de Kort, 1976), we assume that the different results after application of the two types of JH may be due to a differential rate of release from the cuticle to the haemolymph (Goewie and Beetsma, 1976). The persisting high JH-I level in the haemolymph could affect all imaginal discs during the 2 days after application, whereas the assumed short-lived high JH-III-level coincides with the sensitive period of separate imaginal discs. This assumption was confirmed by Goewie and Beetsma (1976). The expression of queen characteristics can be maximised by repeated (3 times) topical applications of 4 μg of JH-III. Repeated applications of

JH-III did not increase mortality, as occurs in the case of application of JH-I, only 4% of the initial number of treated larvae died. It is attractive to suggest that JH-III plays a role not only in the adult bee, but also in larval development. As the radio-immunoassay for the discrimination between the different types of JH appears not to be conclusive, the differences in the content of JH-I and III in larvae of both castes as reported by Lensky *et al.* (1978), must be doubted, Bergot *et al.* (1981).

Hagenguth and Rembold (1978a) detected JH-III in total extracts of spinning larvae, pupae and adult worker bees; traces of JH-I might be present. Rembold *et al.* (1980) applying an improved method conclude that spinning larvae, prepupae, white pupae and newly emerged worker bees contain only JH-III. In this respect, the observation of Czoppelt and Rembold (1978) that topical application of JH-III to young larvae reared *in vitro* on royal jelly has no significant effect on caste differentiation (in contrast to application of JH-I) should be regarded in the light of the experiments by Goewie and Beetsma (1976) mentioned above.

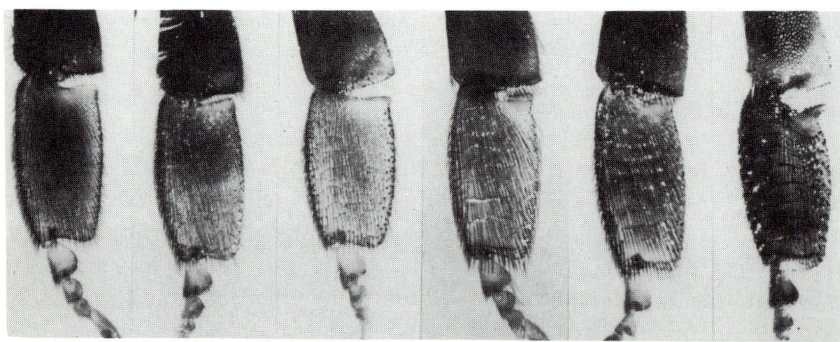

Fig. 15 Effect of JH on the expression of queen characteristics in the hind leg of the honey-bee. The leg of a normal worker is shown on the right that of a normal queen on the left. The 4 examples in the middle represent different degrees of queen expression in the basitarsus after JH-treatment as mentioned in Fig. 14. After JH-treatment the basitarsus is elongated, the hair rows of the basitarsal brush disappear. (After Ebert, 1978)

Final evidence of the significance of JH in the process of queen-worker differentiation was obtained after topical application of the CA-inhibitor, precocene-II. Goewie *et al.* (1978) found that after topical application of precocene-II to queen larvae, the CA partially or completely disappear. Doses of 100 to 150 μg of this substance applied shortly before or after the last larval moult caused an extension of the pupal stage by 3–4 days. The resulting adults are either completely developed queens, or queens with slightly fewer ovarioles but with seriously impaired oocyte formation, or large workerlike individuals. In spite of the fact that all larvae probably had consumed large amounts of food, the precocene-II application caused elimination of the CA, and thereby the development of workerlike characteristics. This experiment also demonstrates that the CA are not involved in the regulation of food intake. Application of 5 μg precocene-II to worker larvae did not disturb their normal development. Application of higher doses invariably caused the death of the individuals. According to Czoppelt and Rembold (1978), topical application of precocene-II to young larvae reared *in vitro* on royal jelly has no effect on caste differentiation and doses of 50–150 μg are toxic. We presently cannot explain why this differential result was obtained.

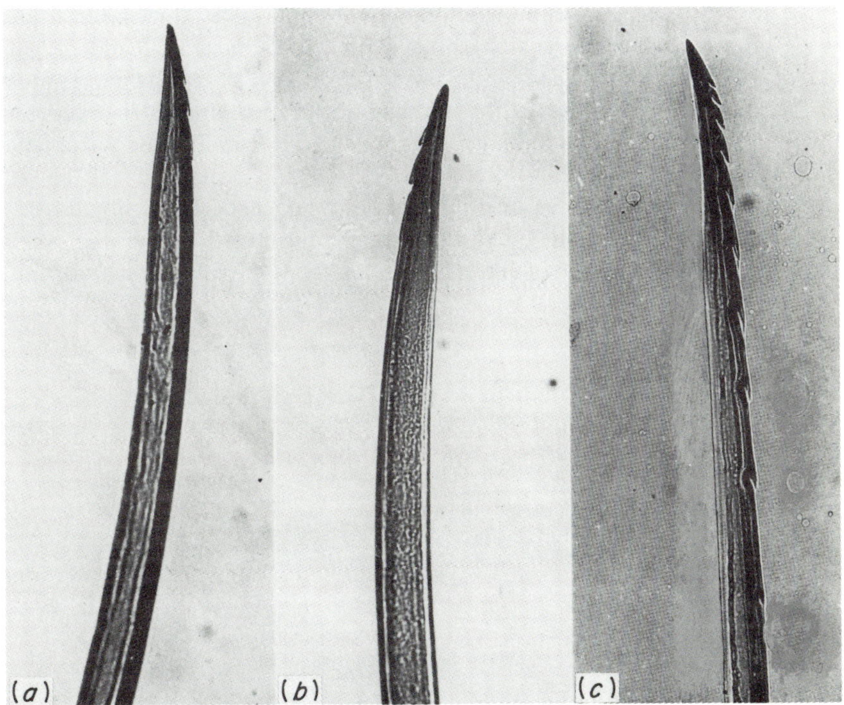

Fig. 16 Effect of JH on the expression of queen characteristics in the shape and structure of the sting of the honey-bee. (*a*) queen; (*c*) worker; (*b*) JH-treated worker. (After Wirtz, 1973)

Naisse (see de Wilde, 1976) detected very interesting effects of JH application on the program of ovarian development in worker larvae. She could demonstrate that JH-I prevents the atrophy of ovariole primordia in worker larvae until the 4th day of development; during this day, a progressively decreasing number of primordia is still left for development (Fig. 17). At day 5 no further change is possible. She also observed reversion of the program in some cases where primordial cysts already showed a developmental arrest and were reactivated by the hormone. But cysts were never observed to reappear where they had already partly atrophied (see de Wilde, 1976). Some of the above morphogenetic effects were obtained with juvenoids by Ždárek and Haragsim (1974) and Copijn *et al.* (1979). The effect of JH on caste programming in the honey bee has been qualified as "speculative" by Rembold (1976) on the ground that no complete and functional queen had been obtained in Wirtz's (1973) experiments. But it is a general experience in endocrinology that it is very difficult, if not impossible, to replace the normal function of an endocrine gland by administering doses of its hormone. Also Lüscher (1977) in discussing the effects of JH-application in termites, found these effects to be different to some degree from increased endogenous activity of their own CA, and ascribed it to the absence of regulation. As most of the research on caste formation in the honey-bee was done by workers not familiar with insect endocrinology, the impact of the findings of Wirtz have met with misunderstanding. Rembold *et al.* (1974) qualified the abovementioned effects of JH as an "indirect morphogenetic effect not directly coupled to caste induction", but were apparently misled by the dual effect of JH in the programming of castes and of instars at different critical periods.

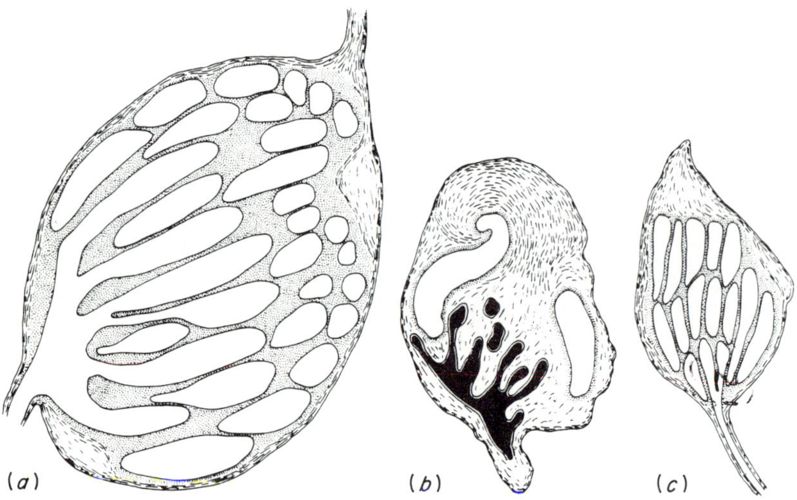

Fig. 17 Schematic drawing of the effect of JH-treatment in the prevention of ovarian atrophy in worker larvae of the honey-bee. (*a*) queen ovary on the 5th day of larval development; (*b*) worker ovary on the 5th day of larval development, showing the reduction in ovariole primordia; (*c*) ovary of a worker larva on the 5th day of larval development after being treated with JH on the 3rd day. (After Naisse, unpublished)

The same authors denied the effect of JH in promoting queen formation, because Dixon and Moser (1972) observed that, implantation of CA of queen larvae 48 h old, had a repressive effect on CA activity in roaches. These findings, however, are easily explained by the fact that an artificially increased titre of JH generally inhibits CA-activity (Schooneveld *et al.*, 1979). This apparently is part of the feedback regulation of this gland. In the honey-bee, JH inhibits pupal and adult development. Application of juvenoids during the prepupal period results in metathetelic pupal and adult characteristics (Žďárek and Haragsim, 1974). The same hormone is able to shift the program of caste development in the direction of queen formation when it is present in a sufficient titre during the third and fourth day of larval development.

After the demonstration of the morphogenetic effects of JH in caste induction, it was found by Ebert (1978, 1980) that the JH-titre also regulates larval orientation behaviour. Having completed the cocoon, queen larvae show a positive geotactic orientation both under natural conditions (queen cells) and when kept under experimental conditions in glass cells. On the contrary, worker larvae orientate themselves in a horizontal position with their heads towards the rough cell capping. Topical application of JH (1 μg of JH-I or 3×5 μg of JH-III) to larvae 78–84 h old in worker cells resulted in a positive geotactic orientation after cocoon spinning, as is found with queen larvae. The same was found with treated worker larvae kept in glass cells. Apparently, JH "switches on" the development of all queen characteristics when present at the critical period for caste induction.

The CA is not the "primo movens" in the above regulatory mechanisms. This gland is generally controlled by the brain via the medial neurosecretory cells (MNSC), and sometimes also by direct reflectory innervation. The evidence of this control in the honey bee is not at all clear. Both Dogra *et al.* (1977) and Goewie (1978) conclude from their histological data that an increased activity of the MNSC occurs *after* the critical period for caste determination. It is suggested that the MNSC inhibit the activity of the CA (Goewie, 1978), but more evidence is needed before we can accept this effect, which would be the inverse from what is known for other adult insects.

According to Ulrich (1979), the pars intercerebralis of honey-bee larvae shows identifiable NSC by the time of the first larval instar. In the queen, a neurosecretory chiasma is visible in the second larval instar, in the worker only in the fourth larval instar. Ritcey and Dixon (1969) described NSC in both castes in the third larval instar. Clearly stainable neurosecretory material was only observed by Ulrich in the queen larva in the 5th larval instar, and in the worker even later.

The CA of worker bees according to Wirtz (1973) are already measurable in the first larval instar, but according to Ulrich (1979) they are only innervated by the cerebral NSC at the end of the fourth larval instar. It is, however, known from other insects (e.g. de Wilde and de Boer, 1969), that neurosecretory control of CA from the brain can take place also in the absence of direct innervation. In this respect, the acceleration of larval and pupal development caused by JH-treatment of worker larvae must be mentioned. The first sign of a successful JH-treatment is the advanced moment of cell capping, which is similar as in queen cells. Secondly, JH-application to young worker larvae shortens the development of the prepupa and the pupa conform to the situation in normal queens. The differential rate of development in the two castes during the last larval and pupal instars is probably related to the respective states of development of the PG. Lukoschus (1952, 1956b) demonstrated that the volume and the secretory activity of PG in last instar queen larvae is enhanced above values found in worker larvae. In addition, Hagenguth and Rembold (1978b) found a considerable difference in ecdysone level during the prepupal stage between queen and worker larvae. Goorden *et al.* (in preparation), found that after application of JH-I to worker larvae, the nuclear size of the PG cells increases to the same degree as in queen larvae. It is a matter of discussion whether the nuclear size of PG cells is affected directly by the increased JH-level or by the neurosecretory material from the MNSC of the brain.

9 Endocrine and pheromonal impact on queen reproductivity and worker functions

In the previous sections, particular stress was laid on caste development. We will now discuss the endocrine impact on caste functioning, and the modification of this endocrine activity by pheromones.

As has been established in adults of several insect orders, it appears that, in *Solenopsis invicta* queens, JH regulates oogenesis, and that JH production by the CA depends on the NSC of the brain (Barker, 1978). In overwintering gynes of *Polistes gallicus*, the ovaries are small and the ecdysone-titre in the haemolymph is high except during a period of 2 weeks in midwinter. The size of the CA is inversely related to the ecdysone-titre. Injection of 2×18 μg of JH into overwintering gynes resulted in the development of oocytes whereas injection of 2×2.5 μg of ecdysterone inhibited the normal development of the ovaries. Strambi *et al.* (1977) suggest that ecdysone and JH play an antagonistic role in ovarian development. The ecdysone-content of trophocytes changes in the same way as in the oocytes leading the authors to suggest that trophocytes are involved either in the synthesis or metabolism of ovarian ecdysone. Röseler *et al.* (1980) confirmed that, even though overwintering gynes have large CA, these glands produce small amounts of JH. After hibernation, egg maturation starts and Röseler *et al.* (1980) found that the

CA of dominant females are larger than those of subordinated ones and in addition that the volume of the CA is well correlated with their synthetic activity. The authors conclude that the social hierarchy affects oogenesis by way of the endocrine system. In 1972, Bohm demonstrated the impact of JH-titre of the haemolymph in relation to ovarian activity in *P. metricus*. When he isolated gynes at 26°C and 16 h photoperiod in June, oogenesis was started. Bohm (1972) suggests that diapause is induced during the growth of immature stages of these females in July and August, because no ovarian activity could be obtained after isolation. However, after topical application of 0.2 or 0.5 μl of synthetic JH, ovarian activity could also be obtained in females isolated in July. Repeated topical applications of JH workers of *P. annularis* caused ovarian maturation and an increase in the frequency of dominance interactions (Barth *et al.*, 1975). *P. gallicus* females parasitised by a Strepsipteron, which causes castration, have large but inactive CA (Röseler *et al.*, 1980). However, Strambi and Strambi (1973) observed small CA in parasitised females and suggested that the size of the CA was affected by their neurosecretory innervation from the brain. The ecdysone titre in the haemolymph was low (Strambi *et al.*, 1977). Oogenesis is induced by topical application of JH to adult workers of *Lasioglossum zephyrum* (Bell, 1973). It is concluded that the queen inhibits CA activity.

After Röseler (1977) demonstrated, in *Bombus terrestris*, that a linear relation exists between the CA-volume and the body size of newly-emerged workers (wing span between 25 and 32 mm), he was able to distinguish differences in CA-volumes between queenright and queenless workers. In 5-day-old queenless workers, the CA-volume was about three times as large as that of queenright workers. Röseler (1977a) also found that the respective average JH-titres were 25 000 and 5000 *Galleria* Units/ml. In addition, Röseler proved that a high JH-titre causes oocyte development. After injection of 50 μg of synthetic JH-I complete oogenesis could be obtained even in queenright workers. Under these experimental conditions the CA-volume remained suppressed. When calculating synthetic activities of CA *in vitro* per unit of body size Röseler and Röseler (1978) found that the average incorporation of ^{14}C-labelled methyl methionine into JH of 4 pairs of CA of queenless workers was about 5 times as high as of those of queenright workers 3 days of age. Average synthetic activity of 1-day-old queenless workers was twice that of queenright workers of the same age. According to Röseler and Röseler, (1978) the JH-titre is mainly regulated by the synthetic activity of the CA. High pressure liquid chromatography showed that JH-III is produced *in vitro*. High CA-activity coincided with a large accumulation of neurosecretory granules in the pars intercerebralis of the brain. JH is broken down by esterases in the hindgut; only traces of JH are excreted (Röseler, 1977). The rates of breakdown and excretion are not affected by the presence or absence of the queen.

Topical application or injection of JH in adult queens of *B. terrestris* and *B. hypnorum* changes their physiology and behaviour into that of a foundress in spring. Increasing the initial low JH-titre inhibits the development of the fat body, decreases the content of lipid and of a queen specific protein in the haemolymph and induces oogenesis. JH-treated queens do not leave the nest, but take part in colony tasks. Such queens however are attractive to drones and are ready to mate (Röseler, 1976). Studies on the nature and amount of pheromones produced by dominant queens and workers in relation to their JH-titre and the induction of dominant behaviour in workers by JH application remain to be done. Special conditions prevail in the honey-bee queen. After allatectomy of either young mated queens (van Laere, 1974), of egg-laying queens 1–3 years old (Hrdý and Sláma, 1963), or virgin queens treated with carbon dioxide (Engels and Ramamurty, 1976), oviposition is continued for several weeks. However, in the last mentioned experiments, vitellogenin synthesis, vitellogenin metabolism and the vitellogenin titre were lowered in comparison to controls. As determinations of JH-titre are lacking in the above experiments, the possibility that a residual JH-titre remained for a sufficiently long period to induce oogenesis, cannot be excluded. Topical application of JH to allatectomised bee queens (Ramamurty, 1977) did not influence their rate of oviposition. In addition, van Laere (1974) suggests that the allatectomised queens still produce pheromones as they evoke retinue behaviour in the workers and inhibit queen cell construction. This is in agreement with the results of Ždárek *et al.* (1976) who topically applied a large dose (1 mg) of 7 different JHA to inseminated honey bee queens. In the majority of cases, this treatment did not affect their survival, onset of oviposition or fecundity. Also, repeated applications of JH-III to queens, from the time of their emergence up to the age of 30 days, did not affect the frequency and duration of mating flights, onset of oviposition, and vitellogenin synthesis (Regel, 1978).

Lüscher and Walker (1963) suggest that the inhibition of ovarian growth of worker bees by the queen's pheromones is brought about by way of the CA. They derive this conclusion from experiments with winter bees. These results were confirmed in experiments with caged summer bees by Gast (1967). Lüscher (1977) assumed that the queen's pheromones must have additional effects, because it has not been possible to induce oocyte development in queenright workers by implanting CA or by treatment with JH. Gast (1967) observed that in a queenright colony, the CA of worker bees increased in two phases. Only in the second phase does a significant increase in volume occur, while the nuclear size of cerebral NSC decreases. Conversely, the size of the CA of queenless workers does not increase during the second phase, whereas the size of the nuclei of the cerebral NSC increases. The initial increase in size of the CA is correlated with oogenesis, the second increase only occurs in the presence of a queen, probably by sensory perception of its pheromone. The activity of the cerebral NSC (inhibited by the queen) is probably necessary for oogenesis. Although Rutz *et al.* (1976) found that vitellogenin synthesis increases after injection of small amounts of JH-III in queenright workers, they never observed oocyte development. Röseler *et al.* (1980) suggest that this is caused by a lack of responsiveness of the ovarioles to JH, their development being arrested at an early stage (differentiation of oogenesis). This suggestion does not account for the occurrence of some ovarian activity in workers of queenright colonies in periods during the active season (Verheijen-Voogd, 1959) when CA activity is high (Fluri *et al.*, 1977). In winter bees, where the ovarian activity is low, there is a much reduced JH-titre (Fluri *et al.*, 1977). The measurement of accurate JH-titres would be of great help in resolving the above contradictions. It has recently been shown that JH controls the functioning of the worker caste in the division of labour. By increasing JH-titre of young workers, it has been demonstrated that nursing and building behaviour, pollen consumption and first flight activities, the development of hypopharyngeal and wax glands, protein and vitellogenin titres and the percentage of dense membrane leucocytes in the haemolymph are all affected (Jaycox *et al.*, 1974; Beetsma and ten Houten, 1975; Fluri *et al.*, 1977; Rutz *et al.*, 1977). Breed (1979) demonstrated that JH-treatment increases the expression of aggressive behaviour in adult workers. The sequence of behavioural activities displayed by worker bees during imaginal life in the colony, now appears to be determined to a large degree by fluctuations in their JH-titre.

Retinue behaviour similar to that evoked by a queen has been observed towards certain worker bees in queenless colonies. Butler (1957) suggested that bees with active ovaries produce a pheromone inhibiting ovarian growth in other workers. Sakagami (1958) observed that, in *Apis cerana*, ovarian size decreased when a laying worker was present. Velthuis (1970b) and Jay and Nelson (1973) confirmed this observation. Velthuis *et al.* (1965) demonstrated that acetone extracts of workers with highly enlarged ovaries gave similar results. Ruttner *et al.* (1976) found that the heads of laying workers of *Apis mellifera capensis* contain trans-9-oxodecenoic acid.

10 Concluding remarks

The physiology of caste determination is an aspect of ecophysiology, and it was, therefore, inevitable to discuss at some length the extrinsic factors involved. As these are often complex, we have sometimes given a detailed picture.

Our understanding of caste differentiation has developed rapidly in recent years. Of particular significance is the role of hormones and pheromones. Of the first category, juvenile hormone stands out. A crucial role in caste differentiation appears to be played by the corpora allata, and the external and internal messages that trigger or inhibit the activity of these glands are now of central interest.

One might even say that all previous explanations of caste differentiation conceived before the impact of juvenile hormone was known, should be subject to reconsideration. Crucial experiments are needed building upon the endocrine facts and testing alternative explanations. This is especially valid for the "Determinator" concept and for genetic theories of caste determination. From the evidence given in section 5, it is clear that in social Hymenoptera, food quantity is a very important element in caste differentiation, and it would seem that factors regulating the intake of food by the larvae, by some phagostimulatory or deterrent action, could be of additional importance. At present, our most detailed understanding of caste differentiation is in the honey-bee where the search for the relevant messages in the extrinsic control of corpus allatum activity is being actively pursued. In addition to the interest in the function and composition of royal jelly, there is now a growing interest in the treatment given to the worker larvae by the nursing bees. As Jung-Hoffmann (1967) has stated, it is not only important *when* certain food components are offered, but also *how* they are offered by the workers.

The factors provided by royal jelly and by the treatment given to the larvae in queen cells deserve continued interest. During a recent visit to the Academy of Agriculture in Peking, Dr Yang (oral communication) informed the senior author of experiments, in which female larvae of *Apis mellifera* (one-day-old) were transferred into queen cells of *Apis cerana*. The resulting *A. mellifera* queens, after careful morphological inspection, showed significant differences in number of hooks of the wing attachment system, the hamuli. Although this is not an essential change, it seems worthwhile noting.

The short-term effect of juvenile hormone in caste differentiation of termites as compared with the long-term effect in honey-bee caste differentiation has given rise to misunderstandings, and sometimes to undue criticism (see Rembold *et al.*, 1974). In the hemimetabolous termites, presoldier differentiation depends on an increased juvenile hormone titre during the preceding intermoult period (soldiers, after all, are larvae). In honey-bees, and most probably in other social Hymenoptera as well, the effect of an increased JH-titre is only visible after several moults and metamorphosis have elapsed. In the holometabolous Hymenoptera, the developmental program is much more complex and includes the differentiation of imaginal discs. In an insect like the honey-bee, effects of juvenile hormone recur during the whole life span, and there are sensitive periods for the determination of stages and castes during postembryonic development, and oogenesis and glandular activities during adult life. This includes the division of labour and the seasonal state of the worker. In this insect, juvenile hormone effects show a remarkable versatility.

The endocrine basis of caste differentiation has important consequences for the application of juvenoid insect growth regulators. Its study has opened up new ways for controlling termites and ants, but also has revealed some risks, especially with regard to pollinating bees.

Developmental programming as well as functioning of castes, modified by hormones, and the underlying trigger mechanisms of endocrine activity, offer a fascinating field of sufficient complexity to provide themes of research for generations of insect physiologists.

Acknowledgements

The authors are indebted to Dr R. H. Leuthold and Dr H. H. W. Velthuis for checking the parts on termites and stingless bees in earlier drafts of the paper.

Many thanks are due to Mrs M. Flinterman and Miss G. P. Bruyn for typing the manuscript with understanding and patience, and to Mr W. C. Th. Middelplaats and Mr J. W. Brangert for preparing and reproducing the figures.

NOTE ADDED IN PROOF

Wheeler and Nijhout (1981) recently demonstrated in the ant *Pheidole bicarinata* the existence of a critical period in the last larval instar during which JH or a JHA induces soldier formation.

References

Abo-Khatwa, N. (1977). Chemistry of "fungus combs" and their role in the nutrition of the termites. Proc. 8th Int. Congr, IUSSI. Wageningen, 253–256

Archer, M. E. (1972). The significance of worker size in the seasonal development of the wasps *Vespula vulgaris* (L.) and *Vespula germanica* (F.). *J. Ent.* **(A) 46,** 175–183

Asencot, M. and Lensky, Y. (1976). The effect of sugars and juvenile hormone on the differentiation of the female honeybee larvae (*Apis mellifera* L.) to queens. *Life Sci.* **18,** 693–700

Banks, W. A., Lofgren, C. S. and Plumley, J. K. (1978). Red imported fire ants: Effects of insect growth regulators on caste formation and colony growth and survival. *J. econ. Ent.* **71,** 75–78

Barbier, M. and Lederer, E. (1960). Structure chimique de la "substance royale" de la reine d'abeille (*A. mellifica* L.). *C.R. Acad. Sci. Paris* **250,** 4467–4469

Barker, J. F. (1978). Neuroendocrine regulation of oocyte maturation in the imported fire ant *Solenopsis invicta. Gen. comp. Endocr.* **35,** 234–237

Barth, R. H., Lester, L. J., Sroka, P., Kessler, T. and Hearn, R. (1975). Juvenile hormone promotes dominance behavior and ovarian development in social wasps (*Polistes annularis*). *Experientia* **31,** 691–692

Beetsma, J. (1979). The process of queen–worker differentiation in the honeybee. *Bee World* **60,** 24–39

Beetsma, J. and Houten, A. ten (1975). Effects of juvenile hormone analogues in the food of honeybee colonies (*Apis mellifera* L.). *Z. angew. Ent.* **77,** 292–300

Beetsma, J. and Schoonhoven. L. M. (1966). Some chemosensory aspects of the social relations between the queen and the worker in the honeybee (*Apis mellifera* L.). *Proc. K. ned. Akad. Wet. C.* **69,** 645–647

Bell, W. J. (1973). Factors controlling initiation of vitellogenesis in a primitively social bee, *Lasioglossum zephyrum* (Hymenoptera: Halictidae). *Ins. Soc.* **20,** 253–260

Bergot, B. J., Schooley, D. A. and Kort, C. A. D. de (1981). Identification of JH-III as the principal juvenile hormone in *Locusta migratoria. Experientia* **37,** 909–910

Ber-Lin Chai and Shuel, R. W. (1970). Effects of supernumerary corpora allata and farnesol compounds on ovary development in the worker bee. *J. apic. Res.* **9,** 19–27

Berndt, K. P. (1975). Physiology of reproduction in the Pharaoh's ant (*Monomorium pharaonis* L.). 1. Pheromone mediated cyclic production of sexuals. Proc. 3rd Symp. med. vet. Acaroentomology, Gdansk, 25–28

Berndt, K. P. and Nitschmann, J. (1979). The physiology of reproduction in the Pharaoh's ant (*Monomorium pharaonis* L.) 2. The unmated queens. *Ins. Soc.* **26,** 137–145

Bertholf, L. M. (1925). The moults of the honeybee. *J. econ. Ent.* **18,** 380–384

Bier, K. (1954). Über den Saisondimorphismus der Oogenese von *Formica rufa rufo-pratensis* minor Gössw. und dessen Bedeutung für die Kastendetermination. *Biol. Zentralblatt* **73,** 170–190

Bigley, W. S. and Vinson, S. B. (1975). Characterization of a brood pheromone isolated from the sexual brood of the Imported Fire Ant, *Solenopsis invicta. Ann. ent. Soc. Am.* **68,** 301–304

Blackith, R. E. (1958). An analysis of polymorphism in social wasps. *Ins. Soc.* **5,** 263–272

Bodot, P. (1969). Composition des colonies de termites: ses fluctuations au cours de temps. *Ins. Soc.* **16,** 39–54

Bodot, P. (1970). La composition des colonies de *Cubitermes subcrenulatus* silvestri (Isoptera, Termitidae). *C.R. Acad. Sci. Paris* **271,** 327–330

Bohm, M. K. (1972). Effects of environment and juvenile hormone on ovaries of the wasp, *Polistes metricus. J. Ins. Physiol.* **18,** 1875–1883

Bollenbacher, W. E., Zvenko, H., Kumaran, A. and Gilbert, L. J. (1978). Changes in ecdysone content during postembryonic development of the wax moth, *Galleria mellonella*: The role of the ovary. *Gen. comp. Endocr.* **34,** 169–179

Bonavita, A. and Passera, L. (1977). On the diet of queen and worker larvae in the ant *Plagiolepis pygmaea* Latr. (Hymenoptera, *Formicidae*). Proc. 8th Int. Congr. IUSSI, Wageningen, 109–112

Bonavita-Cougourdan, A. and Passera, L. (1978). Étude comparative au moyen d'or radioactif de l'alimentation des larves d'ouvrières et des larves de reine chez la Fourmi *Plagiolepis pygmaea* Latr. *Ins. Soc.* **25,** 275–287

Bordereau, C. (1975). Déterminisme des castes chez les termites supérieurs: mise en évidence d'un contrôle royal dans la formation de la caste sexuée chez *Macrotermes bellicosus* Smeethman (Isoptera, Termitidae). *Ins. Soc.* **22,** 363–374

Bordereau, C., Delbecque, J. P., Hirn, M., Lanzrein, B., Imboden, H., Lüscher, M. and O'Connor, J. D. (1977). Ecdysones of the physogastric termite queen. Proc. 8th Int. Congr. IUSSI, Wageningen, 290

Bordereau, C., Hirn, M., Delbecque, J. P. and De Reggi, M. (1976). Présence d'ecdysones chez un insecte adulte: La reine de Termites. *C. R. Acad. Sci. Paris D* **282,** 885–888

Bounhiol, J. J. (1938). Recherches expérimentales sur le déterminisme de la métamorphose chez les Lepidoptères. *Bull. biol. Fr. Belg. Suppl.* **24,** 1

Breed, M. D. (1975). Sociality and seasonal size variation in halictine bees. *Ins. Soc.* **22,** 375–379

Breed, M. D. (1977). Interactions among individuals and queen replacement in an eusocial halictine bee. Proc. 8th Int. Congr. IUSSI, Wageningen, 228–231

Breed, M. D. (1979). Does juvenile hormone regulate dominance and aggression in worker honeybees ? Proc. 16th Int. ethol. Conf., Vancouver

Breed, M. D. and Gamboa, G. J. (1977). Behavioral control of workers by queens in primitively eusocial bees. *Science* **195,** 1694–1696

Breed, M. D., Silverman, J. M. and Bell, W. J. (1978). Agonistic behavior, social interactions, and behavioral specialization in a primitively eusocial bee. *Ins. Soc.* **25,** 351–364

Brian, M. V. (1957). Serial organization of brood in *Myrmica. Ins. Soc.* **4,** 191–210

Brian, M. V. (1965). Studies of caste differentiation in *Myrmica rubra* L. 8. Larval development sequences. *Ins. Soc.* **12,** 347–362

Brian, M. V. (1973). Caste control through worker attack in the ant *Myrmica. Ins. Soc.* **20,** 87–102

Brian, M. V. (1974a). Kastendetermination bei *Myrmica rubra* L. In: *Sozialpolymorphismus bei Insekten* (G. H. Schmidt, ed.) pp. 565–588. Wiss. Verlagsgesellschaft, Stuttgart

Brian, M. V. (1974b). Caste differentiation in *Myrmica rubra*: the role of hormones. *J. Ins. Physiol.* **20,** 1351–1365

Brian, M. V. (1975a). Larval recognition by workers of the ant *Myrmica. Anim. Beh.* **23,** 745–756

Brian, M. V. (1975b). Caste determination through a queen influence on diapause in larvae of the ant *Myrmica rubra. Ent. exp. Appl.* **18,** 429–442

Brian, M. V. (1976). Endocrine control over caste differentiation in a Myrmicine ant. In: *Phase and Caste Determination in Insects* (M. Lüscher, ed.) pp. 63–69, Pergamon Press, Oxford

Brian, M. V. (1977). The synchronization of colony and climatic cycles. Proc. 8th Int. Congr. IUSSI, Wageningen, 202–206

Brian, M. V. (1979). Caste differentiation and division of labour in wasps, bees and ants. In: *Social Insects* (H. R. Hermann, ed.) Vol. 1, pp. 121–222. Academic Press, London

Brian, M. V. and Blum, M. S. (1969). The influence of *Myrmica* queen head extracts on larval growth. *J. Ins. Physiol.* **15,** 2213–2223

Brian, M. V. and Hibble, J. V. (1963). 9-Oxodec-trans-2-enoic acid and *Myrmica* queen extracts tested for influence on brood in *Myrmica. J. Ins. Physiol.* **9,** 25–34

Brothers, D. J. and Michener, C. D. (1974). Interactions in colonies of primitively social bees III. Ethometry of division of labor in *Lasioglossum zephyrum. J. comp. Physiol.* **90,** 129–168

Bruinsma, O. and Leuthold, R. H. (1977). Pheromones involved in the building behaviour of *Macrotermes subhyalinus* (Rambur). Proc. 8th Int. Congr. IUSSI, Wageningen, 257–258

Bühlmann, G. (1977a). The study of caste differentiation in the higher termites. In: *Advances in Medical, Veterinary and Agricultural Entomology in Eastern Africa* (C. P. F. de Lima, ed.) pp. 107–110. 1st. East African Conference on Entomology and Pest Control, Nairobi, Kenya

Bühlmann, G. (1977b). The development of the incipient colony and egg adaption experiments in the termite *Macrotermes subhyalinus*. Proc. 8th Int. Congr. IUSSI, Wageningen, 259–261

Buschinger, A. and Kloft, W. (1973). Zur Funktion der Königin im sozialen Nahrungshaushalt der Pharaoameise *Monomorium pharaonis* (L.). *Forschungsber. Landes Nordrhein-Westfalen* No. 2306, 34

Butler, C. G. (1954). The method and importance of the recognition by a colony of honeybees (*A. mellifera*) of the presence of its queen. *Trans. Roy. Entomol. Soc. London* **105,** 11–29

Butler, C. G. (1957). The control of ovary development in worker honey bees (*Apis mellifera*). *Experientia* **13,** 256–257

Butler, C. G. (1960). The significance of queen substance in swarming and supersedure in honey-bee (*Apis mellifera* L.) colonies. *Proc. Roy. Entomol. Soc. London* **35,** 129–132

Butler, C. G. and Fairey, E. M. (1963). The role of the queen in preventing oogenesis in worker honey bees. *J. apic. Res.* **2,** 14–18

Butler, C. G. and Simpson, J. (1958). The source of the queen substance of the honeybee *A. mellifera* L. *Proc. Roy. Entomol. Soc. London* **33,** 120–122

Butler, C. G., Callow, R. K. and Johnston, N. C. (1961). The isolation and synthesis of queen substance, 9-oxodec-2-enoic acid. *Proc. Roy. Soc.* **155,** 417–432

Butler, C. G., Callow, R. K., Greenway, A. R. and Simpson, J. (1974). Movement of the pheromone, 9-oxodec-2-enoic acid, applied to the body surfaces of honeybees (*Apis mellifera*). *Ent. exp. Appl.* **17,** 112–116

Camargo, C. A. de (1972). Produção "in vitro" de intercastes em Scaptotrigona postica Latreille. *Homenagem a Warwick E. Kerr*, pp. 37–54

Camargo, C. A. de (1977a). Effects of juvenile hormone on diploid drones of *Melipona quadrifasciata* Lep. (Hymenoptera, Apidae). Proc. 8th Int. Congr. IUSSI, Wageningen, 193–194

Camargo, C. A. de (1977b). Properties of the x° gene, sex determining in *Melipona quadrifasciata* Lep. (Hymenoptera, Apidae). Proc. 8th Int. Congr. IUSSI, Wageningen, 191–192

Camargo, C. A. de, Almeida, M. G. de, Parra, M. G. N. and Kerr, W. E. (1976). Genetics of sex determination in bees. IX. Frequencies of queens and workers from larvae under controlled conditions (Hymenoptera: Apoidea). *J. Kansas ent. Soc.* **49,** 120–125

Campos, L. A. de Oliveira, Velthuis-Kluppell, F. M. and Velthuis, H. H. W. (1975). Juvenile hormone and caste determination in a stingless bee. *Naturwiss.* **62,** 98–99

Canetti, S. J., Shuel, W. and Dixon, S. E. (1964). Studies on the mode of action of royal jelly in honey bee development. IV. Development within the brain and retrocerebral complex of female honey bee larvae. *Can. J. Zool.* **42,** 229–233

Carr, C. A. H. (1962). Further studies on the influence of the queen in ants of the genus *Myrmica*. *Ins. Soc.* **9,** 197–211

Castle, G. B. (1934). An experimental investigation of caste differentiation in *Zootermopsis angusticollis*. In: *Termites and Termite Control* (C. A. Kofoid, ed.) pp. 292–310. University of California Press, Berkeley

Chippendale, G. M. (1977). Hormonal regulation of larval diapause. *Ann. Rev. Ent* **22,** 121–138

Cleveland, L. R. (1926). Symbiosis among animals with special reference to termites and their intestinal flagellates. *Quart. Rev. Biol.* **1,** 51–60

Cleveland, L. R., Hall, S. R., Sanders, E. P. and Collier, J. (1934). The wood-feeding roach *Cryptocercus*, its Protozoa, and the symbiosis between Protozoa and roach *Mem. Am. Acad. Arts Sci.* **17,** 185–342

Colombel, P. (1978). Biologie *d'Odontomachus haematodes* L. (Hym. Form.) Déterminisme de la caste femelle. *Ins. Soc.* **25,** 141–151

Copijn, G. M., Beetsma, J. and Wirtz, P. (1979). Queen differentiation and mortality after application of different juvenile hormone analogues to worker larvae of the honey bee (*Apis mellifera* L.) *Proc. K. ned. Akad. Wet.* C **82,** 29–42

Cumber, R. A. (1949). Biology of humble-bees with special reference to the production of the worker caste. *Trans. Roy. Entomol. Soc. London* **100,** 1–45

Cupp, E. W. and O'Neal, J. (1973). The morphogenetic effects of two juvenile hormone analogues on larvae of imported fire ants. *Environ. Ent.* **2,** 191–194

Czoppelt, Ch. and Rembold, H. (1978). Juvenilhormon- und Precocene-Wirkung auf mit Weiselfuttersaft ernährte Arbeiterinlarven der Honigbiene. *Mitt. dtsch. Ges. allg. angew. Ent. Giessen* **1,** 294–295

Darchen, R. (1973). Essai d'interprétation du déterminisme des castes chez les *Trigones* et les Mélipones. *C.R. Acad. Sci. Paris* **276,** 607–609

Darchen, R. and Delage, B. (1970). Facteur déterminant les castes chez les *Trigones* (Hyménoptérès, Apidés). *C.R. Acad. Sci. Paris* **270,** 1372–1373

Darchen, R. and Delage-Darchen, B. (1971). Le déterminisme des castes chez les *Trigones* (Hyménoptères, Apidés). *Ins. Soc.* **18,** 121–134

Darchen, R. and Delage-Darchen, B. (1974). Nouvelles expériences concernant le déterminisme des castes chez les Mélipones (Hyménoptères, Apidés). *C.R. Acad. Sci. Paris* **278,** 907–910

Darchen, R. and Delage-Darchen, B. (1975). Contribution à l'étude d'une abeille du Mexique *Melipona Beecheii* B. (Hyménoptère: Apide). *Apidologie* **6,** 295–335

Darchen, R. and Delage-Darchen, B. (1977). Sur le déterminisme des castes chez les Mélipones (Hyménoptères Apidés). *Bull. biol. France Belgique* **111,** 91–109

Delage-Darchen, B. (1977). Les régulations sociales chez les fourmis. *Année Biol.* **16,** 517–543

Delbecque, J. P., Lanzrein, B., Bordereau, C., Imboden, H. and Hirn, M. (1978). Ecdysone and ecdysterone in physogastric queens and eggs of *Macrotermes bellicosus* and *Macrotermes subhyalinus*. *Gen. comp. Endocr.* **36,** 40–47

Deleurane, E. P. (1948). Le comportement reproducteur est indépendant de la présence des ovaires chez *Polistes* (Hyménoptères, Vespides). *C.R. Acad. Sci. Paris* **227,** 866–867

Dietz, A. and Lambremont, E. N. (1970). Caste determination in honey bees. II. Food consumption of individual honey bee larvae, determined with ^{32}P-labelled royal jelly. *Ann. ent. Soc. Am.* **63,** 1342–1345

Dixon, S. E. and Moser, E. (1972). Duality in function in corpora allata of honey bee larvae. *Can. J. Zool.* **50,** 593–595

Dogra, G. S., Ulrich, G. M. and Rembold, H. (1977). A comparative study of the endocrine system of the honey bee larva under normal and experimental conditions. *Z. Naturforsch.* **32,** 637–642

Ebert, R. (1978). *Die Bedeutung der Schwerkraft bei der longitudinalen Orientierung der weiblichen Honigbienenlarve* (Apis mellifera *L.*) *und deren Beeinflussung durch hormonale Faktoren*. Ph. D. Thesis, University of Würzburg

Ebert, R. (1980). Influence of juvenile hormone on gravity orientation in the female honeybee larva (*Apis mellifera* L.). *J. comp. Physiol. A* **137,** 7–16

Edwards, J. P. (1975a). The use of juvenile hormone analogues for the control of some domestic insect pests. Proc. 8th British Insecticide and Fungicide Conf., Brighton, 267–275

Edwards, J. P. (1975b). The effects of juvenile hormone analogues on laboratory colonies of pharaoh's ant, *Monomorium pharaonis* (L.) (Hymenoptera, Formicidae). *Bull. ent. Res.* **65,** 75–80

Edwards, J. P. (1977). Control of *Monomorium pharaonis* with an insect juvenile hormone analogue. Proc. 8th Int. Congr. IUSSI, Wageningen, 81–82.
Endo, K. (1980). Neuroendocrine regulation of seasonal form development in several species of butterflies. Abstr. 16th Int. Congr. Ent. Kyoto, 187
Engel, M. S. (1979). Is caste determination in *Melipona quadrifasciata*, a stingless bee, influenced by 9-oxo-decenoic acid? *Ins. Soc.* **26,** 273–278
Engels, W. and Ramamurty, P. S. (1976). Initiation of oogenesis in allatectomised virgin honey bee queens by carbon dioxide treatment. *J. Ins. Physiol.* **22,** 1427–1432
Erp, A. van (1960). Mode of action of the inhibitory substance of the honeybee queen. *Ins. Soc.* **7,** 207–211
Evans, H. E. and West-Eberhard, M. J. (1973). *The Wasps.* David and Charles Ltd., Newton Abbot, Devon
Fluri, P., Wille, H., Gerig, L. Lüscher, M. (1977). Juvenile hormone and the determination of winter bee physiology (*Apis mellifera*). Proc. 8th. Int. Congr. IUSSI, Wageningen, 28–29
Free, J. B. (1955). Queen production in colonies of bumblebees. *Proc. Roy. Entomol. Soc. London A* **30,** 19–25
Free, J. B. (1960). The distribution of bees in a honey-bee (*Apis mellifera* L.) colony. *Proc. Roy. Entomol. Soc. London A* **35,** 141–144
Fukaya, M. and Kobayashi, M. (1966). Some inhibitory actions of corpora allata in diapausing larvae of the rice stemborer, *Chilo suppressalis* Walker. *Appl. Ent. Zool.* **1,** 125–129
Gamboa, G. J., Heacock, B. D. and Wiltjer, S. L. (1978). Division of labor and subordinate longevity in foundress associations of the paper wasp, *Polistes metricus* (Hymenoptera: Vespidae). *J. Kansas ent. Soc.* **51,** 343–352
Gast, R. (1967). Untersuchungen über den Einfluss der Königinnensubstanz auf die Entwicklung der endokrinen Drüsen bei der Arbeiterin der Honigbiene (*Apis mellifica*). *Ins. Soc.* **14,** 1–12
Geigy, R. (1931). Erzeugung rein imaginaler Defekten durch ultraviolette Eibestrahlung bei *Drosophila melanogaster. W. Roux Arch. Entw. Mech. Org.* **125,** 406–477
Glancey, R. M., Stringer, C. E., Craig, C. H. and Bishop, P. M. (1973). Evidence of a replete caste in the fire ant, *Solenopsis invicta. Ann. ent. Soc. Am.* **66,** 233–234
Glancey, R. M., Stringer, C. E., Craig, C. H., Bishop, P. M. and Martin, B. B. (1970). Pheromone may induce brood tending in the fire ant, *Solenopsis saevissima. Nature, London* **226,** 863–864
Goetsch, W. (1940). *Vergleichende Biologie der Insektenstaaten.* Akad. Verlagsgesellshaft, Leipzig
Goewie, E. A. (1976). Early queen–worker differentiation in the honey bee (*Apis mellifera* L.). *Proc. K. ned. Akad. Wet. C* **79,** 470–471
Goewie, E. A. (1978). *Regulation of caste differentiation in the honey bee* (Apis mellifera *L.*). Ph. D. Thesis, Agricultural University, Wageningen, Meded. Landb. Hogeschool, 78–157
Goewie, E. A. and Beetsma, J. (1976). Induction of caste differentiation in the honey bee (*Apis mellifera* L.) after topical application of JH-III. *Proc. K. ned. Akad. Wet. C* **79,** 466–469
Goewie, E. A., Beetsma, J. and Wilde, J. de (1978). Wirkung von Precocene II auf die Kastendifferenzierung der Honigbiene (*Apis mellifera* L.). *Mitt. dtsch. Ges. allg. angew. Ent. Giessen* **1,** 304–305

Goorden, A., Goewie, E. A. and Beetsma, J. (in preparation). Effects of juvenile hormone application on development of the prothoracic gland and oenocytes of the honey-bee larva (*Apis mellifera* L.).

Grassé, P. P. (1949). Ordre des Isoptères ou termites. In: *Traité de Zoologie* (P. P. Grassé, ed.) Vol. 9, pp. 408–544. Masson, Paris

Grassé, P. P. and Noirot, Ch. (1946a). La production des sexués neoténiques chez le termite à cou jaune (*Calotermes flavicollis* F.). Inhibition germinale et inhibition somatique. *C.R. Acad. Sci. Paris* **223,** 869–871

Grassé, P. P. and Noirot, Ch. (1946b). Le polymorphisme social du termite à cou jaune (*Calotermes flavicollis* F.) la production des soldats. *C.R. Acad. Sci. Paris* **223,** 929–931

Grassé, P. P. and Noirot, Ch. (1947). Le polymorphisme social du termite à cou jaune (*Calotermes flavicollis* F.). Les faux ouvriers ou pseudergates et les mues regressives. *C.R. Acad. Sci. Paris* **224,** 219–221

Grassé, P. P. and Noirot, Ch. (1960). Role respectif des mâles et femelles dans la formation des sexués neoténiques chez *Calotermes flavicollis. Ins. Soc.* **7,** 109–123

Grassi, B. and Sandias, A. (1896, 1897). The constitution and development of the society of termites. *Quart. J. micr. Sci.* **39,** 245–322; **40,** 1–75

Greenberg, S. L. W. and Stuart, A. M. (1979). The influence of group size on ovarian development in adult and neotenic reproductives of the termite *Zootermopsis angusticollis* (Hagen) (*Hodotermitidae*). *J. Inverteb. Reprod.* **1,** 99–108

Hadorn, E. (1966). Über eine Änderung musterbestimmender Qualitäten in einer Blastenkultur von *Drosphila melanogaster. Rev. Suisse Zool.* **73,** 253–265

Hadorn, E. (1967). Dynamics of determination. In: *Major Problems in Developmental Biology* (M. Locke, ed.) pp. 85–104. Academic Press, New York

Hagenguth, H. and Rembold, H. (1978a) Identification of juvenile hormone 3 as the only juvenile hormone homologue in all developmental stages of the honey bee. *Z. Naturforsch.* **33,** 11–12

Hagenguth, H. and Rembold, H. (1978b). Kastenspezifische Modulation des Ecdysteroid-Titers bei der Honigbiene. *Mitt. dtsch. Ges. allg. angew. Ent. Giessen* **1,** 296–298

Hanser, G. and Rembold, H. (1964). Analytische und histologische Untersuchungen der Kopf- und Thoraxdrüsen bei der Honigbiene *Apis mellifica. Z. Naturforsch.* **19,** 938–943

Haverty, M. J. (1979). Soldier production and maintenance of soldier proportion in laboratory experimental groups of *Coptotermes formosanus* Shiraki. *Ins. Soc.* **26,** 69–84

Haverty, M. J. and Howard, R. W. (1979). Effects of insect growth regulators on subterranean termites (*Reticulitermes flavipes; R. virginicus* and *Coptotermes formosanus*): induction of differentiation, defaunation and starvation. *Ann. ent. Soc. Am.* **72,** 503–508

Haydak, M. H. (1943). Larval food and development of castes in the honey bee. *J. econ. Ent.* **36,** 778–792

Heath, H. (1927). Caste formation in the termite genus *Termopsis. J. Morph.* **43,** 387–425

Herlant-Meeuwis, H. and Pasteels, J. M. (1961). Les glandes de mue de *Calotermes flavicollis* F. (Ins. Isopt.). *C.R. Acad. Sci. Paris* **253,** 3078–3080

Hermann, H. R. (1979). *Social Insects*, Vol. 1, Academic Press, London

Hille Ris Lambers, D. (1966). Polymorphism in *Aphididae. Ann. Rev. Ent.* **11,** 47–78

Howard, R. W. and Haverty, M. J. (1978). Deformation, mortality and soldier differentiation: Concentration effects of methoprene in a termite. *Sociobiol.* **3,** 73–77

Hrdý, I. (1973). Effects of juvenoids on termites and honey bees. Proc. 7th Int. Congr. Zool., London 158–161

Hrdý, I. and Křeček, J. (1972). Development of superfluous soldiers induced by juvenile hormone analogues in the termite *Reticulitermes lucifugus Ins. Soc.* **19,** 105–109

Hrdý, I. and Sláma, K. (1963). The effect of allatectomy upon oviposition by bee queens. 19th Int. Beekeep. Congr. Prague, 49–50

Hrdý, I., Křeček, J., Rupeš, V., Ždárek, J., Chmela, J. and Ledvinka, J. (1977). Control of the pharaoh's ant (*Monomorium pharaonis*) with juvenoids in baits. Proc. 8th Int. Congr. IUSSI Wageningen, 83–84

Hsiao, T. H. and Hsiao, C. (1979). Ecdysteroids in the ovary and the egg of the greater wax moth *J. Ins. Physiol.* **25,** 45–52

Ikan, R., Gottlieb, R., Bergmann, E. D. and Ishay, J. (1969). The pheromone of the queen of the Oriental hornet, *Vespa orientalis*. *J. Ins. Physiol.* **15,** 1709–1712

Imboden, H., Lanzrein, B., Delbecque, J. P. and Lüscher, M. (1978). Ecdysteroids and juvenile hormone during embryogenesis in the ovoviviparous cockroach *Nauphoeta cinerea*. *Gen. comp. Endocr.* **36,** 628–635

Imperatriz-Fonseca, V. L. (1977a). Queen supersedure in *Paratrigona subnuda* (Moure) (Apidae, Meliponinae). Proc. 8th Int. Congr. IUSSI. Wageningen, 78

Imperatriz-Fonseca, V. L. (1977b). Studies on *Paratrigona subnuda* (Moure) (Hymenoptera, Apidae, Meliponinae)–II. Behaviour of the virgin queen. *Bolm. Zool. Universidade de São Paulo* **2,** 169–182

Ishay, J. (1975). Caste determination by social wasps: cell size and building behaviour. *Anim. Beh.* **23,** 425–431

Ishay, J. Ikan, R. and Bergmann, E. D. (1965). The presence of pheromones in the Oriental hornet, *Vespa orientalis* F. *J. Ins. Physiol.* **11,** 1307–1309

Ishizaki, H. (1958). Correlation between colour variation and diapause in the pupa of the swallowtail, *Papilio xuthus* Seiro Seitai. (*Physiol. and Ecol.*) *Kyoto, Japan* **8,** 32–35

Jay, S. C. (1970). The effect of various combinations of immature queen and worker bees on the ovary development of worker honeybees with and without queens. *Can. J. Zool.* **48,** 169–173

Jay, S. C. and Nelson, E. V. (1973). The effects of laying worker honeybees (*Apis mellifera* L.) and their brood on the ovary development of other worker honeybees. *Can. J. Zool.* **51,** 629–632

Jaycox, E. R., Skowronek, W. and Guynn, G. (1974). Behavioral changes in worker honey bees (*Apis mellifera*) induced by injections of a juvenile hormone mimic. *Ann. ent. Soc. Am.* **67,** 529–534

Jeanne, R. L. (1980). Evolution of social behaviour in the Vespidae. *Ann. Rev. Ent.* **25,** 371–396

Joly, P. (1951). Déterminisme endocrine de la pigmentation chez *Locusta migratoria*. *C.R. Soc. biol. Paris* **145,** 1362–1366

Jouvenaz, D. P., Banks, W. A. and Lofgren, C. S. (1974). Fire ants: Attractions of workers to secretions of queens. *Ann. ent. Soc. Am.* **67,** 442–444

Jung-Hoffmann, I. (1967). Die Determination von Königin und Arbeiterin der Honigbiene. *Z. Bienenforsch.* **8,** 296–322

Kaiser, P. (1956). Die Hormonalorgane der Termiten in Zusammenhang mit der Entstehung ihrer Kasten. *Zool. Anz.* **54,** 129–178

Kaissling, K. E. and Renner, M. (1968). Antennale Rezeptoren für Queen substance und Sterzelduft bei der Honigbiene. *Z. vergl. Physiol.* **59,** 357–361

Kamm, D. R. (1974). Effects of temperature, day length and number of adults on the sizes of cells and offspring in a primitively social bee (Hymenoptera: Halictidae). *J. Kansas ent. Soc.* **47,** 8–18

Karlson, P. and Lüscher, M. (1959). Pheromone: Ein Nomenklaturvorschlag für eine Wirkstoffklasse. *Naturwiss.* **46,** 63–64

Kerr, W. E. (1946). Formação das castas no gênero Melipona (Illiger 1806). *Anais Escola Super. Agr. "Luiz de Queiroz"* **3,** 299–312

Kerr, W. E. (1969). Some aspects of the evolution of social bees (Apidae). *Evol. Biol.* **3,** 119–175

Kerr, W. E. (1974). Geschlechts- und Kastendetermination bei stachellosen Bienen. In: *Sozialpolymorphismus bei Insekten* (G. H. Schmidt, ed.) pp. 336–348. Wiss. Verlagsgesellschaft, Stuttgart

Kerr, W. E., Akahira, Y. and Camargo, C. A. de (1975). Sex determination in bees. IV Genetic control of juvenile hormone production in *Melipona quadrifasciata* (Apidae). *Genetics* **81,** 749–756

Knerer, G. and Atwood, C. E. (1966). Polymorphism in some nearctic *Halictine* bees. *Science* **152,** 1262–1263

Knerer, G. and and Plateaux-Quénu, C. (1966). Sur le polymorphisme des femelles chez quelques *Halictinae* (Insectes Hymenopteres) paléarctiques. *C.R. Acad. Sci. Paris* **263D,** 1759–1761

Kort, C. A. D. de and Granger, N. (1981). Regulation of the juvenile hormone titer. *Ann. Rev. Ent.* **26,** 1–28

Kort, C. A. D. de, Wieten, M., Kramer, J. and Goewie, E. A. (1977). Juvenile hormone degradation and carrier proteins in honey bee larvae. Proc. *K. ned. Akad. Wet. C* **80,** 297–301

Kramer, S. J. and Kort, C. A. D. de (1976). Age-dependent changes in juvenile hormone esterase and general carboxyesterase activity in the haemolymph of the Colorado potato beetle, *Leptinotarsa decemlineata. Molec. cell. Endocr.* **4,** 43–53

Krishna, K. and Weesner, F. (1969). *Biology of Termites*. Academic Press, London and New York

Kumar, S. (1975). Relations among bee size, cell size, and caste, in *Lasioglossum zephyrum* (Hymenoptera, Halictidae). *J. Kansas ent. Soc.* **48,** 374–380

Laere, O. van (1974). Physiology of the honeybee corpora allata. 3. A new method for allatectomy of queens. *J. apic. Res.* **13,** 15–18

Lagueux, M., Hetru, C., Galzené, F., Kappler, C. and Hoffmann, J. A. (1979). Ecdysone titre and metabolism in relation to cuticulogenesis in embryos of *Locusta migratoria. J. Ins. Physiol.* **25,** 709–723

Lanzrein, B., Gentinetta, V. and Lüscher, M. (1977). *In vitro* and *in vivo* studies on the endocrinology of the reproductives of the termite *Macrotermes subhyalinus.* Proc. 8th Int. Congr. IUSSI, Wageningen, 265–268

Lanzrein, B., Gentinetta, V. and Lüscher, M. (1978). *In vitro* Juvenilhormon-synthese durch Corpora Allata der Termite *Macrotermes subhyalinus. Rev. Suisse Zool.* **85,** 10

Lawrence, P. A. (1970). Polarity and patterns in the postembryonic development of insects. *Adv. Insect Physiol.* **7,** 197–266

Lebrun, D. (1967). La détermination des castes du termite à cou jaune (*Calotermes flavicollis* Fabr.). *Bull. biol. France Belgique* **101,** 139–217

Lebrun, D. (1969). Glandes endocrines et biologie de *Calotermes flavicollis.* Proc. 6th Int. Congr. IUSSI, Bern, 131–136

Ledoux, A. (1977). Influence de la temperature et de la nourriture sur la formation des castes chez *Aphaenogaster senilis*, (Mayr) (Hym., Formicoidea). *C.R. hebd. Acad. Sci. D* **284,** 1899–1902

Legay, J. M., Caluez, B., Hirn, M. and Reggi, H. L. de (1976). Ecdysone and oocyte morphogenesis in *Bombyx mori*. *Nature* **262,** 489–490

Le Masne, G. (1953). Observations sur les relations entre le couvain et les adultes chez les Fourmis. *Ann. Sci. nat., Zool. Biol. Anim.* **11,** 1–56

Lensky, Y., Baehr, J. C. and Porcheron, P. (1978). Dosages radio-immunologiques des ecdysones et des hormones juvéniles au cours du développement post-embryonnaire chez les ouvrières et les reines d'abeille (*Apis mellifica* L. *var. ligustica*). *C.R. Acad. Sci. Paris* **287D,** 821–824

Lenz, M. (1976). The dependence of hormone effects in termite caste determination on external factors. In: *Phase and Caste Determination in Insects* (M. Lüscher, ed.) pp. 73–89. Pergamon Press, Oxford

Light, S. F. (1943). The determination of castes of social insects. *Quart. Rev. Biol.* **18,** 46–63

Light, S. F. (1944). Experimental studies on ectohormonal control of the development of supplementary reproductives in the termite genus *Zootermopsis* (formerly *Termopsis*). *University of California, Publ. Zool.,* **43,** 413–454

Light, S. and Weesner, F. M. (1951). Further studies on the production of supplementary reproductives in *Zootermopsis* (Isoptera). *J. exp. Zool.* **117,** 397–414

Lindauer, M. (1952). Ein Beitrag zur Frage der Arbeitsteilung im Bienenstaat. *Z. vergl. Physiol.* **34,** 299–345

Lingens, F. and Rembold, H. (1959). Über den Weiselzellenfuttersaft der Honigbiene. III. Vitamingehalt von Königinnen- und Arbeiterinnenfuttersaft. *Hoppe-Seyler's Z. physiol. Chem.* **314,** 141–146

Lukoschus, F. (1952). Über die Prothoraxdrüse der Honigbiene (*Apis mellifera* L.). *Naturwiss.* **39,** 116

Lukoschus, F. (1955). Die Bedeutung des innersekretorischen Systems für die Ausbildung epidermaler Kastenmerkmale bei der Honigbiene (*Apis mellifera* L.). *Ins. Soc.* **2,** 221–236

Lukoschus, F. (1956a). Zur Kastendetermination bei der Honigbiene. *Z. Bienenforsch.* **3,** 190–199

Lukoschus, F. (1956b). Untersuchungen zur Entwicklung der Kastenmerkmale bei der Honigbiene (*Apis mellifera* L.). *Z. Morph. Ökol. Tiere* **45,** 157–197

Lüscher, M. (1949). Continuous observation of termites in laboratory cultures. *Acta tropica* **6,** 161–165

Lüscher, M. (1951). Über die Determination der Ersatzgeschlechtstiere bei der Termite *Kalotermes flavicollis* (Fabr.). *Rev. Suisse Zool.* **58,** 404–408

Lüscher, M. (1952a). New evidence for an ectohormonal control of caste determination in termites. Trans. 9th Int. Congr. Ent., Amsterdam 1951, **1,** 289–294

Lüscher, M. (1952b). Die Produktion und Elimination von Ersatzgeschlechtstieren bei der Termite *Kalotermes flavicollis* Fabr. *Z. vergl. Physiol.* **34,** 123–141

Lüscher, M. (1955). Zur Frage der Uebertragung sozialer Wirkstoffe bei Termiten. *Naturwiss.* **42,** 186

Lüscher, M. (1956a). Die Entstehung von Ersatzgeschlechtstieren bei der Termite *Kalotermes flavicollis*, Fabr. *Ins. Soc.* **3,** 273–276

Lüscher, M. (1956b). Hemmende und fördernde Faktoren bei der Entstehung der Ersatzgeschlechtstieren bei der Termite *Kalotermes flavicollis. Rev. Suisse Zool.* **63,** 261–267

Lüscher, M. (1957). Ersatzgeschlechtstiere bei der Termiten und die Beeinflussung ihrer Entstehung durch die Corpora allata. *Verh. Ber. dtsch. Ges. angew. Ent.* **14,** 144–150

Lüscher, M. (1958a). Experimentelle Erzeugung von Soldaten bei der Termite *Kalotermes flavicollis* (Fabr.). *Naturwiss.* **45,** 1–2

Lüscher, M. (1958b). Über die Entstehung der Soldaten bei Termiten. *Rev. Suisse Zool.* **65,** 372–377

Lüscher, M. (1960). Hormonal control of caste differentiation in termites. *Ann. New York Acad. Sci.* **89,** 549–563

Lüscher, M. (1961). Social control of polymorphism in termites. Symp. Insect Polymorphism. *Proc. Roy. Entomol. Soc. London* 57–67

Lüscher, M. (1964). Die spezifische Wirkung männlicher und weiblicher Ersatzgeschlechtstiere auf die Entstehung von Ersatzgeschlechtstieren bei der Termite *Kalotermes flavicollis* (Fabr.). *Ins. Soc.* **11,** 79–90

Lüscher, M. (1969). Die Bedeutung des Juvenilhormons für die Differenzierung der Soldaten bei der Termite *Kalotermes flavicollis* (Fabr.). Proc. 6th Int. Congr. IUSSI, Bern, 165–170

Lüscher, M. (1973). The influence of the composition of experimental groups on caste development in *Zootermopsis* (Isoptera). Proc. 7th Int. Congr. IUSSI, London, 253–256

Lüscher, M. (1974a). Kasten und Kastendifferenzierung bei niederen Termiten. In: *Sozialpolymorphismus bei Insekten* (G. H. Schmidt, ed.) pp. 695–739. Wiss. Verlagsgesellschaft, Stuttgart

Lüscher, M. (1974b). Die Kompetenz zur Soldatenbildung bei Larven (Pseudergaten) der Termite *Zoothermopsis angusticollis. Rev. Suisse Zool.* **81,** 710–714

Lüscher, M. (1976). Evidence for an endocrine control of caste determination in the higher termites. In: *Phase and Caste Determination in Insects* (M. Lüscher, ed.) pp. 91–103. Pergamon Press, Oxford

Lüscher, M. (1977). Queen dominance in termites. Proc. 8th Int. Congr. IUSSI, Wageningen, 238–242

Lüscher, M. and Karlson, P. (1958). Experimentelle Auslösung von Häutungen bei der Termite *Kalotermes flavicollis* (Fabr.). *J. Ins. Physiol.* **1,** 341–345

Lüscher, M. and Springhetti, A. (1960). Untersuchungen über die Bedeutung der Corpora allata für die Differenzierung der Kasten bei der Termite *Kalotermes flavicollis* (Fabr.). *J. Ins. Physiol.* **5,** 190–212

Lüscher, M. and Walker, I. (1963). Zur Frage der Wirkungsweise der Königinnenpheromone bei der Honigbiene. *Rev. Suisse Zool.* **70,** 304–311

Mandaron, P., Guillermet, C. and Sengel, P. (1977). In vitro development of *Drosophila* imaginal discs: Hormonal control and mechanism of evagination. *Am. Zool.* **17,** 661–670

Mane, S. D. and Rembold, H. (1977). Developmental kinetics of juvenile hormone inactivation in queen and worker castes of the honey bee *Apis mellifera. Insect Biochem.* **7,** 463–468

Markin, G. P. (1970). Food distribution within laboratory colonies of the Argentine ant, *Iridomyrmex humilis* (Mayr). *Ins. Soc.* **17,** 127–158

Mednikova, J. K. (1977). Caste differentiation in the termite *Anacanthotermes ahngerianus* Jacobson (Isoptera, Hodotermitidae). Proc. 8th Int. Congr. IUSSI, Wageningen, 118–120

Melampy, R. M. and Willis, E. R. (1939). Respiratory metabolism during larval and pupal development of the female honeybee (*Apis mellifera* L.). *Physiol. Zool.* **12,** 302–311

Meyer, D. R., Lanzrein, B., Lüscher, M. and Nakanishi, K. (1976). Isolation and identification of a juvenile hormone (JH) in termites. *Experientia* **32,** 773

Michener, C. D. (1974). *The Social Behavior of the Bees*. Harvard University Press, Cambridge, Massachusetts

Michener, C. D. (1977). Aspects of the evolution of castes in primitively social insects. Proc. 8th Int. Congr. IUSSI, Wageningen, 2–6

Michener, C. D., Brothers, D. J. and Kamm, D. R. (1971). Interactions of colonies of primitively social bees: artificial colonies of *Lasioglossum zephyrum*. *Proc. nat. Acad. Sci. USA* **68,** 1241–1245

Miller, E. M. (1969). Caste differentiation in lower termites. In: *Biology of Termites* (K. Krishna and F. M. Weesner, eds) Vol. 1, pp. 283–310. Academic Press, New York and London

Montagner, H. (1963). Études préliminaires des relations entre les adultes et le couvain chez les guêpes sociales du genre *Vespa* au moyen d'un radio-isotope. *Ins. Soc.* **10,** 153–166

Montagner, H. (1966). Sur le déterminisme des castes femelles chez les guêpes du genre *Vespa*. *C.R. Acad. Sci.* **263,** 547–549

Morse, R. A., Dyce, E. J. and Young, R. G. (1966). Weight changes by the queen honey bee during swarming. *Ann. ent. Soc. Am.* **59,** 772–774

Motro, M., Kugler, J., Motro, U. and Ishay, J. S. (1977). Social prerequisites and oocyte development in *Vespa orientalis* workers. Proc. 8th Int. Congr. IUSSI, Wageningen, 153

Motro, M., Motro, U., Ishay, J. S. and Kugler, J. (1979). Some social and dietary prerequisites of oocyte development in *Vespa orientalis* L. workers. *Ins. Soc.* **26,** 155–164

Müller, H. J. (1955). Die Saisonsformenbildung von *Araschnia levana*; ein photoperiodisch gesteuerter Diapause-effekt. *Naturwiss.* **42,** 134–135

Nagin, R. (1972). Caste determination in *Neotermes jonteli* (Banks). *Ins. Soc.* **19,** 39–61

Naisse, J. (1966). Contrôle endocrinien de la différentiation sexuelle chez *Lampyris noctiluca* (Coleoptera, Lampyridae)I I. Phenomènes neurosécrétoires et endocrines au cours du développement postembryonnaire chez le mâle et la femelle. *Gen. comp. Endocr.* **7,** 85–104

Noirot, Ch. (1957). Neurosécrétion et sexualiteé chez le termite à cou jaune *Calotermes flavicollis* F. *C.R. Acad. Sci. Paris* **245,** 743–745

Noirot, Ch. (1969). Formation of castes in higher termites. In: *Biology of Termites* (K. Krishna and F. M. Weesner, eds) pp. 311–350. Academic Press, New York and London

Noirot, Ch. (1974). Polymorphismus bei höheren Termiten, In: *Sozialpolymorphismus der Insekten* (G. H. Schmidt, ed.) pp. 741–765. Wiss. Verlagsgesellschaft, Stuttgart

Okot-Kotber, B. M. (1977). Changes in corpora allata volume during development in relation to caste differentiation in *Macrotermes subhyalinus*. Proc. 8th Int. Congr. IUSSI, Wageningen, 262–264

Osanai, M. and Rembold, H. (1968). Entwicklungsabhängige mitochondriale Enzymaktivitäten bei den Kasten der Honigbiene. *Biochim. biophys. Acta* **162,** 22–31

Pain, J. (1961). Sur la phérormone des reines d'abeilles et ses effects physiologiques. *Ann. Abeille* **4,** 73–152

Pardi, L. (1948). Dominance order in Polistes wasps. *Physiol. Zool.* **21,** 1–13

Passera, L. (1969). Biologie de la reproduction chez *Plagiolepis pygmaea* Latr. et ses deux parasites sociaux *Plagiolepis grassei* Le Masne et Passera et *Plagiolepis xene* Stärcke (Hymenoptera, Formicidae). *Ann. Sci. nat., Zool. et Biol. Anim.* **11, 327–481**

Passera, L. (1972). Déterminisme de la caste de la Fourmi *Plagiolepis pygmaea* Latr. (Hym. Formicidae): étude des protéines totales des larves par électrophorèse en gel de polyacrylamide. *C.R. Acad. Sci. Paris* **D275,** 2013–2016

Passera, L. (1977). Production of soldiers in the colonies coming out of hibernation in the ant *Pheidole pallidula* (Nyl.) (Formicidae, Myrmicinae). *Ins. Soc.* **24,** 131–146

Passera, L. (1978). Demonstration of a pheromone inhibiting oviposition by workers in the ant *Plagiolepis pygmaea* Latr. (Hym. Formicidae). *C.R. hebd. Acad. Sci.* **D286,** 1507–1509

Passera, L. (1980a). La ponte d'oeufs préorientés chez la fourmi *Pheidole pallidula* (Nyl.) (Hymenoptera-Formicidae). *Ins. Soc.* **27,** 79–95

Passera, L. (1980b). La fonction inhibitrice des reines de la fourmi *Plagiolepis pygmaea* Latr.: Rôle des phéromones. *Ins. Soc.* **27,** 212–225

Passera, L. and Suzzoni, J. P. (1978). Sexualisation du couvain de la fourmi *Pheidole pallidula* (Hymenoptera, Formicidae) après traitement par l'hormone juvénile. *C.R. Acad. Sci. Paris* **D286,** 615–618

Passera, L. and Suzzoni, J. P. (1979). Le rôle de la reine de *Pheidole pallidula* (Nyl.) (Hymenoptera, Formicidae) dans la sexualisation du couvain après traitement par l'hormone juvénile. *Ins. Soc.* **26,** 343–353

Petersen-Braun, M. (1977a). Studies on the endogenous breeding cycle in *Monomorium pharaonis* L. (Formicidae). Proc. 8th Int. Congr. IUSSI, Wageningen, 211–212

Petersen-Braun, M. (1977b). Untersuchungen zur sozialen Organisation der Pharaoameise *Monomorium pharaonis* (L.) (Hymenoptera, Formicidae). II. Die Kastendeterminierung. *Ins. Soc.* **24,** 303–318

Pflugfelder, O. (1948). Volumetrische Untersuchungen an den Corpora allata der Honigbiene, *Apis mellifica. Biol. Zblt.* **66,** 211–235

Plateau-Quénu, C. (1967). Tendances évolutives et degré de socialisation chez les *Halictinae* (Hym. Apoidea). *Ann. Soc. ent. de France* (*n.s.*) **3,** 859–866

Ramamurty, P. S. (1977). Vitellogenin synthesis and vitellogenesis in allatectomised honeybee queens. Proc. 8th Int. Congr. IUSSI, Wageningen, 30–31

Raven, C. P. (1961). *Oogenesis: The Storage of Developmental Information.* Pergamon Press, Oxford

Regel, R. (1978). Auswirkungen einer Langzeitbehandlung mit underschiedlichen Dosen von JH-III auf Verhalten und Fertilität von Jung-Königinnen (*Apis mellifica*). Kurzreferate Entomologentagung, Karlsruhe

Rembold, H. (1965). Biologically active substances in royal jelly. *Vitam. Horm.* **23,** 359–382

Rembold, H. (1969a). Biochemie der Kastenbildung bei der Honigbiene. 22nd Int. Beekeep. Congr. Munich, 552–554

Rembold, H. (1969b). Biochemie der Kastenentstehung bei der Honigbiene. Proc. 6th Int. Congr. IUSSI, Bern, 239–246

Rembold, H. (1976). The role of determinator in caste formation in the honey bee. In: *Phase and Caste Determination in Insects* (M. Lüscher, ed.) pp. 21–34. Pergamon Press, Oxford

Rembold, H. and Hanser, G. (1964). Über den Weiselzellenfuttersaft der Honigbiene. VIII. Nachweis des determinierenden Prinzips im Futtersaft der Königinnenlarven. *Hoppe-Seyler's Z. physiol. Chem.* **339,** 251–254

Rembold, H., Czoppelt, Ch. and Rao, P. J. (1974). Effect of juvenile hormone treatment on caste differentiation in the honeybee, *Apis mellifera. J. Ins. Physiol.* **20,** 1193–1202

Rembold, H., Hagenguth, H. and Rascher, J. (1980). A sensitive method for detection and estimation of juvenile hormones from biological samples by glass capillary combined gas chromatography-selected ion monitoring mass spectrometry. *Analytical Biochem.* **101,** 356–363

Richards, O. W. and Richards, M. J. (1951). Observations on the social wasps of South America (Hymenoptera, Vespidae). *Trans. Roy. Entomol. Soc. London* **102,** 1–170

Riddiford, L. M. (1980). Insect endocrinology: Action of hormones at the cellular level. *Ann. Rev. Physiol.* **42,** 511–528

Riddiford, L. M. (1981). Hormonal control of epidermal cell development. In: *Symp. Insect. Systems*, Seattle, Wash. Am. Zool. **21,** 751–762

Ritcey, G. M. and Dixon, S. E. (1969). Postembryonic development of the endocrine system in the female honeybee castes, *Apis mellifera* L., *Proc. ent. Soc. Ontario* **100,** 124–138

Robeau, R. M. and Vinson, S. B. (1976). Effects of juvenile hormone analogues on caste differentiation in the imported fire ant, *Solenopsis invicta. J. Georgia ent. Soc.* **11,** 198–203

Röseler, P. F. (1967). Untersuchungen über das Auftreten der 3 Formen im Hummelstaat. *Zool. Jb. Physiol.* **74,** 178–197

Röseler, P. F. (1970). Unterschiede in der Kastendetermination zwischen den Hummelarten *Bombus hypnorum* und *Bombus terrestris. Z. Naturforsch.* **25,** 543–548

Röseler, P. F. (1974). Vergleichende Untersuchungen zur Oogenese bei weiselrichtigen und weisellosen Arbeiterinnen der Hummelart *Bombus terrestris. Ins. Soc.* **21,** 249–274

Röseler, P. F. (1975). *Die Kasten der sozialen Bienen.* Steiner Verlag, Wiesbaden

Röseler, P. F. (1976). Juvenile hormone and queen rearing in bumblebees. In: *Phase and Caste Determination in Insects* (M. Lüscher, ed.) pp. 55–61. Pergamon Press, Oxford

Röseler, P. F. (1977a). Juvenile hormone control of oogenesis in bumblebee workers, *Bombus terrestris. J. Ins. Physiol.* **23,** 985–992

Röseler, P. F. (1977b). Endocrine control of polymorphism in bumblebees. Proc. 8th Int. Congr. IUSSI, Wageningen, 22–23

Röseler, P. F. and Röseler, I. (1974). Morphologische und physiologische Differenzierung der Kasten bei den Hummelarten *Bombus hypnorum* (L.) und *Bombus terrestris* (L.). *Zool. Jb. Physiol.* **78,** 175–198

Röseler, P. F. and Röseler, I. (1977). Dominance in bumblebees. Proc. 8th Int. Congr. IUSSI, Wageningen, 232–235

Röseler, P. F. and Röseler, I. (1978). Studies on the regulation of the juvenile hormone titre in bumblebee workers, *Bombus terrestris. J. Ins. Physiol.* **24,** 707–713

Röseler, P. F., Röseler, I. and Strambi, A. (1980). The activity of corpora allata in dominant and subordinated females of the wasp *Polistes gallicus Ins. Soc.* **27,** 97–107

Rowell, O. and Frazer, C. H. (1967). Corpus allatum implantation and green-brown polymorphism in three African grasshoppers. *J. Ins. Physiol.* **13,** 1401–1412

Rupeš, V., Hrdý, I., Pinterova, J., Ždárek, J. and Křeček, J. (1978). The influence of methoprene on Pharaoh's ant, *Monomorium pharaonis*, colonies. *Acta Ent. Bohemoslov. Tchecosl.* **75,** 155–163

Ruppli, E. (1969). Die Elimination überzähliger Ersatzgeschlechtstiere bei der termite *Kalotermes flavicollis* (Fabr.). (Inaug. Diss. Bern.) *Ins. Soc.* **16,** 235–248

Ruppli, E. and Lüscher, M. (1964). Die Elimination überzähliger Ersatzgeschlechtstiere bei der Termite *Kalotermes flavicollis* (Fabr.). *Rev. Suisse Zool.* **71,** 626–632

Ruttner, F., Koeniger, N. and Veith, H. J. (1976). Queen substance bei eierlegenden Arbeiterinnen der Honigbiene (*Apis mellifica* L.). *Naturwiss.* **63,** 434–435

Rutz, W., Gerig, L., Wille, H. and Lüscher, M. (1976). The function of juvenile hormone in adult worker honeybees, *Apis mellifera. J. Ins. Physiol.* **22,** 1485–1491

Rutz, W., Imboden, H., Jaycox, E. R., Wille, H. Gerig, L. and Lüscher, M. (1977). Juvenile hormone and polyethism in adult worker honeybees (*Apis mellifera*). Proc. 8th Int. Congr. IUSSI, Wageningen, 26–27

Sakagami, S. F. (1958). The false-queen, fourth adjustive response in dequeened honeybee colonies. *Behaviour* **13,** 280–296

Sannasi, A. and George, C. J. (1972). Termite queen substance: 9-oxodec-trans-2-enoic acid. *Nature* **237,** 457

Sannasi, A. and Sundara Rajulu, G. (1971). 9-oxo-trans-2-decenoic acid in the Indian honeybees. *Life Sci.* **10,** 195–201

Sannasi, A. P. K., Sen-Sarma, C., George, C. J. and Basalingappa, S. (1972). Juvenile hormone activity from various sources of termite castes and their fungus gardens. *Ins. Soc.* **19,** 81–86

Schooneveld, H., Kramer, S. J., Privee, H. and Huis, A. van (1979). Evidence of controlled corpus allatum activity in the adult Colorado potato beetle *J. Ins. Physiol.* **25,** 449–453

Schmidt, G. H. (1974). *Sozialpolymorphismus bei Insekten.* Wiss. Verlagsgesellschaft, Stuttgart

Schneidermann, H. A. (1979). Pattern formation and determination in insects. In: *Mechanisms of Cell Change* (J. Ebert and T. Okada, eds) pp. 243–272. Wiley, Chichester,

Shearer, D. A., Boch, R., Morse, R. A. and Laigo, F. M. (1970). Occurrence of 9-oxodec-trans-2-enoic acid in queens of *Apis dorsata*, *Apis cerana*, and *Apis mellifera. J. Ins. Physiol.* **16,** 1437–1441

Shuel, R. W. and Dixon, S. E. (1959). Studies in the mode of action of royal jelly in honeybee development II. Respiration of newly emerged larvae on various substrates. *Can. J. Zool.* **37,** 803–813

Shuel, R. W. and Dixon, S. E. (1960). The early establishment of dimorphism in the female honey bee, *Apis mellifera* L. *Ins. Soc.* **7,** 265–282

Shuel, R. W., Dixon, S. E. and Kinoshita, G. B. (1978). Growth and development of honeybees in the laboratory on altered queen and worker diets. *J. apic. Res.* **17,** 57–68

Skaife, S. H. (1954). The black-mound termite of the Cape, *Amitermes atlanticus* Fuller. *Trans. R. Soc. S. Afr.* **34,** 251–271

Spradbery, J. P. (1973). *Wasps.* Sidgwick and Jackson, London

Springhetti, A. (1969). Influenza dei reali sulla differenziazione dei soldati di *Kalotermes flavicollis* Fabr. (Isoptera). Proc. 6th Int. Congr. IUSSI, Bern, 267–273

Springhetti, A. (1970). Influence of the king and queen on the differentiation of soldiers in *Kalotermes flavicollis* (Fabr.) (Isoptera). *Monit. Zool. Ital.* (N.S.) **4,** 99–105

Springhetti, A. (1972). The competence of *Kalotermes flavicollis* (Fabr.) (Isoptera) pseudergates to differentiate into soldiers. *Monit. Zool. Ital.* (*N.S.*) **6,** 97–11

Staal, G. B. (1961). *Studies on the physiology of phase induction in* Locusta migratoria migratorioides R & F. Publ. Fonds Landbouw Exportbureau **40** (Ph. D. Thesis, Wageningen)

Strambi, A. and Strambi, C. (1973). Influence du développement du parasite *Xenos vesparum* Rossi (Insecte, Strepsiptère) sur le système neuroendocrinien des femelles de Poliste (Hyménoptère, Vespidae) au début de leur vie imaginale. *Arch. Anat. Micr.* **62,** 39–54

Strambi, A., Strambi, C. and Reggi, M. de (1977). Ecdysones and ovarian physiology in the adult wasp *Polistes gallicus.* Proc. 8th Int. Congr. IUSSI Wageningen, 19–20

Stumpf, H. F. (1967). Über den Verlauf eines schuppenorientierenden Gefälles bei *Galleria mellonella. Roux Arch. Entw. Mech. Org.* **158,** 315–330

Trautmann, K. H., Masner, P., Schuler, A., Suchý, M. and Wipf, H. K. (1974). Evidence of the juvenile hormone methyl (2E, 6E)–10, 11–epoxy–3, 7, 11–trimethyl–2, 6–dodecadienoate (JH-3) in insects of four orders. *Z. Naturforsch.* **29,** 757–759

Troisi, S. J. and Riddiford, L. M. (1974). Juvenile hormone effects on metamorphosis and reproduction of the fire ant, *Solenopsis invicta. Env. Ent.* **3,** 112–116

Tschinkel, W. R. and Howard, D. F. (1978). Queen replacement in orphaned colonies of the fire ant, *Solenopsis invicta. Behav. Ecol. Sociobiol.* **3,** 297–310

Ulrich, G. (1979). *Histologische und biochemische Untersuchungen zur endokrinen Steuerung der Kastenbildung bei der Honigbiene* (Apis mellifera). Ph. D. Thesis, Köln

Varjas, L., Paguia, P. and Wilde, J. de (1969). Juvenile hormone titres in penultimate and last instar larvae of *Pieris brassicae* L. and *Barathra brassicae* L., in relation to the effect of juvenoid application. *Experientia* **25,** 213

Velthuis, H. H. W. (1970a). Queen substances from the abdomen of the honey bee queen. *Z. vergl. Physiol.* **70,** 210–222

Velthuis, H. H. W. (1970b). Ovarian development in *Apis mellifera* worker bees. *Ent. exp. Appl.* **13,** 377–394

Velthuis, H. H. W. (1972). Observations on the transmission of queen substances in the honey bee colony by the attendants of the queen. *Behaviour* **41,** 105–129

Velthuis, H. H. W. (1976). Environmental, genetic and endocrine influences in stingless bee caste determination. In: *Phase and Caste Determination in Insects* (M. Lüscher, ed.) pp. 35–53. Pergamon Press, Oxford

Velthuis, H. H. W. (1977). The evolution of honey bee queen pheromones. Proc. 8th Int. Congr. IUSSI, Wageningen, 220–222

Velthuis, H. H. W. and Velthuis-Kluppell, F. M. (1975). Caste differentiation in a stingless bee, *Melipona quadrifasciata* Lep., influenced by juvenile hormone application. *Proc. K. ned. Akad. Wet. C* **78,** 81–94

Velthuis, H. H. W., Verheijen, F. J. and Gottenbos, A. J. (1965). Laying worker honey bee: similarities to the queen. *Nature, London* **207,** 1314

Verheijen-Voogd, C. (1959). How worker bees perceive the presence of their queen. *Z. vergl. Physiol.* **41,** 527–582

Vierling, G. and Renner, M. (1977). Die Bedeutung des Sekretes der Tergittaschendrüsen für die Attraktivität der Bienenkönigin gegenüber jungen Arbeiterinnen. *Behav. Ecol. Sociobiol.* **2,** 185–200

Vinson, S. B. and Robeau, R. (1974). Insect growth regulator effects on colonies of the imported fire ant. *J. econ. Ent.* **67,** 584–587

Vinson, S. B., Robeau, R. and Dzuik, L. (1974). Bioassay and activity of several insect growth regulator analogues on the imported fire ant. *J. econ. Ent.* **67,** 325–328

Wanyonyi, K. (1974). The influence of the juvenile hormoneanalogue ZR 512 (Zoecon) on caste development in *Zootermopsis nevadensis* (Hagen). (Isoptera). *Ins. Soc.* **21,** 35–44

Wanyonyi, K. and Lüscher, M. (1973). The action of juvenile hormone analogues on caste development in *Zootermopsis* (Isoptera). Proc. 7th Int. Congr. IUSSI, London, 392–395

Watkins, J. F. and Cole, T. W. (1966). Worker Army ants, *Neivamyrmex opacithorax* Emery are attracted to secretions of larvae and pupae. *Texas J. Sci.* **18,** 254

Watson, J. A. L., Barrett, R. A. and Abbey, H. M. (1977). Caste ratios in a long-established, neotenic-headed laboratory colony of *Mastotermes darwinensis* Froggatt. *J. Austr. ent. Soc.* **16,** 469–470

Weaver, N. (1955). Rearing of honeybee larvae on royal jelly in the laboratory *Bee Wld.* **36,** 157–159

Weaver, N. (1958). Rearing honeybee larvae in the laboratory. Proc. 10th Int. Congr. Ent. **4,** 1031–1036

Weaver, N. (1966). Physiology of caste determination. *Ann. Rev. Ent.* **11,** 79–102

Weiss, K. (1971). Über die Ausbildung und Leistung von Königinnen aus Eiern und jungen Arbeitermaden. *Apidologie* **2,** 3–47

Weiss, K. (1974). Zur Frage des Königinnengewichtes in Abhängigkeit von Umlarvalter und Larvenversorgung. *Apidologie* **5,** 127–147

Weiss, K. (1978). Zur Mechanik der Kastenentstehung bei der Honigbiene (*Apis mellifica* L.). *Apidologie* **9,** 223–258

Wendel, L. E. and Vinson, S. B. (1978). Distribution and metabolism of a juvenile hormone analogue within colonies of the red imported fire ant. *J. econ. Ent.* **71,** 561–565

West-Eberhard, M. J. (1969). The social biology of Polistine wasps. *Misc. Publ. Mus. Zool., University Michigan* **140,** 1–101

West-Eberhard, M. J. (1977). The establishment of reproductive dominance in social wasp colonies. Proc. 8th Int. Congr. IUSSI, Wageningen, 223–227

Wheeler, W. M. (1960). *Ants, their Structure, Development, and Behavior.* Columbia University Press, New York

Wheeler, D. E. and Nijhout, H. F. (1981) Soldier determination in ants: New role for juvenile hormone. *Science* **213,** 361–363

White, D. F. (1976). Corpus allatum activity associated with development of wing-buds in cabbage aphid embryos and larvae. *J. Ins. Physiol.* **17,** 761–773

Wigglesworth, V. B. (1961). Insect polymorphism–A tentative synthesis. In: *Insect Polymorphism.* Symp. Roy. Entomol. Soc. London, no. 1

Wilde, J. de (1971). Hormones and diapause. *Mem. Soc. Endocr.* **18,** 487–514

Wilde, J. de (1975). An endocrine view of metamorphosis, polymorphism and diapause in insects. *Am. Zool. Suppl.* 13–28

Wilde, J. de (1976). Juvenile hormone and caste differentiation in the honey bee (*Apis mellifera* L.). In: *Phase and Caste Determination in Insects* (M. Lüscher, ed.) pp. 5–20. Pergamon Press, Oxford

Wilde, J. de (1978). Seasonal states and endocrine levels in insects. In: *Environmental Endocrinology* (J. Assenmacher and D. S. Farner, eds) pp. 10–19. Springer Verlag, Berlin

Wilde, J. de and Boer, J. A. de (1969). Humoral and nervous pathways in photo-periodic induction of diapause in *Leptinotarsa decemlineata* Say. *J. Ins. Physiol.* **15,** 661–675

Wilson, E. O. (1974). *The Insect Societies.* Harvard University Press, Cambridge, Massachusetts

Wirtz, P. (1973). *Differentiation in the honey bee larva.* Ph. D. Thesis, Agricultural University, Wageningen, Meded. Landb. Hogesch., 73–5

Wirtz, P. and Beetsma, J. (1972). Induction of caste differentiation in the honey bee (*Apis mellifera* L.) by juvenile hormone. *Ent. exp. Appl.* **15,** 517–520

Wolpert, L. (1969). Positional information and the spatial pattern of cellular differentiation. *J. theoret. Biol.* **25,** 1–47

Woyke, J. (1971). Correlations between the age at which honeybee brood was grafted, characteristics of the resultant queens, and results of insemination. *J. apic. Res.* **10,** 45–55

Wyss-Huber, M. and Lüscher, M. (1975). Protein synthesis in the "fatbody" and the ovaries of the physogastric queen of *Macrotermes subhyalinus. J. Ins. Physiol.* **21,** 1697–1704

Ždárek, J. and Haragsim, O. (1974). Action of juvenoids on metamorphosis of the honey bee, *Apis mellifera. J. Ins. Physiol.* **20,** 209–221

Ždárek, J., Haragsim, O. and Veselý, V. (1976). Action of juvenoids on the honey bee colony. *Z. angew. Ent.* **81,** 392–401

Chemoreception: The Significance of Receptor Numbers

R. F. Chapman

Centre for Overseas Pest Research, London, UK

1 Introduction

With the advent of the transmission electron microscope and then of the scanning electron microscope there have been great advances in our understanding of the structures of insect chemoreceptor systems with a number of comprehensive reviews (C. T. Lewis, 1970; Slifer, 1970; Altner and Prillinger, 1980; Zacharuk, 1980). At the same time improvements in experimental techniques have given considerable understanding of how the sensilla function (e.g. Kaissling, 1971, 1974; Kijima *et al.*, 1977; Maes, 1980) and an appreciation of the qualitative differences which exist between receptors (e.g. Lacher, 1964; van der Starre, 1972; Blaney, 1974; Ma and Visser, 1978; van der Molen *et al.*, 1978; Schaller, 1978). Attempts have been made to relate the overt behavioural responses of insects to the recorded sensory input. In some cases, with pheromones (Mustaparta *et al.*, 1977, 1979; Priesner, 1980) and with simple food chemicals (Ma, 1972; Blom, 1978), a simple relationship exists, but in other cases, especially where chemically complex food odours or tastes are involved, there is no obvious relationship between sensory input and behavioural response. It has become apparent that chemical discrimination often depends on an analysis by the insect of the overall pattern of sensory inputs from an array of receptors with different qualities (Dethier and Schoonhoven, 1969; Blaney, 1975; Boeckh, 1980; Dethier, 1980a). Yet, despite this appreciation, very little attention has been given to receptor populations. Notable exceptions are the work of Mustaparta (1975) and of Schaller (1978).

It is also evident that great differences exist in the numbers of sensilla on the mouthparts of different insects despite the fact that they have similar feeding habits; compare, for instance, caterpillars (Schoonhoven, 1973) and grasshoppers (Chapman and Thomas, 1978). But no attempt has been made to account for, or even to discuss, these differences.

In this review the existing data on numbers of chemoreceptors on the legs, mouthparts and antennae of insects are brought together and, where possible, the physiological characteristics of the sensilla are briefly described. This is followed by a discussion of the possible significance of numbers of sensilla in relation to insect behaviour and evolution. The picture which emerges is fragmentary, but the fragments should serve to indicate fruitful lines of research which will lead to a better understanding of the physiological significance of chemoreceptor populations and the numerical differences which occur between them.

It is fitting to pay tribute to Dr E. H. Slifer whose pioneering work on insect chemoreceptors laid the foundations for much that is written here and without which the gaps in our knowledge would be even larger than they now are.

2 Numbers of chemoreceptors and their functions

In this section data on numbers of chemoreceptors on different parts of the body are presented and relevant electrophysiological studies are reviewed. Apart from the extensive work on the antenna of *Periplaneta americana* by Schaller and others, there is no detailed study which relates the ultrastructure of sensilla to their physiological responses and numerical abundance. Consequently it is necessary to accept less stringent criteria in classifying the sensilla according to function. A sensillum with a single neurone which ends at the base of a hair or in a tubular body is considered to be a mechanoreceptor (see McIver, 1975). Mechanoreceptors have not been included in this account. A sensillum for which there is histological or ultrastructural evidence of a terminal pore, or which is permeable to crystal violet only at the tip, is considered to be a contact chemoreceptor, notwithstanding the fact that it is known that such sensilla can also respond to odours (Städler and Hanson, 1975). They are also called uniporous sensilla. A sensillum for which there is evidence of the presence of numerous pores in the wall, a multiporous sensillum, is regarded as having an olfactory function, whether or not this function has been proved electrophysiologically. In some cases the function of sensilla is inferred from their outward similarity to other sensilla of known function. Anatomically different types of contact chemoreceptors or olfactory sensilla have not been discussed separately unless there are good physiological reasons for doing so.

Until recently, sensilla containing temperature and humidity receptors have not been differentiated from chemoreceptors because no obvious morphological grounds for separating them existed. They have, therefore, been included in this review, but where possible estimates of their abundance have been made. They have been distinguished on the basis of electrophysiological evidence, or by the presence of a lamellate dendrite beneath unperforated cuticle (Altner and Prillinger, 1980), or by having a similar outward appearance to other receptors with these characteristics in the same group of insects. For instance, styloconic sensilla on the antennae of *Mamestra brassicae* have been shown to be temperature/humidity receptors (Becker, quoted by Altner and Prillinger, 1980); it therefore seems reasonable to regard styloconic sensilla on the antennae of other Lepidoptera as having the same function.

No comprehensive studies of the numbers of sensilla on mouthparts and antennae of any single insect have been carried out, apart from a few larval Endopterygota where the work of Chu-Wang and Axtell (1971, 1972a, b) and of Corbière-Tichané (1971a) is notable. Chapman and Thomas (1978) have made the only extensive survey of sensilla in one group of insects, the Acridoidea, while the various studies on antennal sensilla have mostly been concerned with the perception of pheromones. As used in this review the term "chemoreceptor" refers to a sensillum.

2.1 SENSILLA ON LEGS

There have been few inventories of chemoreceptors on the legs of insects, but the information available shows that nearly all such sensilla are concentrated on the tarsi. In *Phormia regina*, Grabowski and Dethier (1954)

TABLE 1 Numbers of chemoreceptors and associated chemosensitive neurones on the tarsi of various adult insects

Order	Species	Sex	Prothoracic leg		Mesothoracic leg		Metathoracic leg		Ref.
			sensilla	neurones	sensilla	neurones	sensilla	neurones	
Orthoptera	*Schistocerca gregaria*	♂	280	1130	230	930	130	480	1
		♀	240	940	210	910	120	476	1
	Tettigonia viridissima	?	214	–	175	–	154	–	14
Lepidoptera	*Heliothis zea*	♂	70	–	50	–	76	–	2
		♀	67	–	81	–	76	–	2
	Pieris brassicae	♂	65	260	66	264	66	264	3
		♀	95	380	82	328	86	344	3
Diptera	*Anopheles atroparvus*	♀	17	–	18	–	12	–	12
	Aedes aegypti	♂	19	76	10	40	2–3	10	4,5
		♀	25	100	19	76	6–7	25	4,5
	Simulium baffinense	♂			420[a]				11
		♀			373[a]				11
	Simulium arcticum	♂			623[a]				11
		♀			601[a]				11
	Simulium euryadminiculum	♀			801[a]				11
	Simulium rugglesi	♂			448[a]				11
		♀			618[a]				11

	Rhagoletis pomonella	♀	49	–	48	–	48	–	15
	Phormia regina	♂,♀	270	1080	193	772	139	556	8,9
	Phormia terraenovae	♂,♀	232	928	152	608	140	560	10,9
	Calliphora vicina	♂,♀	187	748	–	–	–	–	7,9
	Musca domestica	♂,♀	248	–	208	–	150	–	6
	Glossina palpalis	♂,♀	78	–	68	–	68	–	10
Hymenoptera	*Apis mellifera*	☿	87	348	–	–	–	–	13

[a] For the *Simulium* species the total number of sensilla on all three legs is given; this total includes a few sensilla on the tibiae

[1] Kendall (1970)
[2] Callahan (1969)
[3] Ma and Schoonhoven (1973)
[4] Slifer (1962)
[5] McIver and Siemicki (1978)
[6] Dethier (1955)
[7] van der Wolk (1978)
[8] Grabowski and Dethier (1954)
[9] Dethier (1976)
[10] C. T. Lewis (1954)
[11] Sutcliffe and McIver (1976)
[12] Owen (1971)
[13] Whitehead and Larsen (1976a)
[14] Henning (1974)
[15] Crnjar and Prokopy (in prep.)

found no chemoreceptors on the femora and only 9% of the total numbers on the legs were on the tibiae; in *Musca domestica* the corresponding figures are 2% and 8% on the femora and tibiae respectively (Dethier, 1955). Sutcliffe and McIver (1976) record some chemoreceptors distally on the tibiae of four *Simulium* species, but the majority are on the tarsi, and Owen (1971) observed none on the femora or tibiae of *Anopheles atroparvus*.

Data on numbers of chemoreceptors on the tarsi of adult insects are summarised in Table 1. In the case of *Schistocerca gregaria*, it has been assumed that basiconic sensilla, with six neurones, and canal sensilla, with only one neurone, are chemoreceptors (Kendall, 1970). The numbers of probable chemoreceptors on the tarsi of *Schistocerca gregaria* and *Tettigonia viridissima* are similar to those in most calypterate flies and there are about twice as many on the forelegs as on the hind legs. On each foreleg there are about 1000 chemosensitive neurones compared with 500 on each hind leg. On the other hand, *Glossina palpalis*, which, like *Musca domestica*, belongs to the family Muscidae and is of comparable size, has relatively few sensilla and there are similar numbers on all legs. *Rhagoletis pomonella* also has very few sensilla. Despite their small size, the *Simulium* species have as many sensilla as the larger flies, but *Aedes aegypti* and *Anopheles atroparvus* have only a few.

In *Heliothis zea* the sigmoid sensilla (Callahan, 1969) resemble the known chemoreceptors on the legs of *Pieris brassicae* (Ma and Schoonhoven, 1973) and are assumed also to be chemoreceptors. Approximately similar numbers occur on all the legs. In *Aedes aegypti* and in *Pieris brassicae*, the sexual dimorphism presumably reflects the differences in behaviour between the sexes since there are no marked differences in size.

No chemoreceptors are present on the prothoracic legs of the beetle *Dendroctonus ponderosae* (Whitehead, 1981).

No studies of the sensilla on the legs of larval insects exist, but Henig (1930) gives detailed drawings showing that on the simple thoracic legs of the caterpillar of *Orthosia lota* there are only 12 sensilla in total and only four on the tarsus. Each sensillum is depicted with only a single neurone, suggesting that they may be mechanoreceptors and that no chemoreceptors are present.

Several electrophysiological studies of tarsal chemoreceptors have been carried out on flies, although no surveys of responsiveness to a range of chemicals have been attempted. In *Phormia regina* and *Delia brassicae* the results of McCutchan (1969) and Städler (1978) suggest the presence of separate cells responding to NaCl, sucrose and, in *Delia brassicae*, to water. McCutchan sometimes obtained a response in two cells when a sensillum was stimulated by NaCl, while van der Starre (1972), working with *Calliphora vicina* (=*erythrocephala*), found that both water and sucrose stimulated a number of cells and considered that the tarsal chemoreceptor neurones exhibited a lack of specificity. Behavioural experiments on *P. regina* show that the tarsal sensilla are stimulated by a wide range of chemicals, but do not suggest great sensitivity to any one compound (see Dethier, 1976). On the other hand, oviposition by *Delia brassicae* is stimulated by glucosinolates and Städler (1978) found a cell in the tarsal sensilla of females responding to glucosinolates with a threshold of sensitivity at 10^{-5}–10^{-6} M, while some sensilla on the tarsi of *Rhagoletis pomonella* contain a neurone which is specific for the oviposition deterrent pheromone (Crnjar and Prokopy, in prep.).

Salt (NaCl), water and glucosinolate cells, the latter with a threshold sensitivity at 10^{-4}–10^{-5} M, are also present in the tarsal receptors of *Pieris brassicae* (Ma and Schoonhoven, 1973). No sugar-sensitive cell was found in this species, but one is inferred from the behavioural experiments of Kusano and Sato (1980) on *Pieris rapae*. The beetle *Chrysolina brunsvicensis* feeds only on *Hypericum* which contains hypericin, and it has tarsal receptors which respond to 2×10^{-5} M hypericin; other species of the genus, feeding on other plants, lack this sensitivity (Rees, 1969). There is, then, evidence for the existence of some receptor cells which are highly sensitive (tuned) to key compounds in the tarsal chemoreceptors of these insects. No electrophysiological data are available on tarsal receptors of orthopteroid insects.

2.2 SENSILLA ON MOUTHPARTS

2.2.1 *Sensilla on the mouthparts of Apterygota*

The only detailed studies on Apterygota relate to the terminal segments of the labial palps of various Thysanura (Larink, 1978, 1979) and of the dipluran *Campodea sensillifera* (Bareth and Juberthie–Jupeau, 1977). In *Lepisma saccharina* the number of pointed trichoid sensilla, which are not known to be chemoreceptors, varies with the size of the terminal segment (Fig. 1). In an early instar there are about 40, while in a large male there are over 350. Blunt trichoid sensilla are less abundant and, except in male *L. saccharina*, their number increases only slightly with size. In addition, there are five complex sensilla and six basiconic sensilla. Comparable numbers of sensilla are recorded for other Thysanura. The terminal segment of the labial palp of *C. sensillifera* has 350 sensilla with a total of some 2500 chemosensitive neurones.

2.2.2. *Sensilla on the mouthparts of Odonata*

Sensilla on the mouthparts of *Libellula* spp. are figured and described by Petryszak (1977). She states that all the hair-like sensilla on the mouthparts are innervated by single neurones and it is assumed that they are all mechanoreceptors. The only probable chemoreceptors are "papillae" which are found on the inside of the labrum, on the hypopharynx and mandibles, at the base of the labium and on the tips of the maxillary palps. The total number on all the mouthparts is about 220 with a total of about 900 neurones.

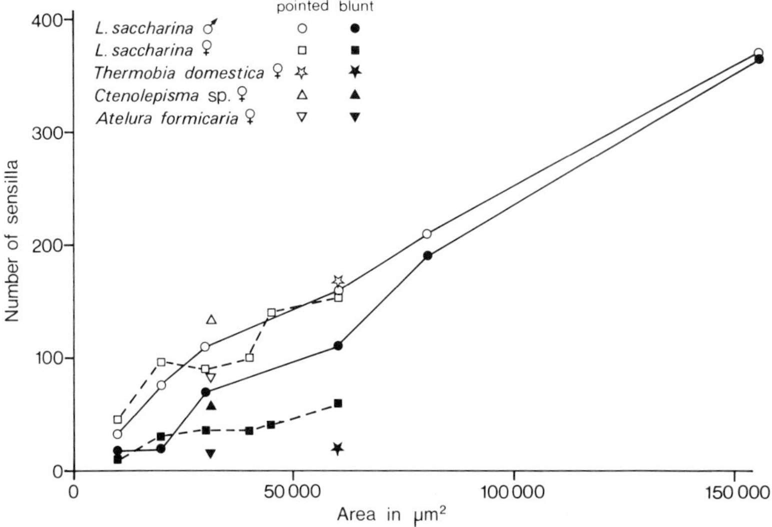

Fig. 1 Numbers of blunt and pointed trichoid sensilla in relation to the surface area of the terminal segment of the labial palp of various species of Thysanura. (After Larink, 1978, 1979)

The mouthparts of larval Odonata are specialised for catching prey and the labium forms an extensible mask. Zwarzin (1912) regarded all the hair sensilla on the mouthparts of *Aeschna* spp. as mechanoreceptors and depicted them with only single neurones. Some of the other sensilla which he regarded as chemoreceptors are almost certainly campaniform sensilla and others pressure receptors of the type described by Corbière-Tichané (1971) in the mandibles of the larva of *Speophyes lucidulus* (Coleoptera). Only the sensilla which Zwarzin observed on the inside of the labrum, and which were subsequently described by Pritchard (1965), and those on the tips of the antennae and maxillary palps, later described in other species by Bassemir and Hansen (1980), can confidently be regarded as chemoreceptors. Zwarzin shows 14 of these sensilla in an illustration of one side of the labrum and Pritchard shows 16 in *Aeschna interrupta*. There are four or five sensilla on the tip of the antenna and four to six in each palp in the *Aeschna* species as well as in *Agrion puella* and *Ischnura elegans* (Bassemir and Hansen, 1980). Hence there are about 50 chemoreceptors on the larval head. The labral sensilla each have two to five neurones, while of those on the palp two have two neurones and two have five (Bassemir and Hansen, 1980). Thus there are probably about 200 chemosensitive neurones on the mouthparts and antennae.

2.2.3 *Sensilla on the mouthparts of orthopteroid insects*

Amongst the Orthopteroidea the most complete studies on the sensilla on the mouthparts are on Acridoidea. The work of Chapman and Thomas (1978) and of Viscuso (1974) shows that the distribution of sensilla is similar throughout the group. Scattered sensilla with the structure of chemoreceptors are probably present on all the mouthparts (e.g. Blaney and Chapman, 1969; Thomas, 1966), but because they cannot be recognised with certainty in whole mounts the following discussion is largely restricted to the sensilla which occur in discrete groups and which are known to be chemoreceptors by virtue of their fine structure or from electrophysiological evidence. These groups are the A1, A2, A3 and A10 groups of the epipharyngeal face of the labrum, and others on the hypopharynx, the inner face of the galea and the outer face of the paraglossa (Chapman and Thomas, 1978; Thomas, 1966). All these are conical contact chemoreceptors barely projecting above the general surface of the cuticle (Cook, 1972, 1979; Louveaux, 1973). In addition, each of the maxillary and labial palps has a discrete terminal group of trichoid contact chemoreceptors (Blaney and Chapman, 1969, 1970; Blaney *et al.*, 1971).

Within any one species, the number of chemoreceptors in most groups is proportional to the size of the insect (see Table 2 in Chapman and Thomas, 1978), but the change in numbers with insect size varies in different groups of sensilla. For instance, the number of sensilla in the A1 group increases from 50 in a first instar larva of *Chortoicetes terminifera* to about 150 in the adult. By contrast the number of A3 sensilla increases from 50–60 to about 80. In general, the numbers of scattered sensilla increase roughly in proportion to the surface area of the labrum, while the sensilla in groups show smaller increases (Fig. 2).

For a given size of insect, Chapman and Thomas (1978) concluded that grass-feeding species tended to have slightly fewer sensilla than species feeding on broad-leaved plants and that species with a specialised diet had slightly lower numbers than other closely related, but polyphagous species.

In Fig. 3 the numbers of sensilla in the A1 epipharyngeal groups of various species of Gomphocerinae and Pyrgomorphidae are compared. Data from six species of Gomphocerinae, feeding primarily on a variety of grass species, are compared with counts from *Anablepia granulata*, which feeds only on *Brachyaria* spp. (Hummelen and Gillon, 1968; Gillon, 1972), and from *Xenocheila zarudnyi*, which is said to specialise on *Ephedra* (Chapman, 1966).

Individuals of the latter two species have fewer A1 sensilla than any "polyphagous" species of comparable size. Numbers of A2 sensilla are similar irrespective of feeding habit, but *X. zarudnyi* has fewer A3 sensilla than the other species (Chapman and Thomas, 1978).

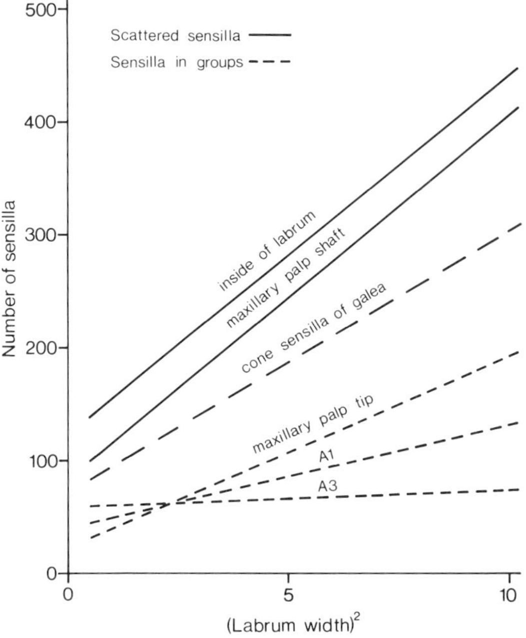

Fig. 2 Relationship between numbers of sensilla on the mouthparts and insect size [expressed as (labrum width)^2] in *Chortoicetes terminifera*. (After Chapman and Thomas, 1978)

Poekilocerus species feed primarily on Asclepiadaceae although they will eat a variety of other plants (Abushama, 1968; Fishelson, 1960; Joyce, 1952); other Pyrgomorphidae are much more catholic in their normal diets (Akhtar, 1971; Bernays *et al.*, 1975; Kevan, 1954). *Poekilocerus* species have fewer A1 sensilla than other Pyrgomorphidae of similar size (Fig. 3), but have similar numbers of A3 sensilla.

The only other species of Acridoidea which are known to be relatively specific in their food selection and of which the sensilla have been examined are *Cornops longicorne* and *Paulinia acuminata*. *C. longicorne* normally feeds only on water hyacinth (*Eichhornia*) and is further specialised by virtue of its oviposition habits (Bennett, 1970; Perkins, 1973). No data are available for other species of Leptysminae, the subfamily to which *C. longicorne* belongs, but the numbers of sensilla on all the mouthparts are very small for the size of insect (Chapman and Thomas, 1978). *P. acuminata* feeds mainly on the aquatic fern *Salvinia* (Bennett, 1966). No other species of Pauliniidae have been examined, but *P. acuminata* has similar numbers of sensilla to polyphagous Pyrgomorphidae of similar size (Chapman and Thomas, 1978).

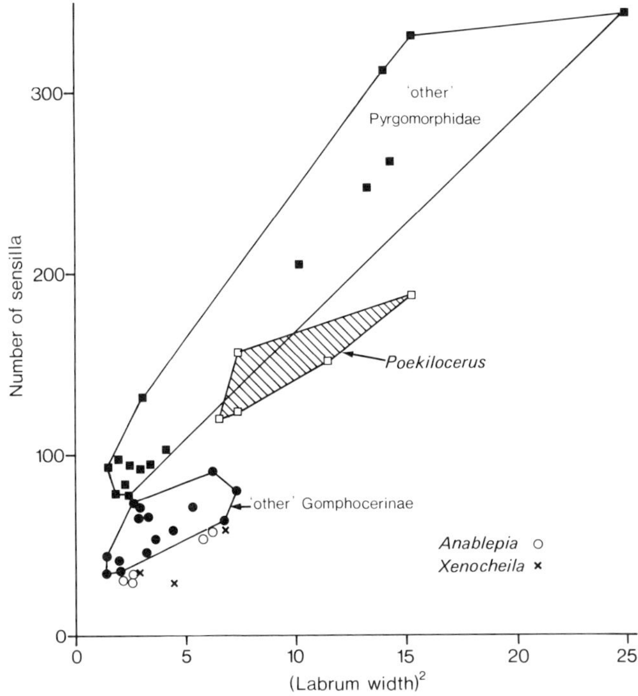

Fig. 3 Numbers of A1 sensilla (both sides combined) of "polyphagous" feeders belonging to the Pyrgomorphidae and Gomphocerinae compared with the "specialists", *Poekilocerus* (Pyrgomorphidae) and *Anablepia* and *Xenocheila* (Gomphocerinae). The "envelopes" contain all the points relating to the group shown. (After Chapman and Thomas, 1978)

The smallest numbers of sensilla in a group are found in the smallest Acridoidea. Amongst a sample of over 20 small Gomphocerinae, with labrum width less than 2 mm, there were rarely less than 20 sensilla in each of the A2 or A3 groups, though smaller numbers occurred more frequently in the A1 groups. The lowest number in a "polyphagous" species was 16; one group in *Anablepia granulata* contained only eight sensilla. Numbers were higher in other sensillum-groups and taxa (Chapman and Thomas, 1978).

Studies on the ultrastructure of these sensilla indicate they each contain between two and ten chemosensitive neurones. The total number of chemosensitive neurones in the main sensillum groups on the mouthparts of *L. migratoria* is about 3000 in the first instar nymph and about 15000 in the adult male. In addition there are numerous scattered chemoreceptors; on

TABLE 2 Numbers of chemoreceptors and neurones in groups on the mouthparts of *Locusta migratoria* of different ages. (References in parentheses)

Sensillum group	Number of neurones/sensillum	Instar I[2]		Instar V, male		Adult, male[2]	
		number of sensilla	total neurones	number of sensilla	total neurones	number of sensilla	total neurones
A1	5[3]	64	320	286[3] 221[6]	1430	403	2015
A2	5[4]	79	395	132[4] 144[6]	660	153	765
A3	5[4]	85	425	95[4] 77[6]	475	79	395
A10	2[5]	22	44	80[4]	160	107	214
Hypopharynx	3–6[7]	49	245	70[6]	350	101	505
Galea	4–5[6]	34 × 2	340	165[6] × 2	825	358 × 2	3580
Maxillary palp tip	6, 10[1]	41 × 2	492	370[1] × 2	4440	401 × 2	4812
Paraglossa	4[5]	37 × 2	296	92 × 2	736	149 × 2	1192
Labial palp tip	6?	50 × 2	600	280 × 2	3360	326 × 2	3260
Total		623	3157	2477	12 436	3311	15 036

[1] Blaney *et al.* (1971)
[2] Chapman and Thomas (1978)
[3] Cook (1972)
[4] Cook (1979)
[5] Louveaux (1972)
[6] Louveaux (1973)
[7] Louveaux (1975)

the maxillary palp of *Schistocerca gregaria*, Blaney and Chapman (1969) found over 250 scattered contact chemoreceptors contributing at least another 1250 chemosensitive neurones to the total. Including all the scattered sensilla, the total number of chemosensitive neurones associated with the mouthparts of an adult *L. migratoria* probably approaches 20 000 (Table 2).

Comprehensive studies have not been carried out on other orthopteroid groups of insects, but Rohr (1978) in a detailed account of the sensilla of the epipharynx of *Acheta domestica* shows that the numbers increase with the stage of development of the insect (Fig. 4). Estimates of the numbers of sensilla on the mouthparts of *A. domestica* (instar XIII nymphs) and *Periplaneta americana* (adults) can be obtained from the illustrations of Rościszewska and Fudalewicz–Niemczyk (1974) and Petryszak (1975) respectively. Malz and Hintze–Podufal (1979a, b) give a complete analysis of the mouthpart sensilla of *Gromphadorhina brunneri*.

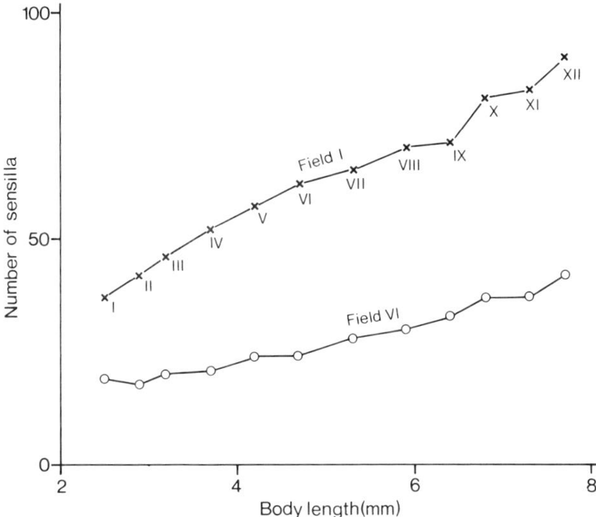

Fig. 4 Numbers of sensilla in fields I and VI of the epipharynx of *Acheta domestica* in relation to body length and nymphal instar (indicated by Roman numerals). Field I is equivalent to the A1 sensilla of acridids and field VI to the A3 sensilla. (After Rohr, 1978)

Within the Acrididae comparisons between species have been based on the width of the labrum. This is considered reasonable because the structure of the mouthparts and the shape of the labrum is relatively uniform throughout the group. However, other groups, such as the cockroaches, have a differently shaped labrum and there is no obvious objective basis on which comparisons of size can be made. Only big differences in numbers of sensilla can be considered important.

TABLE 3 Numbers of probable chemoreceptors occurring in groups on the mouthparts for maxilla and labium are for one side only. Data refer to adults or late instar nymphs

Species	Epipharynx		Hypopharynx		Galea	
	sensilla	neurones	sensilla	neurones	sensilla	neurones
Acheta domestica	250	1200	–	–	170	250
Gryllus bimaculatus	–	–	–	–	–	–
Periplaneta americana	224	650	132	400	125	500
Blaberus craniifer	–	–	110	450	–	–
Gromphadorhina brunneri	214	550	60	250	60	300
Forficula auricularia	35	175	10	50	–	–

[1] Rohr (1978)
[2] Rościszewska and Fudalewicz-Niemczyk (1974)
[3] Klein (1979)
[4] Petryszak (1975)
[5] Wieczorek (1978)
[6] Moulins (1968, 1971)

The number of conical sensilla on the epipharyngeal surface of the labrum is rather lower in gryllids and blattids than in the acridids, although the illustrations of Viscuso *et al.* (1978) show that relatively large numbers of sensilla are present on the epipharynx throughout the Orthopteroidea. The numbers of sensilla on the hypopharynx, the galea and the paraglossa are similar throughout the groups (Table 3), but the terminal sensilla of the palps are much more numerous in *G. bimaculatus* (Klein, 1979), in *P. americana* (Wieczorek, 1978; Altner and Stetter, 1980) and in *G. brunneri* (Malz and Hintze–Podufal, 1979b) than they are even in the biggest acridid. In *G. bimaculatus* there is considerable diversity in the form of the sensilla on the palps, and the number of chemosensitive neurones on the mouthparts of these species is very large, over 50 000 from the maxillary palps alone in *G. bimaculatus* (Klein, 1981).

In *Forficula auricularia* (Dermaptera) the numbers of sensilla are much smaller; this is only partly a reflection of the smaller size of the insects. The adult has 30 to 40 epipharyngeal sensilla in two paired groups and there are a further 10 on the hypopharynx (Lhoste, 1957; Moulins, 1969). The hypopharyngeal sensilla have 4–5 neurones (Moulins, 1969) so the total number of chemosensory neurones from epipharynx and hypopharynx probably is about 225. This number would be reduced to about 100 in the first instar larva.

The palps of *F. auricularia* have terminal sense organs which differ in outward form from those of other orthopteroids studied. The maxillary palp, ends in a papilla bearing 12 (17 in plate Ic of Popham, 1979) conical sensilla, f various orthopteroid insects with estimates of the total numbers of neurones. Numbers

Maxillary palp		Glossa + Paraglossa		Labial palp		Total		Ref.
sensilla	neurones	sensilla	neurones	sensilla	neurones	sensilla	neurones	
–	–	50	130	–	–	>700	>2000	1,2
3245	25 882	–	–	–	–	>6500	>50 000	3
2650	15 000	96	400	–	–	>6100	>32 000	4,5
–	–	–	–	–	–	–	–	6
4080	>10 000	90	450	260	2600	11 000	>28 000	7,8
12	60	–	–	63	620	>200	>1500	9,10,11

Malz and Hintze-Podufal (1979a)
Malz and Hintze-Podufal (1979b)
Lhoste (1957)
[10] Moulins (1969)
[11] Popham (1979)

while the tip of the labial palp is covered by about 60 scales each with a cushion-like receptor. Popham (1979) regards these as pressure receptors, but by analogy with other related insects it is more likely that they are chemoreceptors. Assuming that this is so and that they each have five neurones, the total sensory array on the palp tips is similar to that of acridids of comparable size (Table 3).

The only detailed neurophysiological studies on the mouthpart sensilla of Orthopteroidea are by Blaney (1974, 1975, 1980, and Winstanley and Blaney, 1978). These and the earlier studies of Haskell and Schoonhoven (1969) have shown that the sensilla on the tips of the palps and those in the epipharyngeal groups respond to a range of salts, sugars and other substances. Haskell and Schoonhoven (1969) suggested that specific neurones were sensitive to each class of compound, including one with maximal sensitivity to "grass chemicals" and one to repellent compounds. However Blaney (1974, 1975) concluded that most compounds tested caused a number of neurones in each sensillum to fire and that each neurone probably responded to a range of chemicals. He suggested that the sensillum is the sensory unit and showed that a measure of specificity occurred amongst sensilla. Winstanley and Blaney (1978) suggest that the sensilla on the palp tips of *S. gregaria* which respond positively to nicotine might be particularly sensitive to deterrent chemicals and so might provide labelled lines to the brain. Haskell and Mordue (1969) had concluded, on the basis of behavioural experiments, that the normal role of the A3 receptors of the epipharynx was to act as deterrent receptors. Nevertheless Winstanley and Blaney (1978) conclude that there is no unique pattern of response associated with acceptance or rejection (and see p. 319). Klein (1979) showed that the sensilla on the maxillary palps of *Gryllus bimaculatus* responded to a range of salts, but not to a mixture of sugars.

In addition to a large number of contact chemoreceptors, each maxillary palp of *P. americana* has a group of about 200 olfactory sensilla of four physiologically different types. These have principal responses to pentanol/hexanol, heptanol, decanol and butyric acid but, unlike the sensilla on the antennae, none of them responds to the sex pheromone (Altner and Stetter, 1980).

2.2.4 *Sensilla on the mouthparts of hemipteroid insects*

In general, Hemiptera have only a few sensilla on the mouthparts: a small group at the tip of the labium, up to five pairs in the maxillary and mandibular stylets, and some in the food canal. Wensler (1977) and Tjallingii (1978) show that in aphids the sensilla at the tip of the proboscis are mechanoreceptors and only the sensilla on the epipharynx and hypopharynx have the structure of chemoreceptors. In *Brevicoryne brassicae* and *Tuberolachnus salignus* there are 14 sensilla with 60 neurones on the epipharynx, and a number on the hypopharynx (Wensler and Filshie, 1969). The total number of chemosensitive neurones in the mouthparts of aphids is probably about 100. The planthopper, *Nilaparvata lugens*, has approximately 70 chemoreceptor neurones associated with sensilla on the labial lobes and 86 neurones from receptors in the cibarial cavity (S. Foster, unpublished).

Small numbers of sensilla are present on the distal lobes of the labium in many Heteroptera (Cobben, 1978) and in some species they are known to be chemoreceptors. On each labial lobe, *Dysdercus* spp. have 10 or 12 chemoreceptors, each with two to five chemosensitive neurones, with a total of 32 neurones in *D. intermedius* (Peregrine, 1972; Schoonhoven and Henstra, 1973; Gaffal, 1981). In *Lygus lineolaris* there are 11 uniporous sensilla on each side with a total, for the two sides, of 74 presumed chemosensitive neurones (Hatfield and Frazier, 1980). A general chemoreceptor function for these sensilla has been proved electrophysiologically (Avé *et al.*, 1978). Chemoreceptors in the wall of the food canal have been seen in a number of species (Kraus, 1957; Cobben, 1978) and they are probably present in most, or even all Heteroptera. In *Triatoma infestans* there are 13 sensilla in this position (Bernard, 1974). Assuming the presence of similar numbers in other species, the total number of chemosensitive neurones on the mouthparts of *Dysdercus* spp. and *L. lineolaris* is about 150.

Cenocorixa bifida differs from the other Hemiptera studied. It is a detritus feeder, ingesting particulate matter as well as fluids, and the rostrum is much broader. On the outer face of the labrum there are bands of sensilla, about

2500 in all. Only 190 of these have been shown to be permeable to dyes and have only one neurone each (Lo and Acton, 1969). This figure must be regarded as the minimal number of chemosensitive cells associated with feeding: the number may well be greater.

No detailed electrophysiological studies have been carried out on the chemosensilla of hemipteroids, but 3.8×10^{-6} M ATP and other nucleotides induce gorging in *Rhodnius prolixus*, apparently acting via the cibarial sensilla (Friend and Smith, 1977).

2.2.5 *Sensilla on the mouthparts of adult Endopterygota*

No comprehensive studies have been made on the numbers of sensilla on the mouthparts of Lepidoptera, but a general impression can be gained from information on different species. Städler *et al.* (1974) record about 122 chemoreceptors with 210 neurones on each galea of *Choristoneura fumiferana*, while in the cibarial cavity of *Trichoplusia ni* and *Manduca sexta* there are 24 sensilla with 116 neurones (Eaton, 1979). On the labial palp of *Heliothis zea* Callahan (1969) records only 4–8 chemoreceptors. These data suggest that the total number of chemosensitive neurones on the mouthparts of any one species would be about 600, though this should probably be regarded as a conservative estimate. In some galeal sensilla of *C. fumiferana* one cell responds to sucrose (Städler and Seabrook, 1975).

Various groups of blood-sucking Diptera Nematocera have been studied. Only small numbers of chemoreceptors occur on the maxillary palps, in the cibarium and on the labrum of mosquitoes (Table 4); no sensilla are present on the other stylets (Lee, 1974). About 50 chemoreceptors are also present on each labellum (Owen *et al.*, 1974; Pappas and Larsen, 1976). Labral sensilla are absent in males and in the nectar-feeding females of *Toxorhynchites* spp., but they are apparently present in the female of *Wyeomyia smithii* (Gernet and Buerger, 1966) which is also nectar feeding (McIver and Hudson, 1972). In *Toxorhynchites* there is also a tendency for the number of cibarial receptors to be reduced (Table 4).

The sensilla on the mouthparts are sensitive to several sugars and inorganic salts (Salama, 1966; Sinitsyna, 1971; Owen *et al.*, 1974) and to adenine nucleotides which are phagostimulants. The behavioural threshold to ATP is about 5×10^{-4} M (Friend and Smith, 1977) so there is a suggestion of a specialised receptor for this and related compounds (AMP and ADP). Kellogg (1970) demonstrated that some sensilla on the palps of mosquitoes were sensitive to CO_2, and one neurone in the multiporous pegs of the palps has a lamellate structure (McIver, 1972) associated with thermo- or hygroreception (Altner and Prillinger, 1980). Female mosquitoes have more chemoreceptors on the palps than males of the same species even where, as in *W. smithii*, both sexes have the same feeding habits.

TABLE 4 Numbers of chemoreceptors on the mouthparts of various mosquitoes. (– indicates no data)

Species	Sex	Number of sensilla on				Food	Ref.
		labrum	cibarium	maxillary palp	labellum		
Aedes aegypti	♂	0	10	17	–	nectar	1,2,11
	♀	4	10	29	–	blood	3,11
Aedes albopictus	♂	0[a]	10	–	–	nectar	4
	♀	4[a]	10	–	–	blood	4
Aedes atropalpus	♀	4[a]	–	24	–	blood	2
Culex pipiens	♂	0	10	15	–	nectar	4,2,7
	♀	4	10	78	–	blood	4,3,7
Culex territans	♂	0	10	15	–	nectar	2,7
	♀	4	10	37	–	blood	3,7
Culiseta spp.	♂	0	10	–	–	nectar	7
	♀	4	10	–	56	blood	7,9,10
Mansonia perturbans	♂	0	10	–	–	nectar	7
	♀	4	10	–	–	blood	7
Toxorhynchites brevipalpis	♂	0	8–10	–	–	nectar	5
	♀	0	8–10	–	–	nectar	5
Toxorhynchites rutilus	♂	0	6–10	–	–	nectar	5
	♀	0	6–8	–	–	nectar	5
Toxorhynchites splendens	♂	0	10	–	–	nectar	7
	♀	0	10	–	–	nectar	5
Wyeomyia aporonoma	♀	–	–	31	–	blood	6
Wyeomyia smithii	♂	0	10	20	–	nectar	6,7
	♀	4	10	34	–	nectar	6,7
Anopheles gambiae	♂	0	10	–	–	nectar	7
	♀	4	10	67	–	blood	7,8
Anopheles stephensi	♂	–	–	9	–	nectar	8
	♀	–	–	98	–	blood	8

[a] Numbers assumed from Lee (1974)

[1] Lee (1974)
[2] McIver (1971)
[3] McIver and Charlton (1970)
[4] Uchida (1979)
[5] Lee and Davies (1978)
[6] McIver and Hudson (1972)
[7] Gernet and Buerger (1966)
[8] McIver and Siemicki (1975)
[9] Larsen and Owen (1971)
[10] Owen *et al.* (1974)
[11] McIver and Siemicki (1981)

Simuliidae have approximately 12 chemoreceptors at the tip of the labral food canal and four cibarial receptors (see Colbo *et al.*, 1979, for details); no information is available on the labial receptors. On the maxillary palps there are 52–68 uniporous sensilla, but the number of multiporous sensilla varies with sex and feeding habit, being greatest in blood-sucking females (Table 5) (Mercer and McIver, 1973). Comparably small numbers of sensilla are present on the mouthparts of *Culicoides* spp. (Ceratopogonidae) with more sensilla on the palps of bird-feeding than of mammal-feeding species (Braverman and Hulley, 1979) (Table 5).

TABLE 5 Numbers of chemoreceptors on the mouthparts of some *Simulium* spp. and *Culicoides* spp.

Species	Sex	Numbers of sensilla on				Food	Ref.
		labrum	cibarium	maxillary palp			
				uniporous	multiporous		
S. baffinense	♂	–	–	55	20	nectar	1
	♀	–	–	52	30	nectar?	1
S. euryadminiculum	♀	–	–	62	75	blood	1
S. rugglesi	♂	–	–	–	10	nectar	1
	♀	–	–	58	78	blood, birds	1
S. venustum	♂	12	4	–	–	nectar	2
	♀	12	4	68	55	blood, mainly mammals	1,2
C. brucei	♂	–	–	0	6	nectar	3
	♀	–	–	0	20	blood, mainly mammals	3
C. furens	♂	–	–	0	5	nectar	4
	♀	–	–	0	10	blood, mammals	4
C. variipennis	♂	4	–	0	3	nectar	5,6
	♀	4	–	0	16	blood, mammals	5,6
C. crepuscularis	♂	–	–	0	10	nectar	6
	♀	–	–	0	64	blood, birds	6
C. nivosus	♂	–	–	0	1	nectar	3
	♀	–	–	0	>50	blood, birds assumed	3

[1] Mercer and McIver (1973)
[2] Colbo *et al.* (1979)
[3] Braverman and Hulley (1979)
[4] Chu-Wang *et al.* (1975)
[5] Buerger (1967)
[6] Rowley and Cornford (1972)

Although the numbers of sensilla on the labium are not known in most of these groups it appears likely that the total number of chemoreceptors on the mouthparts of a female mosquito is less than 250, while in the male there are less than 75. In Simuliidae and Ceratopogonidae the probable figures are 300 and 200, and 150 and 50 respectively. The total number of neurones is unknown, but probably ranges from about 1000 in blood sucking females to about 200 in male Ceratopogonidae.

Adult Cyclorrhapha have chemoreceptor hairs on the outer surface of the labellar lobes and interpseudotracheal papillae on the ventral surface. *Phormia regina* and *Calliphora vicina* have over 250 hairs on the labellum, while *Drosophila melanogaster* has only 80 (Table 6). This smaller number might simply be a reflection of the much smaller size of *D. melanogaster*, but might also reflect the tendency of the former species to feed on a wider range of materials compared with *Drosophila. D. melanogaster* also has fewer interpseudotracheal sensilla than *P. regina*. Cibarial sensilla have only been described in *C. vicina* of these species, but Rice (1973) suggests that similar small numbers are present in other Diptera. *D. melanogaster* has only about 25% of the number of chemosensitive neurones present on the proboscis of its larger relatives. On the other hand, *Glossina austeni*, which is directly comparable in size to *Calliphora vicina*, has a total of only 20 chemoreceptors with 28 chemosensitive neurones on the proboscis.

All these species have maxillary palps. Peters (1963) depicts sensilla with a chemoreceptor-like structure terminally on the palps, but no indication is given of the number present.

Extensive neurophysiological investigations of the labellar sensilla of Cyclorrhapha have been carried out, notably on *P. regina* by Dethier (1974, 1976, 1980). The sensilla do fall into different physiological categories, but each of the cells responds to a spectrum of chemicals, with different cells generally responding to different groups of chemicals. Typically present are: the "sugar" cell which responds to sugars, amino acids, glycosides and some other substances; the "salt" cell which responds to salts, acids, amino acids and glycosides; and the "water" cell whose activity is suppressed by most compounds, but which responds to tropaeolin; and an "anion" cell which responds to salts and some acids. The cells are not highly sensitive to any one substance and a single compound may cause two or three cells to fire.

In *Glossina* spp., on the other hand, the adenine nucleotides, and especially ATP, are the only important phagostimulants (Friend and Smith, 1977). The threshold concentration of ATP to which the sensilla in at least some individuals respond is about 10^{-6} M, and the levels for AMP and ADP are only slightly higher. The sensilla respond to other nucleotides and to sodium chloride, but at much higher concentrations (Mitchell, 1976).

TABLE 6 Numbers of chemoreceptors on the mouthparts of adult Diptera Cyclorrhapha. The numbers include sensilla from both labellar lobes. (– indicates no data; references in parentheses)

	Drosophila melanogaster(1)			*Phormia regina* (2,3,4)			*Calliphora vicina*= *erythrocephala* (5,6,7)			*Glossina austeni* (8,9)		
	number of sensilla	neurones/sensillum	total neurones	number of sensilla	neurones/sensillum	total neurones	number of sensilla	neurones/sensillum	total neurones	number of sensilla	neurones/sensillum	total neurones
Labellar hairs	80	2–4	300	271	3–4	1070	256–264	4	1024–1056	16	1–2	24
Interpseudotracheal papillae	40	1	40	135	3	405	–	–	–	0	0	0
Cibarial sensilla	–	–	–	–	–	–	4	1	4	4	1	4
Total	120		340	406		1475	> 260		> 1030	20		28

[1] Falk *et al.* (1976)
[2] Wilczek (1967)
[3] Felt and Vande Berg (1976)
[4] Tominaga *et al.* (1969)
[5] Maes and Vedder (1978)
[6] Peters (1963)
[7] Rice (1973)
[8] Rice *et al.* (1973a)
[9] Rice *et al.* (1973b)

The worker bee, *Apis mellifera* has contact chemoreceptors on the mouthparts which are called sensilla chaetica and basiconica by Whitehead and Larsen (1976a). They record about 150 of the former which are uniporous sensilla with four chemosensitive neurones. Their sensilla basiconica, about 180 in total, also appear to be uniporous. If these also have four neurones, the total number of chemosensitive neurones on the outside of the mouthparts is about 1300. In addition there are 180 epipharyngeal sensilla, each with one neurone, which are probably contact chemoreceptors (Galic, 1971), and 90 hypopharyngeal sensilla with four neurones each. Bees are well known to "taste" a number of sugars and Whitehead and Larsen (1976b) and Whitehead (1978) have shown electrophysiologically that single cells respond to several sugars and salts; only a limited range of chemicals has been tested.

The only extensive study on the sensilla of the mouthparts of adult Endopterygota which chew their food is on *Dendroctonus ponderosae* (Coleoptera) (Whitehead, 1981). There are 37 probable chemoreceptors with a total of 126 chemosensitive neurones on the maxillary and labial palps and the galea of each side. This includes three sensilla which Whitehead calls "campaniform", but which from their appearance and position may be chemoreceptors. The labrum and hypopharynx were not studied, but the total number of chemosensitive neurones on the mouthparts is clearly not much over 250. Sutcliffe and Mitchell (1980) record 8–12 contact chemoreceptors, each with four neurones, on the galea of *Entomoscelis americana*. *Trichogramma dendrolimi* (Hymenoptera) is not typical since, although it uses its mandibles to bite its way out of the host egg, it is essentially a fluid-feeder. The work of Bangying (1979) suggests that there are only about 14 chemoreceptors on the mouthparts.

2.3 SENSILLA ON THE HEAD OF LARVAL ENDOPTERYGOTA

In larval Endopterygota the antennae and mouthparts are much more closely associated than in the adults and it is logical to consider them together although it is still generally true that the chemoreceptors of the mouthparts have a gustatory function and those of the antennae are olfactory.

The sensory array in caterpillars is similar in all the species studied (Ma, 1972, 1976; Schoonhoven, 1973; de Boer *et al.*, 1977; van Drongelen, 1979; Albert, 1980; Dethier, 1980a). On the epipharyngeal surface of the labrum there are commonly two placoid sensilla, though none is present in *Mamestra brassicae* (Blom, 1978). These sensilla have three chemosensitive neurones each (Ma, 1972). On each galea there are two styloconic sensilla with four chemosensitive neurones each and on the tip of the maxillary palp there are eight chemoreceptors with four neurones each. The antennae have three sensilla with 16 neurones. This gives a total of about 118 chemosensitive neurones. Most of the insects studied so far are oligophagous in different groups of plants, but similar numbers of sensilla are present in *Heliothis zea* (Avé, 1980) which is polyphagous.

Fairly extensive electrophysiological studies have been carried out on the contact chemoreceptors of lepidopterous larvae. All the species examined respond to a wide range of sugars, amino acids and salts (e.g. Dethier, 1973; Dethier and Kuch, 1971; van Drongelen, 1979; Ma, 1972; Schoonhoven, 1973; Wieczorek, 1976). It is probably common for each compound to stimulate more than one cell (Dethier and Kuch, 1971), but there is good evidence for some degree of specificity within a sensillum, one cell being most sensitive to sugars, one to salts and one to amino acids.

In a number of oligophagous species one cell in certain sensilla has been shown to be particularly sensitive to a compound which is characteristic of the host plant. These are shown in Table 7 which does not include responses to compounds which are widely distributed amongst plants. There is also behavioural evidence for the initiation of feeding by certain host-plant-specific chemicals (see Table 32). While such behavioural evidence does not prove that specifically tuned receptor cells exist, this would be the simplest explanation of the behaviour. The evidence suggests that such tuning is common in oligophagous larval Lepidoptera.

A large number of species have also been shown to respond electrophysiologically to compounds which deter feeding. In the larva of *Danaus plexippus* Dethier (1980) observed that most deterrent compounds elicited responses in a number of receptor cells. On the other hand, Ma (1972) working with *Pieris brassicae*, and Schoonhoven (1973), with a number of species, indicate that certain cells fire preferentially in response to deterrent substances. These findings are not in conflict: it is probable that deterrents may stimulate several cells, one of which is more sensitive than the others.

In larvae of *Pieris brassicae*, Blom (1978) and Ma (1972) obtained straight line relationships between the impulse frequency of cells in various sensilla in response to sucrose or sinigrin and the amount of feeding induced by the same concentrations of the same chemicals. In *Mamestra brassicae* the relation between sensory input and food intake was less regular, but a general correlation was still present (Blom, 1978). In these instances, the insects were fed on semi-synthetic diets lacking deterrent chemicals; the sensory input resulting from stimulation by such chemicals increased with concentration, though the amount of feeding went down when they were added to the diet. Hence, in an insect feeding on a natural diet containing both phagostimulants and deterrents, the overt behaviour is the outcome of central integration of the various inputs.

TABLE 7 Insects with chemosensitive cells in the mouthparts or tarsi known to be highly sensitive to compounds which are characteristic of the host plant

Species	Stage	Host plant	Compound	Threshold concentration (molar)	Ref.
Lepidoptera					
Pieris brassicae	larva	Cruciferae	glucosinolates	10^{-7}	Schoonhoven (1969)
	adult[a]			10^{-4}	Ma and Schoonhoven (1973)
Mamestra brassicae	larva	Cruciferae	1-naphthyl-β-glucoside	10^{-4}	Wieczorek (1976)
Manduca sexta	larva	Solanaceae	tomatin	10^{-4}	Schoonhoven (1969)
Laothoe populi	larva	Salicaceae	salicin	–	Schoonhoven (1972)
Papilio polyxenes	larva	Umbelliferae	apiin	10^{-3} or less	Dethier (1973)
Diptera					
Delia brassicae	adult[a]	Cruciferae	sinigrin	10^{-6}	Städler (1978)
Coleoptera					
Chrysolina brunsvicensis	adult[a]	Hypericaceae	hypericin	10^{-4}	Rees (1969)

[a] Tarsal sensilla

Olfactory processes in larval Lepidoptera have been considered by Dethier (1980) and Dethier and Schoonhoven (1969) with respect to *Malacosoma americana* and *Manduca sexta*. The olfactory cells of both species are generalists, each responding to a spectrum of odours, and it is apparent that discrimination depends on the central integration of the input from a number of receptor cells. Sense cells in "gustatory" receptors may also respond to odours at high concentrations (Städler and Hanson, 1975).

The larva of *Limnophilus rhombicus* (Trichoptera) is phytophagous and has few sensilla on the mouthparts (Barbier, 1961). It is probable that the sensilla described by Barbier as ampullaceous, placode, basiconic and "batonnets tactile" are chemoreceptors; there are only 32 of these on the mouthparts so that the number of chemosensitive neurones is probably about 100.

The larvae of species belonging to four groups of Coleoptera have been studied in detail. The larva of *Leptinotarsa decemlineata* has 79 probable chemoreceptor sensilla on the mouthparts with a further 22 on the antennae (Chin, 1950) (Table 8). In *Speophyes lucidulus* only 45 sensilla on the mouthparts have the structure of chemoreceptors (Corbière-Tichané, 1969, 1970, 1971a), but there are six other sensilla on the tip of each palp which have not been characterised with certainty so that the total number of chemoreceptors may be greater. In addition, there are 14 contact chemoreceptors, six olfactory receptors and eight sensilla with lamellate neurones on the antennae. *Tribolium* larvae have similar numbers of sensilla, about 67, on the mouthparts (Ryan and Behan, 1973a), not including trichoid sensilla which have been excluded on the assumption that they are all mechanoreceptors. However, the shapes of a few of those shown in Ryan and Behan's illustrations suggest that they may be chemoreceptors. There are also 18 contact chemoreceptors and two olfactory sensilla on the antennae. Finally, in larval elaterids, Zacharuk (1962) indicates that the number of sensilla on each of the palps may be as high as 43, though not all these are necessarily chemoreceptors. The total number of probable chemoreceptors in the mouthparts of these larvae is between 60 and 188, with a further 18 contact chemoreceptors plus two olfactory receptors on the antennae.

In all these insects, with feeding habits ranging from phytophagous to virtually omnivorous in some elaterids, the number of chemoreceptors on the mouthparts is commonly less than 100 with about 20 more on the antennae. Although the number of chemosensitive neurones is only known in detail in two species the total is certainly generally less than 500. There is no information on differences between instars, only the final larval instar having been studied. In no case have chemoreceptors been positively identified on the mandibles.

TABLE 8 Number of probable chemoreceptor sensilla and chemosensitive neurones on the mouthparts and antennae of various larval Coleoptera. (References in parentheses)

	L. decemlineata (1)	*S. lucidulus* (2)		*Tribolium spp.* (3)		Elateridae (4)	
Width of labrum	–	0.25 mm		0.2 mm		0.7 mm	
	no. of sensilla	no. of sensilla	no. of neurones	no. of sensilla	no. of neurones	no. of sensilla	no. of neurones
Labrum/epipharynx	21	25	55	13	–	10	40
Maxilla	–	+?	–	?	–	3 × 2	24
Maxillary palp	15 × 2	5 × 2	34	13 × 2	–	(11–43) × 2	44–172
Labium	2 × 2	0	–	4	–	–	–
Labial palp	11 × 2	5 × 2	36	12 × 2	–	(11–43) × 2	44–172
Antenna							
contact chemoreceptors	} 11 × 2	7 × 2	56	9 × 2	50	9 × 2	72
olfactory receptors		7 × 2	86	1 × 2	260	1 × 2	16–60
Total	101	65	267	87	–	80–208	240–540

[1] Chin, 1950
[2] Corbière-Tichané (1969, 1970, 1971a); Corbière-Tichané and Bermond (1971)
[3] Ryan and Behan (1973a); Behan and Ryan (1978)
[4] Zacharuk (1962); Scott and Zacharuk (1971)

TABLE 9 Number of basiconic sensilla on the mouthparts of different larval instars of *Macrodytes* spp. (After Hamon, 1961)

Structure	Instar		
	I	II	III
Labrum	0	0	0
Mandible	3 × 2	93 × 2	195 × 2
Maxilla	3 × 2	23 × 2	26 × 2
Maxillary palp	7 × 2	7 × 2	8 × 2
Labium	3 × 2	10 × 2	10 × 2
Labial palp	0	5 × 2	5 × 2
Total	32	276	488

This contrasts with the carnivorous larvae of *Macrodytes* spp. (Dytiscidae) (Hamon, 1961). In these insects the number of basiconic sensilla increases in the second and third instars with very large numbers present on each mandible (Table 9). It is not certain that these are chemoreceptors.

Very few electrophysiological studies of chemoreception by beetle larvae exist. The larva of *Entomoscelis americana* has a single cell in a sensillum on the galea which responds to a range of sugars and amino acids (Mitchell, 1978; Mitchell and Gregory, 1979). Glucosinolates stimulate feeding by *E. americana* and Mitchell (1978) obtained an electrophysiological response to glucosinalbin at a concentration of 10^{-3} M, but the response to sinigrin was very variable. In the larva of *Leptinotarsa decemlineata* two cells commonly respond to sucrose (Mitchell and Schoonhoven, 1974). This insect has neurones which are highly sensitive to amino acids (Mitchell, 1974) and respond to sodium chloride, though the existence of a specific salt receptor has not been established (Mitchell and Schoonhoven, 1974).

A number of chemicals known or assumed to be feeding deterrents have been investigated electrophysiologically with larval *L. decemlineata* and *E. americana* (Mitchell and Schoonhoven, 1974; Mitchell, 1978). No response was obtained with most compounds and there is no evidence for the occurrence of a general deterrent receptor in Coleoptera.

No comprehensive study of the sensilla of nematocerous larvae has been made, but the available data indicate that the total number of chemoreceptors on the mouthparts and antennae does not generally exceed 50 (Table 10) irrespective of the habitat or feeding habit. In Simuliidae and Chironomidae, at least, the number of sensilla on the maxillary palp is the same in all instars (Craig and Borkent, 1980) and it is evident from the illustrations of species from other families by other authors that there is, in general, little change in

TABLE 10 Sensilla on the mouthparts and antennae of larvae of Nematocera, omitting probable mechanoreceptors

Family/subfamily	Antenna	Labrum	Galea	Maxillary palp	Labium (both sides)	Total (approx)	Feeding habit	Ref.
Culicidae	3–6	–	5	4–7	8	40	filtering	1,2,3,4,9,10
Thaumaleidae	–	–	–	13	–	?	browsing	5
Chironomidae							filtering	
Diamesinae	–	–	–	12[a]	–	?		5
Orthocladiinae	4	4+8[b]	4	9	10	56		6,11
Chironominae	6[a]	–	4	9[a]	10	48		6,7
Tanyderidae	9	12+2[c]	–	13	–	58	?	5,12
Blepharoceridae	–	–	–	12	–	?	?	5
Simuliidae	–	–	2	11	–	?	filtering	5
Mycetophilidae							fungus feeding	
Bolitophilinae	8[a]	6	0	7	–	36		8
Diadocidiinae	0	10	0	8	–	26		8
Keroplatinae	0	10–12	0	9	–	30		8
Sciophilinae	0	8–10	0	8	–	26		8
Mycetophilinae	0	8	0	8	–	24		8
Ditomyiinae	3[a]	2	0	9[a]	–	26		8
Sciaridae	0	10	0	5[a]	–	20	fungus feeding	8

[a] Counted on photographs in reference cited; [b] hypopharyngeal sensilla; [c] epipharyngeal sensilla

[1] Harbach and Knight (1977)
[2] Jez and McIver (1980)
[3] Zacharuk and Blue (1971a)
[4] Zacharuk and Blue (1971b)
[5] Craig and Borkant (1980)
[6] Mozley (1971)
[7] Sublette and Sublette (1973)
[8] Plachter (1979)
[9] Gardner *et al.* (1973)
[10] Pao and Knight (1970)
[11] Strenzke (1960)
[12] Exner and Craig (1976)

number from instar to instar although the arrangement may vary (e.g. Sublette and Sublette, 1973). The chemoreceptors on the antenna of larval *Aedes aegypti* have about 20 neurones in total (Zacharuk and Blue, 1971a, b; Zacharuk *et al.*, 1971), while on the antenna of the predaceous larva of *Toxorhynchites brevipalpis* there are three chemoreceptors with 16 neurones (Jez and McIver, 1980). The total number of chemosensitive neurones in the head is unlikely to exceed 200 in any of the insects listed in Table 10.

The arrangement of sensilla in three paired groups is similar in all the species of larval Cyclorrhapha studied although the only recent detailed work is on *Musca domestica*. In this species each dorsal organ has 12 receptors with a total of 29 probable chemoreceptor neurones, each terminal organ has 16 sensilla with 37 neurones, and the ventral organ has a single chemoreceptor with two neurones (Chu and Axtell, 1971; Chu-Wang and Axtell, 1972a, b). This gives a total of 136 neurones which are probably sensitive to chemicals. In addition, Bolwig (1946) records six pairs of sensilla in the cibarial cavity and anterior pharynx. Richter (1962) observed 21 neurones in the dorsal organ of *Calliphora vicina*. The larva of *Hylemya antiqua* has only about 30 sensilla in the cephalic region (Yamada *et al.*, 1981) while the rather smaller larva of *Delia brassicae* has a total of 24 sensilla of which six are regarded by Ryan and Behan (1973b) as campaniform sensilla, and in *Psila rosae* there are 24 including 10 campaniform sensilla (Ryan and Behan, 1973c). Both *D. brassicae* and *P. rosae* larvae are known to respond to specific chemicals from their host plants (Jones and Coaker, 1978), while saprophagous species respond to a range of widely occurring compounds such as CO_2, ammonia and acetic acid.

2.4 SENSILLA ON THE ANTENNAE

2.4.1 *Sensilla on the antennae of Apterygota*

In the collembolan *Folsomia candida* the sensilla on the two basal segments are probably all mechanoreceptors (Slifer and Sekhon, 1978a). On the third antennal segment is a simple antennal organ, comprising two sensilla, and six wall-pore sensilla, while on segment four there are approximately 150 wall-pore sensilla. Each of the latter has 2–4 chemosensitive neurones, while in *Onychiurus* the sensilla of the antennal organ have three and five neurones (Altner and Thies, 1972). Hence the total number of chemosensory neurones in the antenna of *Folsomia* is probably about 600.

A peg is present in *Hypogastrura socialis* and *Allacma fusca* forming part of a sensory complex at the tip of the antenna with 10 and 13 neurones respectively (Altner and Thies, 1973, 1978). *Megalothorax minimus* has a large sensillum subapically (Massoud and Deboutteville, 1969) which is probably similar in structure. The antennal organ consists of two sensilla in

M. minimus (Massoud and Deboutteville, 1969) and *Hypogastrura copiosa* (Slifer and Sekhon, 1978a) and four sensilla, with a total of 16 neurones, in *Onychiurus* sp. (Altner and Thies, 1972). All species have, like *F. candida*, scattered chemoreceptors on segments three and four. These sensilla generally appear to have four chemosensitive neurones (Altner and Ernst, 1974; Altner and Thies, 1978) so that a total of a few hundred chemosensitive neurones per antenna is probably common in Collembola.

The amputation experiments of Waldorf (1976) on *Sinella curviseta* show that in this species at least some sensilla on segments three and four of the male antenna are sensitive to a female pheromone.

The Collembola possess postantennal organs on the head, one close to the base of each antenna. The outward form of the organ varies, but in all cases studied there are pores in the cuticle and a single associated neurone (Altner *et al.*, 1970, 1971; Dallai, 1971; Karuhize, 1971; Altner and Thies, 1976).

Ctenolepisma lineata (Thysanura) has a total of about 980 contact chemoreceptors and 150 olfactory sensilla on one antenna. They are associated with more than 4500 neurones (Slifer and Sekhon, 1970).

The pseudoculus of Protura may be homologous with the Collembolan post-antennal organ and may have a chemosensory function. In *Acerentulus trägårdhi* there are six neurones in each sensillum (Bedini and Tongiorgi, 1971), while in *Eosentomon transitorium* there are only two neurones (Haupt, 1972).

2.4.2 *Sensilla on the antennae of Odonata*

Only coeloconic sensilla are normally present externally on the antennae of Odonata, although occasional trichoid sensilla do occur (Slifer and Sekhon, 1972). In the six species of Zygoptera which have been studied there are three, four or five pit openings, while in nine species of Anisoptera there are 10–48 pit openings. The larger pits are compound and contain several pegs, some of which are multiporous. Each peg has between one and three neurones so the total number of chemosensitive neurones per antenna probably never exceeds 200 and may be less than 50 in Zygoptera.

2.4.3 *Sensilla on the antennae of Orthopteroidea*

(*a*) *Dictyoptera* By far the most comprehensive studies on insect antennal sensilla have been on *Periplaneta americana*. The number of contact chemoreceptors on the antenna of an adult male varies from 6000 to 17 000, depending on the strain of insects and the length of the flagellum (Lambin, 1973; Schafer and Sanchez, 1973, 1976; Schafer, 1977; Schaller, 1978). Each sensillum has 3–5 neurones, one of which is a mechanoreceptor, so that there are approximately 24 000 to 68 000 contact chemoreceptor neurones in each antenna. Females sometimes have rather fewer contact chemoreceptors than males (Schafer and Sanchez, 1976).

TABLE 11 Numbers of chemoreceptor sensilla and neurones on the antennae of different species of cockroach and a mantis[a]

Species	Ratio of ♂:♀ length	Number of contact chemoreceptors		Number of olfactory sensilla		Total number of chemosensitive neurones (in 1000s)		Ref.
		♂	♀	♂	♀	♂	♀	
Periplaneta americana	1.0	10 200	7 600	25 400	18 100	140	100	Schafer and Sanchez (1976)
		13 700	10 200	40 800	22 600	210	130	Schaller (1978)
P. australasiae	0.9	4800	4300	25 200	16 800	120	80	Schafer and Sanchez (1976)
P. brunnea	0.9	2800	3300	19 100	11 800	90	60	
P. fuliginosa	0.9	4500	4900	22 800	16 700	105	85	
P. japonica	1.6	4000	2800	18 800	10 900	85	50	
Blaberus craniifer	1.0	4800	4800	45 000	45 000	137	137	Lambin (1973)
Leucophaea maderae	1.0	5400	5400	26 700	26 700	93	93	Schafer (1971)
Arenivaga sp.	1.6	500	400	8100	3800	32	24	Hawke and Farley (1971)
Tenodera angustipennis[a]	1.5	699	402	37 600	10 200	191	53	Slifer (1968b)

TABLE 12 Total numbers of chemoreceptors on one antenna in different stages of development of *L. maderae* and *P. americana*. (References in parentheses)

Instar	*L. maderae* (1)			*P. americana* (2)			*P. americana* (3)
	surface area (mm^2)	olfactory pegs	contact pegs	surface area (mm^2)	olfactory pegs	contact pegs	axons in nerve
1	5.2	2410	1180	2.3	2700	1000	14 000
2	6.3	2500	1370	3.6	3100	1100	–
3	7.7	2910	1430	4.0	3200	1100	18 000
4	9.8	4100	1720	6.4	3800	1300	–
5	16.6	6100	2340	7.9	3900	1400	24 000
6	23.4	10 460	3070	8.9	4900	1700	–
7	42.7	15 890	4090	15.9	6500	2600	–
8				20.0	11 700	3500	47 000
9				22.3	11 800	3800	–
10				30.3	14 500	4100	–
11				34.2	16 300	6900	–
12							84 000
13							142 000
Adult ♂	43.8	27 800	5250	41.0	39 000	10 200	270 000
Adult ♀	–	–	–	37.1	22 100	7600	–

[1] Schafer (1973) [2] Schafer and Sanchez (1973) [3] Schaller (1978), includes mechanoreceptor axons

Olfactory sensilla have a variety of forms and it is difficult to equate with certainty all the forms described by different authors (see especially Altner *et al.*, 1977; Toh, 1977), but the total number in the male varies from 25 000 to 52 000, again depending on the strain and length of flagellum (Schafer and Sanchez, 1973, 1976; Schaller, 1978). Schafer (1973) believed that these sensilla contained two neurones, but the subsequent work of Altner *et al.* (1977) and Schaller (1978) indicates that many have four neurones. Hence a male with 25 000 olfactory sensilla may have nearly 100 000 olfactory neurones, while one with 40 000 sensilla would have approaching 160 000. Schaller (1978) estimates that there are 180 000 neurones from olfactory sensilla in one antenna. Female *P. americana* have 17 000 to 23 000 olfactory sensilla (Schafer and Sanchez, 1976).

On each antenna there are a few sensilla responding to moisture and temperature each of which has three or four neurones (Altner *et al.*, 1977; Yokohari, 1978). There are relatively few of these sensilla, and they are absent from many annuli (Loftus, 1976). In addition, a cold receptor cell is also sometimes present in typical olfactory sensilla associated with a hexanoic acid-sensitive cell (Sass, 1978). The total number of temperature- and moisture-sensitive neurones is small, probably only two or three hundred per antenna.

The total numbers of chemosensilla on the antennae of other cockroach species are comparable with those in *Periplaneta*, though the desert living *Arenivaga* sp. has rather fewer (Table 11). The numbers of contact chemoreceptors are generally similar in the two sexes, the excess in male *Arenivaga* and *P. japonica* being correlated with sexual dimorphism in the length of the antenna. On the other hand, olfactory sensilla are generally more abundant in males. This is true for *Blatella germanica* and *Blatta orientalis* (Lambin, 1973) as well as for most of the species shown in Table 11. In *Arenivaga* sp. the large grooved pegs, which account for the difference in numbers, are only present in adult males (Hawke and Farley, 1971) and this is also true of long wavy sensilla, on the antennae of male *Gromphadorhina* spp. (Hintze-Podufal and Otto, 1975; Slifer, 1968a) although these are not proven olfactory sensilla. In all the species examined, except for *Blaberus craniifer* and *Leucophaea maderae*, the male has at least 20 000 olfactory neurones more than the female and it is likely that these are associated with specific male sexual activity, probably the perception of a female attractant pheromone.

Numbers of antennal receptors on nymphal cockroaches have been investigated by Schafer (1973) in *L. maderae* and Schafer and Sanchez (1973) in *P. americana*, while Schaller's (1978) counts of axons in the antennal nerve provide further data on the latter. Neither species exhibits sexual dimorphism in sensillum numbers during nymphal development. In first instar nymphs of *P. americana* there are 1000 contact chemoreceptors and 2700 olfactory receptors; in *L. maderae* there are 1200 and 2400 respectively (Table 12). The number increases in each instar. In *P. americana* the proportions of contact chemoreceptors and olfactory receptors do not alter greatly, but there are 6 to 7 times as many in the final, eleventh, nymphal instar as in the first. Schaller (1978) observed a comparable increase to the twelfth instar, but a further increase to 10 times the first instar number by the final nymphal stage. In *L. maderae* olfactory receptors become relatively more abundant, increasing to 16 000 in the seventh instar, while the contact chemoreceptors only number 4000. During development there is a progressive decrease in the density of sensilla in both species, but the density of olfactory sensilla increases sharply at the final moult (Fig. 5).

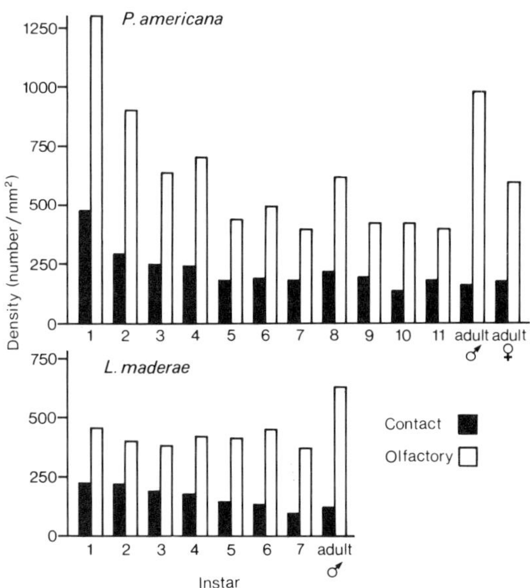

Fig. 5 Density of chemoreceptors on the antennae in different stages of development of *P. americana* and *L. maderae*. (After Schafer and Sanchez, 1976)

The functions of the antennal chemosensilla of *P. americana* have been investigated fairly extensively. In the contact chemoreceptors each cell responds to a spectrum of chemicals. Three categories are recognised with maximum responsiveness to sugars, fatty acids and alcohols respectively (Ruth, 1976).

The sexual dimorphism in numbers and, in some species, types of olfactory receptors suggests that males have some sensilla specialised for the perception of female pheromones. This is supported by the results of Boeckh *et al.*

(1970b) and Washio and Nishino (1976) who obtained a large electro-antennogram (EAG) in male *P. americana* stimulated by the female pheromone, but no response in females, and by Nishino and Takayanagi (1979) and Washio *et al.* (1976) who obtained a bigger EAG response in males stimulated with the pheromone mimic germacrene D. The results of Schafer (1977) and Sass (1980) confirm that some receptor cells in the male are specific to the female pheromone. Male *P. americana* have, depending on the strain, 8000 to 18 000 more olfactory sensilla on one antenna than the females and the evidence presented strongly suggests that these sensilla are involved in pheromone perception. Each sensillum has four neurones (Altner *et al.*, 1977) two of which respond to known components of the female pheromone so up to 36 000 neurones may be concerned in pheromone perception. Sass (1980) puts the figure even higher with 37 000 neurones responding to periplanone A and the same number to periplanone B.

The remaining antennal olfactory sensilla of the male, as well as all those of females and nymphs, are presumably involved in the perception of food odours as well as other pheromones. Cockroaches produce an aggregation pheromone and possibly also an aggression-stimulating pheromone (Persoons and Ritter, 1979) and sensilla of both sexes and all developmental stages will be involved in their perception. In the male, receptors of the aggregation pheromone are known to be on the antennae (Block and Bell, 1974) but there is no information about the numbers involved.

Sass (1976, 1978) recognises seven classes of cells responding to food odours. Each class of cell responds to a spectrum of food odours, but each class responds maximally to a different chemical. For instance, the so-called pentanol cell responds to a number of alcohols and a few other compounds, but maximum response is to pentanol, butanol and 3-methyl-butanol; the hexanol cell also responds to pentanol, but has a different range of responsiveness to other chemicals. Sass (1980) states that each antenna has 2000 cells in each class, that is about 14 000 cells responding to food odours. This is, of course, a minimum value since there may be other classes of receptor which are maximally sensitive to compounds not yet tested.

On the basis of this information it is possible to construct a balance sheet for the functions of sensory neurones in the flagellum of a male *P. americana*. This is based primarily on the work of Schaller (1978) (Table 13).

In the deutocerebrum the axons from the antennal sensilla converge on to about 250 interneurones (Ernst *et al.*, 1977). Each interneurone responds to antennal stimulation by a range of chemicals, but the ranges do not correspond with the cell classes identified by Sass (Waldow, 1975; Sass, 1976; Selzer, 1979) and the results suggest that the information relayed to higher centres in the brain is of a very generalised nature. These interneurones do not respond to cockroach odours but the axons from the pheromone receptors converge on to a separate macroglomerulus which is only present in the male (Boeckh *et al.*, 1977). Waldow (1977) found that the interneurones from this macroglomerulus were sensitive to a range of odours applied to the antennae, but suggested that specific odours might be identified from, for instance, the time course of the response. Yamada (1971), however, identified neurones which were effectively specific for sex attractant and aggregation pheromones, but found similar units in the female. He worked with unpurified pheromone and this may have affected his findings. On balance it seems likely that the specific information from the pheromone receptors is transmitted to the higher neuronal centres in an identifiable form. A distinct macroglomerulus is also present in male *Blaberus carniifer* (Chambille *et al.*, 1980). In this species about one third of the 107 glomeruli on one side of the brain are readily identifiable from their structure and position, and the overall arrangement is constant (Chambille and Rospars, 1981).

TABLE 13 Number of neurones on one antenna of a male *P. americana* responding to different sensory modalities

Effective stimulus	Number of neurones (Schaller, 1978)	Sensillum type
Mechanical	420	Campaniform and marginal sensilla
Mechanical	17 170	Terminal pore sensilla
Temperature/moisture	240	Wet, dry, cold sensilla
Chemical contact	68 700	Terminal pore sensilla
Sex attractant pheromone	74 190	Wall-pore sensilla
Possibly other pheromones[a]	74 190	Wall-pore sensilla
Food and other odours[b]	35 720	Wall-pore sensilla
Total neurones in antenna	270 630	
Total annuli	170	

[a] Neurones in same sensilla as those responding to sex attractants and not known to respond to food odours
[b] Includes a few neurones sensitive to temperature stimuli

The male mantid, *Tenodera angustipennis*, has much larger antennae than the female and the number of contact chemoreceptors is roughly proportional to the antennal length (Table 11, Slifer, 1968b). However, the male has many more probable olfactory receptors than the female with a very large number of sensilla. Slifer (1968b) concludes that the antennal sensilla are not concerned in prey selection, but presumably have a sexual function.

TABLE 14 Numbers of different types of chemoreceptors and associated neurones on the antennae of Acrididae. Numbers in parentheses show the number of neurones/sensillum on the basis of which the total number of neurones is calculated

Species	Instar	Sex	Size of antenna (area or length)	Contact chemo-receptors	Olfactory sensilla			Total neurones $\times 10^3$	Ref.
					large basiconic	small basiconic	coeloconic		
Melanoplus differentialis	Adult	♂	6 mm²	487 } (4)	2424 } (37) (large and small basiconic together)		1015 } (4)	96	1
	Adult	♀	8 mm²	702	2366		888	94	1
Melanoplus bivittatus	1	–	1.8 mm[a]	90	207 (large and small basiconic together)		115	8	4
	2	–	2.5 mm	77	224		138	9	4
	3	–	3.3 mm	152	257		167	11	4
	4	–	6.0 mm	241 } (4)	417 } (37)		287 } (4)	118	4
	5	–	8.4 mm	314	725		455	30	4
	Adult	♂	13.9 mm	394	1020		435	41	4
	Adult	♀	12.4 mm	386	920		482	38	4
Schistocerca gregaria	1	♂	2.2 mm	153	162	83	339	8	2
	1	–	2.2 mm	156	95	39	1145	5	3
	4	–	7.7 mm	442 } (4)	177 } (36)	97 } (6)	416 } (3)	10	3
	5	♂	9.6 mm	631	688	208	350	30	2
	Adult	♂	14.3 mm	769	2674	496	1149	106	2
Truxalis nasuta	Adult	♂	13.0 mm	3738 } all trichoid and basiconic sensilla			1718	–	5
	Adult	♀	17.0 mm	2330			1362	–	5

[a] Lengths of *M. bivittatus* antennae from Chapman, unpublished

[1] Slifer *et al.* (1959) [2] Chapman (unpublished) [3] Aziz (1958) [4] Riegert (1960) [5] Röhler (1906)

(*b*) *Orthoptera* Within the Orthoptera the main studies on antennal sensilla have been on Acridoidea. The total number of chemoreceptors on an antenna of either sex of *Melanoplus differentialis* is about 4000 (Slifer *et al.*, 1959). Since the most abundant sensilla have, on average, 37 neurones, the total number of chemosensitive neurones is large, almost 100 000 (Table 14). The adult male *Schistocerca gregaria* has similar numbers, about 5000 sensilla and 106 000 neurones (Chapman, unpublished), but adults of *Melanoplus bivittatus* have only 1800 sensilla and about 40 000 neurones (Riegert, 1960). Male *Truxalis nasuta* have about 5400 sensilla and females 3700 (Röhler, 1906). Ernst *et al.* (1977) record about 50 000 axons in the antennal nerve of adult male *Locusta migratoria*.

In *S. gregaria* the numbers of sensilla approximately quadruple between the first and fifth nymphal instars, in proportion to the increase in size of the antennae (Table 14). The small number of coeloconic sensilla in the fifth instar compared with those recorded by Aziz in the fourth, probably arose from some abnormality in this antenna (only one was examined). In general, the counts of Aziz (1958) are low, except for the more conspicuous contact chemoreceptors.

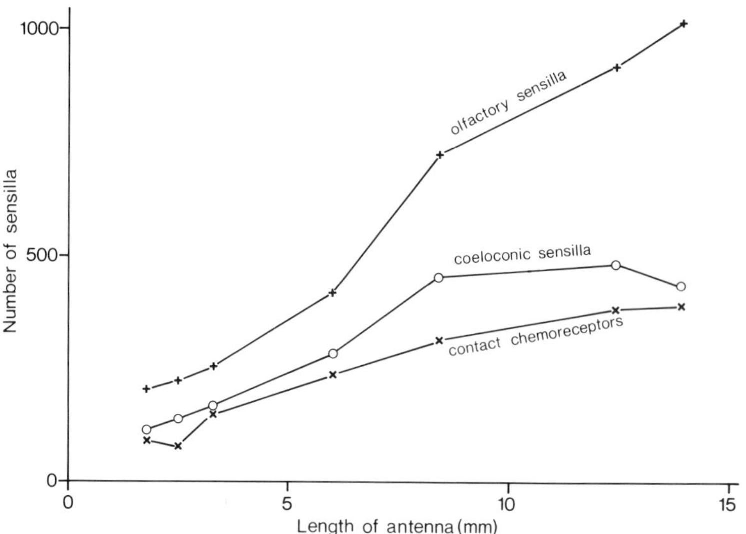

Fig. 6 Numbers of sensilla on the antenna of *Melanoplus bivittatus* in relation to antennal lengths. (After Riegert, 1960; lengths from Chapman, unpublished)

In *M. bivittatus* the numbers of coeloconic sensilla increase approximately in proportion to the length of the antenna up to the final nymphal instar, but no further increase occurs in the adults (Fig. 6) so that the density declines sharply. Numbers of contact chemoreceptors and olfactory receptors are proportional to the length of the antenna, but their density decreases progressively. In *M. differentialis*, however, the male has many more basiconic and coeloconic sensilla in relation to the size of the antennae than does the female. This is also true in *Truxalis nasuta*, while in *S. gregaria* the adult male has relatively many more basiconic sensilla than the final instar nymph.

The disproportionately large number of basiconic sensilla in adult males suggests that these sensilla are involved in sexual recognition. This might also be true of the coeloconic sensilla in *M. differentialis* and *T. nasuta*. In *S. gregaria* it is probably also true that, in all stages of development, some sensilla perceive social pheromones (see e.g. Gillett, 1975). However, in the nymphs of *M. bivittatus* it is likely that most of the sensilla and neurones are concerned in food recognition, the number of neurones increasing from 8000 to 30 000 from the first to the fifth instar. A similar number is present in nymphs of *S. gregaria*.

Extensive studies by Kafka (1970, 1971) have shown that each cell in the coeloconic pegs of *L. migratoria* responds to a range of odours of organic compounds. Different cells respond to different spectra of odours. A few coeloconic sensilla, which differ in lacking any pores (Altner and Prillinger, 1980), are sensitive to moisture and temperature (Waldow, 1970).

In the brain the antennal axons converge on to about 800 interneurones (Ernst *et al.*, 1977). The responses of these interneurones to stimulation of the antenna by odours show that peripheral nerve cells with different response characteristics converge on to single interneurones (Boeckh, 1973), so at least with respect to food odours there is no evidence for specially tuned pathways.

No comprehensive studies have been made on Tettigonioidea, but in *Neoconocephalus ensiger* the study of Slifer (1974) indicates that there must be well over 1000 chemoreceptors and four or five times as many neurones. Probably the figure is much higher than this. In *Grylloblatta campodeiformis* (Grylloblattodea), Pritchard and Scholefield (1978) observed 70 multiporous and 150 coeloconic sensilla on each antenna. They also recorded about 10 000 trichoid sensilla, but regarded them all as mechanoreceptors, although the data of Slifer (1976a) indicate that uniporous sensilla are present.

(*c*) *Other orders* The adult *Carausius morosus* has 600–700 basiconic and 50–70 coeloconic sensilla on the antenna; in addition, there are over 4000 trichoid sensilla, some of which are probably contact chemoreceptors (Weide, 1960). The density of the receptors decreases with age.

The male *Ptilocerembia* sp. (Embioptera) has antennae 9.4 mm long. These bear 294 contact chemoreceptors and 1435 probable olfactory receptors with a total of 6743 neurones (Slifer and Sekhon, 1973).

Other studies do not give accurate counts, but indicate that orthopteroid insects normally have some hundreds of sensilla on the antennae. *Forficula auricularia* has "numerous" olfactory receptors and contact chemoreceptors (Slifer, 1967) and, since the former have about 30 associated neurones, the total number of chemosensitive neurones probably runs into thousands.

Zorotypus hubbardi (Zoraptera) has antennae less than 1 mm long; nevertheless they each bear at least 200 chemoreceptors as indicated by the description of Slifer and Sekhon (1978b).

2.4.4 *Sensilla on the antennae of Hemipteroidea*

(*a*) *Homoptera* Aphids have three types of sensilla on the flagellum which are possible chemoreceptors: trichoid sensilla with a terminal pore, coeloconic sensilla, and plate organs. No information is available on numbers of neurones associated with the trichoid sensilla, but by analogy with other insects there are probably about five. These sensilla are probably contact chemoreceptors. The coeloconic sensilla of the flagellum have two neurones. Since they have no cuticular pores it is possible that they are thermoreceptors (Bromley *et al.*, 1979). Plate organs are almost certainly olfactory receptors. The neurones associated with them occur in groups of 1 to 3 and there are commonly about 10 neurones per sensillum (Bromley *et al.*, 1979).

The total number of sensilla is small (Table 15). *Megoura viciae* probably has about 10 contact chemoreceptors irrespective of the morph, while *Acyrthosiphon pisum* has about 25 in the first instar nymph and 36–40 in all the later stages of development. Coeloconic pegs are associated with plate organs in the primary rhinaria and are constant in number in nymphs. A few plate organs are present on nymphal antennae, rather more in apterous adult females, with more in later generations of virginoparae whether they are apterous or alate (see Fig. 7, and Hardie, 1980). The relatively large number in alate aphids is a feature of all the species examined (Bromley *et al.*, 1979; Dunn, 1978; Krzywiec, 1968; Nault *et al.*, 1973) and males have more than females. These figures indicate that there are about 200 chemosensitive neurones on the antennae of apterous aphids, and about 500 in alatae.

The reduction in male responsiveness to the female following ablation of the antennae in *Megoura viciae* has been interpreted as indicating that the additional plate organs (secondary rhinaria) of male aphids are the principal receptors of female pheromones (Pettersson, 1971; Marsh, 1975), while the greater sensitivity of alate females to alarm pheromones is attributed to their having a greater number of plate organs compared with apterous females (Nault *et al.*, 1973). On the other hand it is the alate form which is responsible for host-finding and work by Chapman *et al.* (1981) on *Cavariella aegopodii* indicates that attraction to specific chemicals may be important. They suggest that the secondary rhinaria are involved in the perception of these chemicals.

TABLE 15 Chemoreceptors on the flagellum of aphids

Species	Morph	Length of antenna (mm)	Numbers of sensilla			Total no. of neurones (approx.)	Ref.
			contact chemoreceptors	plate organs	coeloconic pegs		
Acyrthosiphon pisum	nymph 1	1.3	25	4	4	170	1
	nymph 2	2.0	40	4	4	240	1
	nymph 3	2.4	36	4	4	220	1
	nymph 4	3.8	38	4	4	230	1
	apterous ♀	4.8	38	9	4	230	1
	alate ♀	4.8	39	24	4	440	1
Acyrthosiphon solani	apterous ♀	–	–	3	–	–	5
Aphis fabae	apterous ♀	–	–	8	–	–	6
	alate ♀ (virginopara)	–	–	16–34	–	–	6,7
	alate ♀ (gynopara)	–	–	48	–	–	7
	alate ♂	–	–	81	–	–	6
Aphis nerii	apterous ♀	1.9	–	15	4	–	3
Megoura viciae	nymph 4 ♂	–	10	2	6	80	2
	apterous ♀	–	10	19	6	250	2
	?	3.5	12	17–42	6	370	4
	alate ♂	3.5	10	48	6	540	2
Myzus persicae	apterous ♀	–	–	2	–	–	5
Tuberolachnus salignus	?	1.7	7	22	6	260	4

[1] Shambaugh *et al.* (1978)
[2] Marsh (1975)
[3] Wang and Huber (1976)
[4] Slifer *et al.* (1964)
[5] Nault *et al.* (1973)
[6] Jones (1944)
[7] Hardie (1980)

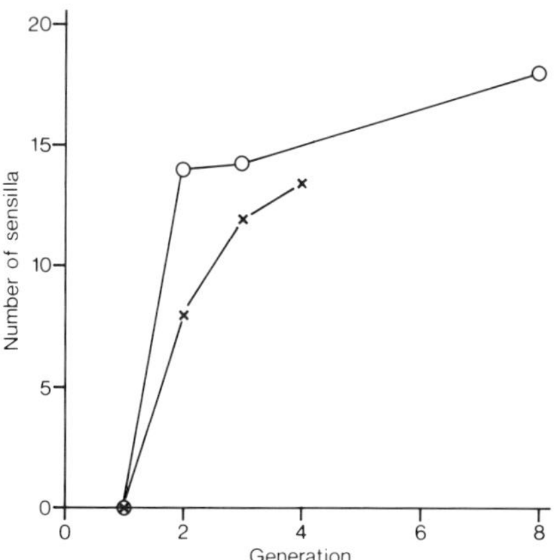

Fig. 7 Numbers of plate organs on the second flagellar annulus of successive generations of apterous *Megoura viciae* virginoparae. None is present in the first generation, the fundatrix. The two lines show the results of two separate experiments. (After Lees, 1966)

TABLE 16 Numbers of sensilla on the antennae of different stages of *Psylla pyricola*. (From Singleton-Smith *et al.*, 1978)

Type of sensillum	Numbers of sensilla in each stage					
	1st instar	2nd instar	3rd instar	4th instar	5th instar	adult
Campaniform sensilla	0	0	0	1	0	0
Trichoid sensilla	1	2	3	3	16	18
Sensilla chaetica	2	2	2	2	2	2
Basiconic sensilla	1	1	0	0	0	0
Placoid sensilla	1	2	2	3	4	4

Psyllidae have very small numbers of sensilla on the antennae; the principal chemoreceptors are plate organs. In *Psylla pyricola* there is very little change in numbers with the stage of development, although trichoid sensilla, which are presumed by Singleton-Smith *et al.* (1978) to be mechanoreceptors, increase markedly in the final nymphal instar (Table 16). Adult *Trioza erytreae* of both sexes have antennae about 1.0 mm long bearing 6 plate organs and a basiconic sensillum at the tip. Other sensilla are few in number and are presumed to be mechanoreceptors (Moran and Brown, 1973). Ablation experiments on the latter species indicate that the antennal sensilla are capable of perceiving the odours of host and non-host plants.

Amongst the Fulgoroidea most antennal sensilla are on the enlarged pedicel. Complex plaques are almost certainly olfactory receptors. The function of the trichoid sensilla is unknown, but some are possibly chemoreceptors. Numbers of plaques per antenna range from 12 in *Ossoides lineatus* to 100 or more in *Pyrops candelaria* (Marshall and Lewis, 1971) and in the latter species each plaque has between 200 and 300 neurones associated with it (C. T. Lewis and Marshall, 1970). Consequently there are at least 20 000 neurones in the plaques of each antenna and the number may be double this. In addition, there are about 200 basiconic sensilla scattered between the plaques.

(*b*) *Heteroptera* The antenna in most Heteroptera consists of a relatively elongated scape and pedicel with a flagellum of two similar annuli. Most chemoreceptors occur on the flagellum. On the antenna of an adult female *Dysdercus fasciatus* there are over 8000 sensilla (Madge, 1965), but how many of these are chemoreceptors is not known. *Lygaeus kalinii* adults have several hundred probable olfactory receptors on the terminal annulus. These each have 50–60 neurones, and in addition there are presumed contact chemoreceptors each with four chemosensitive neurones (Slifer and Sekhon, 1963). These figures suggest that there are at least 20 000 chemosensitive neurones in each antenna which is about 5 mm long. On the terminal annulus of *Oncopeltus fasciatus* there are 990 smooth pegs with about 40 neurones each and over 1000 other chemoreceptors with between two and eight neurones each (Harbach and Larsen, 1976). Thus there are about 45 000 chemosensitive neurones in each antenna.

The adult of *Cimex lectularius*, which has antennae only 1.8 mm long, has 12 contact chemoreceptors and 44 olfactory receptors on each antenna. H.Z. Levinson *et al.* (1974) note that the total number of olfactory neurones does not exceed 200 so that the total number of chemosensitive neurones is less than 250. Steinbrecht and Müller (1976) recorded about 190 axons from probable chemoreceptors in the antennal nerve. The adult *Triatoma infestans* has about 2900 probable olfactory receptors (Bernard, 1974). Each sensillum has from one to many neurones so that the total number of chemosensitive neurones in an antenna might be about 8000. In *Rhodnius prolixus*, Wigglesworth (1959) estimates that there are 4000–5000 neurones, including mechanoreceptors, in the terminal annulus.

In *C. lectularius* the olfactory sensilla are restricted to the terminal annulus in both nymphs and adults, but in adult *T. infestans* they are more widely distributed on the adult antennae than in the nymph. There are also many fewer sensilla on the antennae of nymphal *T. infestans* compared with the adults (Bernard, 1974). These differences suggest that higher numbers and wider distribution of sensilla are associated with greater mobility, and that the increase is related to host-finding, mate-finding, or both. An absence of sexual dimorphism in the antennae is recorded for *C. lectularius*, *T. infestans* and *L. kalinii.*

Electrophysiological studies of the functions of these sensilla have only been undertaken with the blood-sucking species. Trichoid sensilla, which comprise over half the total number of olfactory receptors on the antennae of *C. lectularius*, respond to the alarm pheromone of the bug (H.Z. Levinson *et al.*, 1974), while in *T. infestans* similar receptors failed to respond to a range of host odours. These odours did, however, stimulate other types of sensilla which collectively include a total of about 900 neurones.

TABLE 17 Chemoreceptors on the antennae of Thysanoptera. (After Slifer and Sekhon, 1974)

Species	Antennal length in μm	Numbers of sensilla			Total neurones
		contact chemo-receptors	olfactory receptors	coeloconic peg	
Bagnalliella yuccae	220	6 (4)[a]	11 (8–14)	1 (2)	140
Frankliniella tritici	170	4 (4)	7 (8–14)	0	90

[a] Numbers in parentheses show the approximate number of neurones per sensillum

(*c*) *Other orders* There are very few studies on the antennal sensilla of other hemipteroid insects which are all very small. Thysanoptera have less than 20 chemoreceptors on the antennae (Table 17); male *B. yuccae* have a few less olfactory sensilla than the females (Slifer and Sekhon, 1974). The total number of chemosensitive neurones is about 100 to 150.

In the bird louse (Mallophaga) *Craspedorrhynchus americanus* the antenna has about 9 contact chemoreceptor hairs, 4 coeloconic pegs and 2 olfactory pegs. The total number of chemosensitive neurones is small (Slifer, 1976b). The sucking lice (Anoplura) have comparably low numbers. *Pediculus humanus*, with antennae about 400 μm long, has two multiporous tuft organs, two pore organs and ten chemoreceptor pegs as well as a multiporous area at the tip of the antenna (Slifer and Sekhon, 1980), while *Polyplax serrata* has only 13 probable chemoreceptors (Miller, 1970). None of these sensilla has very large numbers of neurones so that the total number of chemosensitive neurones per antenna is almost certainly less than 50.

2.4.5 *Sensilla on the antennae of adult Endopterygota*

(*a*) *Lepidoptera* The sensilla on the antennae of Lepidoptera generally fall into the categories shown in Table 18, though different authors have used different names. Apart from the trichoid sensilla of some species, notably *Bombyx mori*, little is known about the fine structure or physiology of these sensilla and their functions are consequently not known with certainty. However, it is probable that the sensilla chaetica are contact chemoreceptors (Slifer, 1979a), the trichoid and basiconic sensilla are certainly olfactory receptors (Schneider *et al.*, 1964; den Otter, 1977; Kaissling *et al.*, 1978) and the other sensilla probably are also.

Most species examined have about 300 sensilla chaetica in both sexes, irrespective of the length of the antenna (Table 18). If these sensilla are similar to contact chemoreceptors in other insects, they probably have four chemosensitive neurones, with a total of about 1200 to 1500 per antenna. Long trichoid sensilla are the most abundant sensillum type in males of all species studied except for the two species of butterfly, *Danaus gilippus* and *Colias eurytheme*. Commonly there are over 5000, while in *B. mori* there are 17 000 (Steinbrecht, 1970) and in *Antheraea polyphemus* about 55 000 (Boeckh *et al.*, 1960). Jefferson *et al.* (1970) found only a small difference between the sexes of *H. zea*, but Callahan (1969) records that long trichoid sensilla are completely absent from the female. This is certainly the case in female *Lymantria dispar*, *L. monacha* (Schneider *et al.*, 1977), *A. polyphemus* (Boeckh *et al.*, 1960) and *Manduca sexta* (Sanes and Hildebrand, 1976). The long trichoid sensilla commonly have two neurones, as in *B. mori* (Steinbrecht, 1973) and *Adoxophyes orana* (den Otter *et al.*, 1978), but have three in *Argyrotaenia velutinana* (O'Connell, 1972) and possibly five in *C. fumiferana* (Albert *et al.*, 1974). Shorter trichoid sensilla are also relatively abundant with roughly similar numbers in the two sexes. They vary from about 1200 in female *O. nubilalis* to 10 000 in female *B. mori* in which each sensillum has from one to three neurones (Steinbrecht, 1970). Other types of sensilla are far less numerous with no obvious sexual dimorphism. There are commonly between 100 and 200 sensilla auricillica, less than 100 styloconic sensilla and about 500 coeloconic sensilla, though the last are totally absent from *H. anisocentra* (Slifer, 1979a).

There is no correlation of numbers of contact chemoreceptors with antennal length; on the contrary, the overall similarity of numbers of sensilla chaetica in all the species studied is striking. The very large number of trichoid sensilla in the male *B. mori* and *A. polyphemus* is related to the complexity of the antennae in these species. In all the other species studied the antennae of the two sexes are similar in appearance.

TABLE 18 Numbers of different types of sensilla on the antennae of Lepidoptera

Species	Antennal length (mm)	Numbers of sensilla			
		chaetica		long trichoid	
		♂	♀	♂	♀
Noctuidae					
Trichoplusia ni	13.0	295	303	6469[a]	5448[a]
Heliothis zea	13.0	405	435	8604[a]	7532[a]
H. zea		492	492	2592	0
Prodenia ornithogalli	12.6	396	383	9060[b]	7555[b]
Spodoptera exigua	8.6	274	281	4969[b]	4257[b]
Sphingidae					
Manduca sexta	20.0		–	98 455[b]	–
Bombycidae					
Bombyx mori	8.0	120	120	17 000	6000
Saturniidae					
Antheraea polyphemus	18.0, 13.0	244	278	55 000	0
Danaidae					
Danaus gilippus	15.0	220[c]	220[c]	2500	2500
Pieridae					
Colias eurytheme	7.0	400	400	1800	1800
Tortricidae					
Choristoneura fumiferana	4.9, 5.1	136	172	2346[b]	1375[b]
Cydia nigricana	3.5	327	318	12 180	7200
Pyralidae					
Ostrinia nubilalis	7.5	300	300	5716	3640
Gelechiidae					
Pectinophora gossypiella	–	172	204	1979	1468
Glyphipterygidae					
Homadaula anisocentra	3.5	365	365	700	700

[a] Values estimated from data in Jefferson *et al.* (1970); [b] Types of trichoid and basiconic sensilla

[1] Jefferson *et al.* (1970)
[2] Callahan (1969)
[3] Wall (1978)
[4] Cornford *et al.* (1973)
[5] Slifer (1979a)
[6] Steinbrecht (1970)
[7] Schneider and Kaissling (1957)
[8] Boeckh *et al.* (1960)

Numbers of sensilla								
short trichoid and basiconic		auricillica		coeloconic		styloconic		Ref.
♂	♀	♂	♀	♂	♀	♂	♀	
2772	2684	130	134	381	411	69	69	1
3023	3710	128	171	690	717	76	76	1
10 084	12 382	328	328	492	492	82	82	2
–	–	175	208	724	702	70	70	1
–	–	123	87	301	286	55	55	1
–	–	0	–	1889	–	76	–	12
7500	10 000	0	0	800	800	90	90	6,7,13
10 000	12 000	0	0	1380	720	67	76	8
12 700	12 700	0	0	>600	>600	present		9
5000	5000	0	0	0	0	?	?	11
–	–	occasionally present		259	177	36	30	10
4060	3600	>200	>200	570	514	46	45	3
1563	1212	200	200	220	290	66	66	4
782	933	occasional	–	146	151	33	33	14
?	?	100	100	0	0	17	17	5

not differentiated; [c] Uniporous sensilla, estimated from Myers (1968)

[9] Myers (1968)
[10] Albert and Seabrook (1973)
[11] Grula and Taylor (1980)
[12] Sanes and Hildebrand (1976)
[13] Schneider and Kaissling (1957a)
[14] Yin and Li (1980)

Largely as a consequence of the big differences in numbers of trichoid and basiconic sensilla there is a considerable range of variation in the number of chemosensitive neurones in an antenna. The female of *C. fumiferana* has about 10 000 neurones, compared with 13 000 in the male which has slightly shorter antennae. In the two species of Bombycoidea, in which the males have strongly pectinate antennae, the total number of neurones is much greater and there is a big difference between the sexes: in *B. mori* 52 000 and 35 000 in male and female respectively, and in *A. polyphemus* 138 000 and 64 000 respectively. The two sexes of *D. gilippus* have about 47 000 neurones, while the male *Manduca sexta* has over 260 000 (Sanes and Hildebrand, 1976).

The neurones which are sensitive to the attractant pheromone of the female are in the long trichoid sensilla of male moths. This is true in *B. mori* (Kaissling *et al.*, 1978; Kaissling, 1979), *Antheraea polyphemus* and *A. pernyi* (Boeckh and Boeckh, 1979; Kaissling, 1979), *Choristoneura fumiferana* (Albert *et al.*, 1974), *Adoxophyes orana* (den Otter, 1977), *Lymantria dispar* and *L. monacha* (Schneider *et al.*, 1977), *Argyrotaenia velutinana* (O'Connell, 1975) and *Yponomeuta* spp. (van der Pers and den Otter, 1978). These neurones are relatively insensitive to other odours so that for all practical purposes they are specific pheromone receptors. In *Argyrotaenia velutinana*, *Adoxophyes orana*, *B. mori* and *Antheraea* spp., where the female pheromone contains two component chemicals, each sensillum contains separate cells which respond selectively to each of the components (O'Connell, 1975; den Otter, 1977; Kaissling, 1979; Boeckh and Boeckh, 1979). Hence in each antenna of *B. mori* about 17 000 neurones are specific to each of the two components; in *A. polyphemus* about 55 000 neurones are tuned to each component. By contrast in *C. fumiferana* there are only about 300 trichoid sensilla of the type responding to the female pheromone. Each sensillum has five neurones, but not all of them respond to the pheromone (Albert *et al.*, 1974) so there are almost certainly less than 1000 pheromone-sensitive cells per antenna. A third cell in the trichoid sensilla of male *A. pernyi* respond to a number of fruit and flower odours (Schneider *et al.*, 1964).

The males of many species of Lepidoptera produce pheromones which elicit sexual behaviour in the female (see e.g. Ellis and Brimacombe, 1980). These pheromones are perceived by sensilla on the antennae, but it is not known which sensilla are involved. In *Danaus gilippus* and *Pseudaletia unipuncta* the male pheromone produces a similar EAG in both sexes and even in closely related species (Schneider and Seibt, 1969; Grant, 1971a; Grant *et al.*, 1972). This may indicate that these pheromones are perceived by unspecialised odour cells.

Cells in the basiconic sensilla on the antennae of *A. pernyi* respond to a wide range of odours, but no two cells are alike in their ranges of response (Schneider *et al.*, 1964). Behan and Schoonhoven (1978) and den Otter *et al.*

(1980), recording largely from basiconic (=small trichoid) sensilla on the antenna of *Pieris brassicae*, showed that the neurones responded to a wide range of leaf and flower odours and found no evidence of tuning to specific chemicals. However, Behan and Schoonhoven (1978) found that only 10% of the sensilla they recorded from responded to the plant odours tested. Electro-antennogram studies of several species show that antennal receptors in both sexes respond to odours of both host and non-host plants (Grant, 1971b; Schoonhoven, 1974; Behan and Schoonhoven, 1978; den Otter *et al.*, 1978). In *P. brassicae* some cells respond to the odour of eggs of the same species, but these cells are not specifically tuned to the odour (Behan and Schoonhoven, 1978).

Hence, apart from the neurones in the male which are selective for the female attractant pheromone, the olfactory cells in the antennae of Lepidoptera appear to be generalised odour receptors mainly involved with food-finding and oviposition site selection. In the large Bombycoidea there are more of these generalised neurones in the females than in males, but in *C. fumiferana* the converse is true (Table 19).

TABLE 19 Numbers of pheromone specialist, odour generalist and humidity/temperature receptor cells in the antennae of some Lepidoptera

Species	Sex	Pheromone specialist cells	Odour generalist cells	Humidity/ temperature receptor cells	Ref.
Bombyx mori	♂	34 000	15 000	270	1,2,3,4
	♀	0	32 000	270	
Antheraea polyphemus	♂	110 000	20 000	201	4,5,6,7
	♀	0	24 000	228	
Choristoneura fumiferana	♂	~1000	10 000	108	8
	♀	0	6500	90	

[1] Schneider and Kaissling (1957)
[2] Steinbrecht (1970)
[3] Kaissling *et al.* (1978)
[4] Kaissling (1979)
[5] Boeckh *et al.* (1960)
[6] Schneider *et al.* (1964)
[7] Boeckh and Boeckh (1979)
[8] Albert *et al.* (1974)

Becker (cited in Altner and Prillinger, 1980) found the styloconic sensilla on the antennae of *Mamestra brassicae* to contain neurones sensitive to temperature and humidity. Small numbers of styloconic sensilla are present in both sexes of all the moths studied and the number is correlated with antennal length (Tables 18, 19). It is probable that they are temperature and humidity receptors in all the species.

TABLE 20 Antennal chemoreceptors of different castes of *Apis mellifera* (from Esslen and Kaissling, 1976)

Type	No. of neurones/ sensillum	Drone		Worker		Queen	
		no. of sensilla	no. of neurones	no. of sensilla	no. of neurones	no. of sensilla	no. of neurones
Plate organs	18	19 035	334 692	2672	46 818	1388	24 984
Trichoid sensilla (type A)	5–10	377	2793	2048	16 035	No data	
Trichoid sensilla (type D)	5–7	130	780	100	696		
Basiconic sensilla	?	0	0	61	?		
Ampullaceous sensilla	1	81	81	111	111		
Coeloconic sensilla	1	83	83	63	63		

In the male of *Antheraea pernyi* and of *Manduca sexta* the axons of the cells responding to the female pheromone converge on to a "macroglomerulus" which is separate from the normal glomerular structure associated with other antennal axons (Boeckh and Boeckh, 1979; Hildebrand *et al.*, 1980). The interneurones in the macroglomerulus of *A. pernyi* which respond to female pheromones do not respond to other chemicals while no response to these pheromones was observed in other parts of the deutocerebrum. In *M. sexta* separate interneurones connect the macroglomerulus to the mushroom bodies. This indicates that information concerning the female pheromones is relayed to the brain via a specialised, probably exclusive pathway separate from all other olfactory information from the antennae.

(*b*) *Hymenoptera* The most comprehensive studies on sensillum numbers in Hymenoptera are on *Apis mellifera* and the results are summarised by Esslen and Kaissling (1976). Six types of sensilla are almost certainly chemoreceptors (Table 20), the plate organs, type A trichoid sensilla and basiconic sensilla probably having an olfactory function while the type D trichoid sensilla are contact chemoreceptors. Apart from the plate organs, which are more or less evenly distributed along the flagellum, the numbers tend to increase distally.

The flagellum of drone bees is longer than that of workers, about 7.5 mm compared with 5.5 mm, and the numbers of type D trichoid sensilla and coeloconic sensilla are proportional to the length of the flagellum. Plate organs, however, are much more abundant in the drones than in workers, while the converse is true of type A trichoid sensilla.

Plate organs have about 18 neurones per sensillum and at least some of the cells are specialised receptors of queen substance, *trans*-9-oxo-2-decanoic acid, in all three castes, drone, worker and queen (Kaissling and Renner, 1968). Drones are much more sensitive to queen substance than workers or queens (Adler *et al.*, 1973), probably indicating a much bigger number of specialised neurones. Kaissling and Renner (1968) found five cells responding to queen substance for every one responding to the Nasanov pheromone in drones, while in workers and queens the ratio was 1 : 3. Other cells in the plate organs respond to a wide range of odours (Lacher, 1964; Vareschi, 1971). Each cell responds to a number of chemicals, but different cells may be grouped into some seven categories with little overlap. Cells in Vareschi's class I responded to *trans*-9-oxo-2-decanoic acid among a number of other chemicals and he did not record cells tuned specifically to this substance even though most of his results were obtained with drones. Vareschi (1971) considers that they cannot be regarded either as generalist or as specialist cells.

The major activity of drones is that of finding a mate on the nuptial flight; they do not forage or find food for themselves. The dominance of sexual behaviour, together with the electrophysiological data, may be taken as evidence that a very large number of the 330 000 neurones in the plate organs of the drone are specialised pheromone receptors. The workers have a total of about 60 000 olfactory neurones in the plate organs and type A trichoid sensilla, and odours play a major part in governing their social behaviour (e.g. Ferguson and Free, 1981) as well as in foraging for food. Hence it may be that a proportion of these cells is tuned to receive these socially significant odours and the number capable of responding to flower odours may be less than 60 000. On the other hand, floral odours are complex mixtures of chemicals which would probably stimulate cells in more than one of the categories defined by Vareschi (1971) so that a relatively large number of cells is likely to be involved in food selection.

Ablation experiments by Kuwabara and Takeda (1956) suggest that the sensilla ampullacea are hygroreceptors and later electrophysiological studies show that they and the coeloconic sensilla are receptors of CO_2, humidity and temperature (Lacher, 1964). The total number of cells involved in these functions is about 160 in drones and 170 in workers.

TABLE 21 Antennal chemoreceptors of different castes of *Lasius fuliginosus*. (After Dumpert, 1972a)

Type of sensillum	No. of neurones/ sensillum	Male		Worker	
		no. of sensilla	no. of neurones	no. of sensilla	no. of neurones
Sensilla chaetica	1	1863	1863	2404	2404
Trichoid sensilla	3	35	105	131	393
Curved trichoid sensilla	20–30	278	6950	298	7450
Basiconic sensilla	10	25	250	92	920
Coeloconic sensilla	3	16	48	18	54
Ampullaceous sensilla	1	13	13	18	18

The only complete survey of sensilla on the antennae of ants is on *Lasius fuliginosus* (Dumpert, 1972a). Sensilla chaetica comprise the bulk of the sensillum population; they are probably contact chemoreceptors with a single chemosensitive neurone. Trichoid and basiconic sensilla are more abundant in the worker (Table 21) despite the fact that the antennae are slightly shorter than in the male. Coeloconic and ampullaceous sensilla are present in small numbers and this was also recorded in a number of other ant species by Jaisson (1969) and Masson and Friggi (1971a). The number ranged from 7 coeloconic and 11 ampullaceous sensilla in *Aphaenagaster gibbosa* to 24 and

22 respectively in *Camponotus vagus*. The total numbers of chemosensitive neurones in males and workers of the latter species are about 9000 and 11 000 respectively.

The neurones associated with curved trichoid sensilla on the antennae of *L. fuliginosus* responded to a wide range of organic compounds including the alarm substances of other ant species. Different cells had different reaction spectra and some were relatively specific for undecane, the ant's own alarm substance (Dumpert, 1972b). Although relatively few of these sensilla are present on the antennae they are associated with about 7000 neurones in both male and worker (Table 21).

On the basis of ablation experiments, Jaisson (1969) considered the sensilla ampullacea of *Myrmica laevinodis* to be important in trophallaxis. The electrophysiological results of Masson and Friggi (1971b) indicate responses to a number of odours, but the placement of the electrodes was, perhaps, too imprecise to be sure of the sensilla involved. By analogy with other insects the sensilla ampullacea and coeloconica are likely to be temperature, humidity and CO_2 receptors. Dumpert (1972a) recorded a neurone responding to CO_2 in the coeloconic sensilla of *L. fuliginosus*.

TABLE 22 Numbers of plate organs on the antennae of various parasitic Hymenoptera

Species	Sex	Number of plate organs	Neurones/ sensillum	Total neurones	Ref.
Braconidae					
Aphidius smithii	♂	401	37	14 837	Borden *et al.* (1978b)
	♀	140	37	5180	
Coeloides brunneri	–	~650	13?	8450	Richerson *et al.* (1972)
Ichneumonidae					
Itoplectis conquisitor	♂,♀	700	27	18 900	Borden *et al.* (1978a)
Pteromalidae					
Nasonia vitripennis	♂	43	?	–	Slifer (1969)
	♀	81	?	–	
Trichogrammatidae					
Trichogramma evanescens	♀	5	?	–	Voegelé *et al.* (1975)

Within the deutocerebrum the fibres from the antennal nerve separate into bundles which run to the glomeruli. In *C. vagus* the glomeruli are all of similar size, but those in *Megaponera caffraria* vary in size, the largest being about four times as big as the smallest. This led Masson (1972) and Masson and Strambi (1977) to suggest that different glomeruli are associated with different sensory modalities, some with mechanoreception, others with olfaction. It also suggests the possibility of different glomeruli being associated with the perception of specific chemicals.

Some fragmentary information on the numbers of plate organs in a number of parasitic Hymenoptera is given in Table 22. The male *Aphidius smithi* has many more plate organs than the female, indicating a function in mate finding, but in *Nasonia vitripennis*, with a much smaller number, the female has more. These sensilla are generally associated with large numbers of neurones; in chalcids there are over 50 per sensillum (Barlin and Vinson, 1981). Only in Trichogrammatidae have complete counts of sensilla been made (Voegelé *et al.*, 1975). In *Trichogramma evanescens* there are five plate organs, 7 coeloconic sensilla, 36 uniporous sensilla, and two other probable chemoreceptors – 50 in all.

(*c*) *Diptera* Several comprehensive studies have been made of the chemoreceptors on the antennae of mosquitoes (Table 23). In Culicini, only trichoid and basiconic sensilla are present, but Anophelini have, in addition, coeloconic sensilla with thin-walled, presumably olfactory, pegs. These chemoreceptors are present on all the flagellar annuli in females, but in males are restricted to the two terminal annuli. Both sexes possess a small number of thick-walled coeloconic sensilla (Boo and McIver, 1975) which probably function as temperature and humidity receptors (Davis and Sokolove, 1975).

The number of sensilla varies with the genus and, within the Culicini, is correlated with antennal length. Males have fewer sensilla than females except for *Wyeomyia smithii* in which the female is not known to take blood meals (McIver and Hudson, 1972). The implication is that a large number of the female sensilla are concerned with host finding. Four or five neurones are associated with each of the trichoid sensilla (Steward and Atwood, 1963), while McIver (1974) shows that there are usually three in the basiconic sensilla. The total number of chemosensitive neurones on one antenna thus ranges from about 2000 to 6000 in female mosquitoes and is between one and two thousand in males.

Trichoid sensilla on the antennae respond to a range of odours (Lacher, 1967, 1971), but different morphological types may exhibit some degree of specialisation since Davis (1976) has shown that the blunt-tipped trichoid sensilla on the antennae of female *Aedes aegypti* respond selectively to a range of odours associated with oviposition sites. They are, in general, insensitive to a range of host-related or plant-related compounds, while other types of sensilla are insensitive to the oviposition site odours.

TABLE 23 Chemoreceptors on the antennae of mosquitoes

Species	Sex	Length (mm)	No. of chemoreceptors			Total chemosensitive neurones	No. of coeloconic (thick walled)	Ref.
			trichoid	basiconic	coeloconic (thin walled)			
Culicini								
Wyeomyia smithii	♂	1.1	189	30	0	1035	7	McIver and Hudson (1972)
	♀	1.1	218	35	0	1195	9	McIver and Hudson (1972)
Wyeomyia aporonoma	♀	1.7	360	70	0	2010	9	McIver and Hudson (1972)
Aedes aegypti	♂	1.6	184	35	0	1025	9	McIver (1971)
	♀	2.0	788	105	0	4255	6	Ismail (1964)
Aedes atropalpus	♂	–	208	30	0	1130	6	McIver (1971)
Culex fatigans	♀	2.2	1006	233	0	5729	6	Ismail (1964)
Culex pipiens	♂	–	249	90	0	1515	11	McIver (1971)
	♀	2.3	935	265	0	5470	6	Ismail (1964)
Culex restuans	♀	2.4	902	348	0	5552	7	McIver (1970)
Culex tarsalis	♀	2.2	624	272	0	3936	7	McIver (1969)
Culex territans	♂	–	223	170	0	1625	10	McIver (1971)
	♀	2.0	537	275	0	3510	7	McIver (1970)
Anophelini								
Anopheles gambiae	♀	1.4	685	84	33	3677[b]	7	Ismail (1964)
Anopheles maculipennis	♂	2.0 (0.8[a])	300	28	11	1584[b]	6	Ismail (1962)
	♀	1.6	618	62	28	3276[b]	6	Ismail (1962)
Anopheles stephensi	♀	1.3	708	107	31	3861[b]	7	Ismail (1964)

[a] Length of two terminal (sensillum bearing) annuli; [b] Not including thin walled coeloconic sensilla

Davis and Sokolove (1976) conclude that two neurones in the basiconic sensilla of *Aedes aegypti* are specific for lactic acid since they obtained no response with most other chemicals tested. Lacher (1967) obtained responses to several n-fatty acids and to citral from the same class of sensillum. The tuning of these neurones is clearly not very sharply defined. Kellogg (1970) has suggested that the basiconic sensilla contained receptors for water vapour and carbon dioxide, but Davis and Sokolove (1976) argue that the response to water vapour is non-specific.

The only complete study of antennal sensilla in Ceratopogonidae is of the female *Culicoides furens* by Chu-Wang *et al.* (1975). In this insect the antenna is 0.58 mm long with approximately 40 blunt-tipped trichoid sensilla, each with 2 to 12 neurones, and 85 sharp-tipped trichoid sensilla with one, two or three neurones. In addition there are about 40 basiconic sensilla with three or four neurones and 10 coeloconic sensilla with five neurones each. Hence there is a total of about 600 chemosensitive neurones.

In other species of Ceratopogonidae the number of coeloconic sensilla, which have previously been called pits or campaniform sensilla, is between 7 and 10 in eight species known to feed on mammals and between 17 and 35 in four species feeding on birds (Braverman and Hulley, 1979). The implication is that these sensilla are concerned in host-finding, but, at least in some species, similar numbers are present on the antennae of males (Cornet *et al.*, 1974) which do not feed on blood. The number of basiconic sensilla in the females of twelve species investigated by Braverman and Hulley (1979) varied from 14 to 36, and the total number of chemosensitive neurones probably does not vary greatly from 600.

On the antennae of male *Contarinia sorghicola*, as in other Cecidomyiidae, there are two circumfila per annulus. These sense organs are derived from a total of about 280 multiporous sensilla, probably with only a single neurone each (Slifer and Sekhon, 1971). In addition there are 20–30 multiporous sensilla, with five neurones each, and 10–20 "stubby pegs" which may be chemoreceptors. The total number of chemosensitive neurones in one antenna is almost certainly less than 500.

In Cyclorrhapha, chemoreceptors are restricted to the funiculus, the basal annulus of the flagellum (Bay and Pitts, 1976; Greenberg and Ash, 1972; C. T. Lewis, 1971). Some are present on the surface of the funiculus, others are sunk into pits below the surface. In the females of three species of *Hippelates* (Chloropidae) there are about 500 chemoreceptors on the surface, with a total of about 1500 neurones, and from 14 to 38 sensilla in pits, with 42 to 114 neurones (Dubose and Axtell, 1968). Thus the total number of chemosensitive neurones on one antenna is about 1600.

Stomoxys calcitrans has about 4800 chemoreceptors, with over 12 000 neurones, on the surface of the funiculus. Other chemoreceptors are present in two pits (C. T. Lewis, 1971). In other calypterate Diptera the number of antennal pits varies from two in male *Musca autumnalis* (Bay and Pitts, 1976) to 261 in female *Sarcophaga argyrostoma* (Slifer and Sekhon, 1964). This is not simply a matter of antennal size since the male *S. argyrostoma* has only 52 pits and *Glossina* spp., with similar sized antennae, have only three. In general, female flies have more antennal pits than males. The number of sensilla per pit varies from six in *M. autumnalis* to 300 in *S. argyrostoma* and is related to the size of the pit. These sensilla are variable in form, but generally have only one or two neurones each so in *Musca autumnalis*, for example, the total number of neurones associated with the pit sensilla in the male is less than 50 compared with about 250 in the female (Bay and Pitts, 1976), while in *S. argyrostoma* the number probably runs to several thousands (Slifer and Sekhon, 1964). There are about 7300 axons, most of them presumably from olfactory sensilla, in the antennal nerve of *Musca domestica* (Strausfeld, 1976).

Liebermann (1925) carried out an extensive survey of the sensilla on the antennae of cyclorrhaphous Diptera. Two of the species which he studied have also been the subject of more recent investigations: *Musca autumnalis* (= *corvina*) by Bay and Pitts (1976) and *Stomoxys calcitrans* by C. T. Lewis (1971). Unfortunately the results differ widely, probably because of the limitations of the equipment available to Liebermann, and detailed consideration of his results seems unjustified.

Boeckh *et al.* (1965) and Kaib (1974) have investigated some antennal sensilla of *Calliphora* spp. They found that each cell responded to a limited range of chemicals. Some cells responded to meat odours, others to flower edours, but they were not highly specialised in the perception of specific compounds. In *Lucilia sericata* each member of a series of alkanes and alkyl alcohols stimulates cells of the pit sensilla, but sensitivity varies over the range with octane and heptyl alcohol being the most stimulating of these series (Kay, 1971).

Within the brain of *Musca domestica* the axons of the antennal nerve disperse to the glomeruli of the antennal lobe (Strausfeld, 1976). Four different types of glomeruli are recognisable on the basis of the arrangement of the axon terminals. A few "long olfactory fibres" byepass the antennal lobe and run directly to the calyces. In *Calliphora vicina* and *Phormia regina* some of the antennal axons pass to the antennal lobe of the contralateral side (Boeckh *et al.*, 1970a; Dethier, 1976).

(*d*) *Coleoptera* The chemoreceptors of a number of beetles with different feeding habits have been examined (Table 24). The Carabidae, *Aphaenops erypticola* and *Nebria brevicollis*, are predaceous; they have less than 1000 olfactory sensilla on each antenna. *Necrophorus vespilloides* feeds on

TABLE 24 Numbers of chemoreceptors on one antenna of various Coleoptera

Species	Sex	Length (mm)	Olfactory		Contact chemoreceptor		Hygroreceptors	Ref.
			sensilla	neurones	sensilla	neurones		
Carabidae								
Nebria brevicollis	♂,♀	0.6[a]	812	1624	244	976	?	1
Aphaenops crypticola	♂	3.5	902	–	15	–	107	13
	♀	3.5	932	–	15	–	107	13
Silphidae								
Necrophorus vespilloides	♂,♀	1.0[a]	23 500	30 000	none?		?	2
Tenebrionidae								
Tenebrio molitor	?	3.5	3188	6376	280	1289	76	3
Cerambycidae								
Monochamus notatus	♂	69.0	4200	16 800	365	2555	?	4
	♀	29.0	3200	12 800	321	2247	?	4
M. scutellatus	♂	43.0	3400	13 600	400	3200	?	4
	♀	26.0	2500	10 000	300	2400	?	4
Curculionidae								
Curculio caryae	♂	–	1421	2000[b]	24	120[b]	10	5
	♀	–	1452	2000[b]	24	120[b]	9	5
Hylobius abietis	♂,♀	1.0[a]	6000	8500	40	200	?	6
Hypera meles	♂	0.2[a]	440	600[b]	6	30[b]	17	7
	♀	0.2[a]	481	700[b]	6	30[b]	14	7

Scolytidae								
Dendroctonus frontalis	♂	0.1[a]	780	1100	none?		10	8
D. ponderosae	♀	–	430	860	60	240	15	12
Ips confusus	♂	0.2[a]	394	1376	none?		?	9,10
	♀	0.2[a]	429	1716				9,10
Scolytus multistriatus	?	0.4[a]	315	730	240	1440[c]	?	11
Trypodendron lineatum	♂	0.3[a]	612	1224	22	132	?	14
	♀	0.3[a]	581	1162	27	162		14

[a] Length of club only; [b] Numbers of neurones based on Mustaparta (1973); [c] Based on six neurones per sensillum

[1] Daly and Ryan (1979)
[2] Waldow (1973)
[3] Harbach and Larsen (1977)
[4] Dyer and Seabrook (1975)
[5] Hatfield *et al.* (1976)
[6] Mustaparta (1973)
[7] Smith *et al.* (1976)
[8] Dickens and Payne (1978)
[9] Borden (1968)
[10] Borden and Wood (1966)
[11] Borg and Norris (1971)
[12] Whitehead (1981)
[13] Juberthie and Massoud (1977)
[14] Moeck (1968)

carrion and flowers and oviposits at carrion; it has 23 000 olfactory sensilla on each antenna. Similar relatively large numbers of olfactory sensilla are present on the lamellate clubs of Scarabaeidae: 30 000 in male *Melolontha melolontha*, 24 000 in *Bolbelasmus* sp. 15 000 in *Potosia* sp. (Meinecke, 1975). These insects are relatively unspecific flower and leaf feeders, though their oviposition habits may be more specific. The larvae of *Potosia* species feed in the nests of *Formica* spp. The Curculionidae are also phytophagous, but are much more specific in their host plants; they have 400–6000 olfactory sensilla on the antennal club. They are mostly small insects, but the club of *Hylobius abietis* is the same length as that of *N. vespilloides* which has four times as many sensilla. The Scolytidae are all wood-borers with limited host ranges; they have 300–800 olfactory sensilla on one antenna.

Sexual dimorphism in the numbers of sensilla is negligible in the species studied except for *Monochamus* spp. Here the males have many more olfactory sensilla than the females of the same species, although relative to the lengths of the antennae the females have more olfactory sensilla and contact chemoreceptors.

The number of contact chemoreceptors is small especially in the Curculionidae and Scolytidae (Table 24). *Scolytus multistriatus* is apparently exceptional in this group. It has a relatively large number of trichoid sensilla and the figures of Borg and Norris (1971; fig. 18) show them to have the structure of contact chemoreceptors.

In some species a number of small sensilla, called grooved pegs, fluted sensilla, styloconic sensilla or, in *A. crypticola*, coeloconic pegs, have been described. By analogy with similar-looking sensilla in other insects these may be hygroreceptors and perhaps also thermoreceptors (Altner and Prillinger, 1980). The larger number in *A. crypticola* may be related to the cave-dwelling habit of this species.

Recordings from single cells in *Ips pini* and *I. paraconfusus* (Mustaparta *et al.*, 1977, 1979) and in *Hylobius abietis* (Mustaparta, 1975) indicate that some cells in both sexes of these species are sharply tuned to their aggregation pheromone. Electroantennogram studies on *Dendroctonus frontalis* (Dickens and Payne, 1977) suggest similar tuning in this species with a greater response in the males. EAG studies of *Trogoderma granarium* show that the males have tuned receptors to female sex attractant pheromone (A.R. Levinson *et al.*, 1978). The bark beetles also have some cells which respond only to the odours of host trees, but they are much less sensitive than the "pheromone cells".

The most comprehensive examination of the olfactory response of beetles has been on *H. abietis* by Mustaparta (1975) who used a wide range of pheromones, host odours and other compounds. Her results indicate that each antenna has about 300 specialist pheromone cells, about the same number of cells responding to a limited number of host odours, and about 5000 cells responding to a wide range of chemicals (Table 25).

In a more limited study of *Ips pini*, Mustaparta *et al.* (1979) investigated the responses of 95 olfactory cells to a number of pheromones and related compounds. Almost 50% were specialist cells for the attractant pheromone, ipsdienol, with a sensitivity 100 to 1000 times greater than that of other cells. Some cells were also specialised receptors of ipsenol, which inhibits the response to ipsdienol, while the remainder exhibited a more general response. Assuming that *I. pini* has similar numbers of receptors to *I. paraconfusus*, the proportions observed by Mustaparta *et al.* indicate about 600 ipsdienol specialist cells, 200 ipsenol specialist cells and 600 other olfactory cells. In *I. paraconfusus*, where both ipsdienol and ipsenol are attractant, the cells sensitive to these two substances are present in equal numbers (Mustaparta, 1979). On the basis of experiments in which they blocked acceptor sites on the antennae, Dickens and Payne (1977) suggest that in *Dendroctonus frontalis* there is a lack of labelled lines except for the pheromone frontalin.

TABLE 25 Specificity of olfactory cells on one antenna of *Hylobius abietis*. (After Mustaparta, 1973, 1975)

	Trichoid sensilla			Basiconic sensilla	
	pheromone specialist	host odour specialist	limited spectrum	limited spectrum	broad spectrum
Number of cells tested	6	6	6	15	30
Estimated number on antenna	300	300	300	1700	3300

The responses of chemoreceptors to host odours have also been investigated in other species of beetle. The cells in these cases respond to ranges of chemicals (Boeckh, 1962; Waldow, 1973; Ma and Visser, 1978) though in *Monochamus notatus* tuned cells might exist (Dyer and Seabrook, 1978). Only 25 out of about 100 cells on the antennae of *Leptinotarsa decemlineata* responded with excitation to general plant odours (Ma and Visser, 1978). Eight of these cells responded to a wide range of chemicals, while the remainder exhibited much narrower spectra. Two cells were specialist methyl salicylate receptors, but the biological significance of this is unknown. *L. decemlineata* feeds on potato which lacks specific odours, but has a characteristic combination of general plant odours to which the beetle responds (Visser and Avé, 1978).

In *Dytiscus marginalis*, an aquatic predator, some neurones associated with the antennal plate organs respond to a range of organic and amino acids, while other cells respond to amines. Similar responses occur in air and water (Behrand, 1971).

(*e*) *Other orders* Slifer (1979b) estimates that there are about 10 000 sensilla on one antenna of *Chrysops carnea* (Neuroptera) of which about half are tactile hairs. Most of the remainder are thin-walled chemoreceptors, each with a few neurones. There are probably about 20 000 chemosensitive neurones in total.

The sensilla on the antennae of two species of flea (Siphonaptera) have been examined. In both sexes of *Cediopsylla simplex* there are eight pits each containing four sensilla, eight pits with a single sensillum, one coeloconic sensillum and three other sensilla. The male has, in addition, 13 basiconic pegs while the female has only two (Amrine and Lewis, 1978). Thus there are 57 probable chemoreceptors on the male antenna and 46 on the female. In the female *Ctenocephalides canis* there are five contact chemoreceptors and about 40 sensilla in pits (Slifer, 1980). The number of neurones is unknown.

2.5 SENSILLA ON THE OVIPOSITOR

Thomas (1965) indicates large numbers of sensilla on the ovipositor valves of *Schistocerca gregaria.* It is not certain which of these are chemoreceptors, but probably her types F, I and K have a chemoreceptor function. There are about 80, 300 and 200 of these types on the dorsal and ventral valves of each side. Rice and McRae (1976) record about 50 papillae, equivalent to Thomas's type I, on the "ovipositor" of *Locusta migratoria.*

The numbers of sensilla on the ovipositors of Lepidoptera and Diptera are in marked contrast to this. *Chilo partellus* has two chemoreceptors, each with four neurones, on each valve, while *Spodoptera littoralis* has three sensilla, with five neurones, on each valve (Chadha and Roome, 1980). These give totals of 16 and 30 chemosensitive neu ones respectively.

In the calypterate flies the ovipositor comprises the terminal abdominal segments plus a pair of anal, or lateral, leaflets. Chemoreceptors are probably restricted to the latter although Hooper *et al.* (1972) observed cuticular pits on the ventral plates of segment 9 of *Musca autumnalis.* On each anal leaflet this species has two multiporous pegs, with a single neurone each, and two uniporous pegs with four and three dendrites extending to the tips. In *Lucilia cuprina* there are two multiporous and five uniporous sensilla (Rice, 1976), in *Phormia regina* there are five (possibly eight) multiporous and two uniporous sensilla (Wallis, 1962), while in *Delia brassicae* there are two multiporous and two "styloconic" pegs (Behan and Ryan, 1977). The totals of chemosensitive neurones from the anal leaflets in *M. autumnalis*, *L. cuprina* and *P. regina* are, respectively, about 18, 34 and 22. No obvious chemoreceptors were recorded by Behan and Ryan (1977) on the ovipositor of *Psila rosae*, but they suggest that some of the trichoid sensilla might be chemoreceptors. This might also be true in *D. brassicae*.

The braconid parasite, *Orgilus lepidus*, has 10 chemoreceptors with five neurones each on the lateral stylets. The median stylet is probably of paired origin and, by analogy, probably has about 20 chemoreceptors which each have only four chemosensitive dendrites (Hawke *et al.*, 1973). Hence there are probably about 180 chemosensitive neurones in total in the ovipositor.

3 The significance of numbers of chemoreceptors

It is apparent from the foregoing account that the numbers of chemosensilla present in insects vary from a few tens up to 100 000 with over 300 000 receptor cells. The numbers of sensilla and receptor cells which are present on any organ of any insect are consequent upon a variety of interacting selection pressures among which size, sex and feeding habits will be of considerable significance. In this section the likely effects of these different pressures are discussed.

3.1 INSECT SIZE

The primary function of the mouthparts is the mechanical transport of food into the body. Monitoring the chemical quality of the food is essential but, except for largely sensory structures such as the maxillary and labial palps, the arrangement of sensilla performing this function, will be governed largely by the mechanical functions of the mouthparts. This may to some extent limit the numbers of sensilla. In contrast, the primary and often the only function of the antennae is as sense organs. It might, therefore, be expected that their size, and the number of sensilla they bear, is a direct reflection of the sensory needs of the insect.

Within a taxon comparisons of size can be obtained from some relevant single parameter; the width or area of the labrum is appropriate for the mouthparts of biting and chewing insects. Comparisons between taxa must be relatively subjective because of differences of form even where the feeding mechanism is basically similar.

All Orthopteroidea have biting and chewing mouthparts. Within a species the number of scattered contact chemoreceptors, on the shaft of the maxillary palp for instance, is proportional to the size of the insect (Fig. 2). A similar relationship between insect size and number of contact chemoreceptors holds throughout the superfamily Acridoidea (Chapman and Thomas, 1978). The numbers of sensilla which occur in discrete groups also generally increase with the size of the insect, but less so than the scattered sensilla. Often the density of sensilla in groups is very high; in the fifth instar of *Locusta migratoria*, for instance, there are about 1400 mm^{-2} in one of the A1 groups of the labrum-epipharynx (Cook, 1972), and presumably the groups are positioned to perform particular functions. Those on the tips of the palps of acridids come into contact with the surface of the leaf when the insect palpates (Blaney and Chapman, 1970); the A1 and A2 sensillum groups are at the ends of tracts of hydrophilic hairs which convey fluid directly to them from the food as it is crushed by the mandibles (Cook, 1977). This diminishes the need for an increase in the numbers of sensilla in relation to the size of the insect and Fig. 2 shows that, for a 10-fold increase in the size of the labrum, the numbers of A1 sensilla in *Chortoicetes terminifera* only increase between two and three times. The numbers of A3 sensilla, which are at the edge of the labrum, hardly increase at all (and see Table 2, and Chapman and Thomas, 1978). Groups of sensilla, which probably act as functional units (see p. 320), are present in similar positions in other Orthopteroidea; the number of groups does not alter with the size of the insect.

There are no comparable studies on changes in numbers of sensilla with size in other groups of insects with biting and chewing mouthparts. Only the later instars of caterpillars have been examined and different species have similar, small numbers of sensilla irrespective of their size, ranging from the larva of *Yponomeuta* about 1 cm long to that of *Manduca sexta* which is over 5 cm long. In the larvae of various beetle species there is no obvious relationship between sensillum number and insect size (Table 8).

Amongst fluid feeders, no data are available for comparisons within species of Hemipteroidea, but in all of the Hemiptera studied there are 10–12 sensilla on each lobe of the labium. In aphids these are all mechanoreceptors (Tjallingii, 1978), but in some Delphacidae and in several Heteroptera they are chemoreceptors (Schoonhoven and Henstra, 1973; Avé *et al.*, 1978; Hatfield and Frazier, 1980; Foster, unpublished results). The number of sensilla is the same for a planthopper 3 mm long as for *Dysdercus* 15 mm long. Ten to 15 sensilla also occur on the wall of the food canal, again irrespective of the size of the insect, so that even in the biggest bugs there are only about 150 chemosensitive neurones associated with the mouthparts.

Adult Endopterygota exhibit relatively little variation in size within a species, but in *Calliphora vicina* Maes and Vedder (1978) found no correlation between the length of the labellum and the number of sensilla it bears. However, there are fewer sensilla on the labellum of *Drosophila melanogaster* compared with other, bigger flies.

Aquatic fly larvae have about 50 sensilla on the mouthparts with little variation from instar to instar (Sublette and Sublette, 1973).

It can be concluded that the number of chemoreceptors on the mouthparts varies in relation to size in the Orthopteroidea, but not in Hemipteroidea or larval Endopterygota, and the Orthopteroidea have many more sensilla than insects of comparable size in the other groups.

These generalisations also appear to hold in respect of contact chemoreceptors on the antennae. Within species of Dictyoptera and Orthoptera the number increases with length of the antennae (Tables 12 and 14), although it is common for the density to decrease. The numbers of contact chemoreceptors in the adults of different species of *Periplaneta* are also correlated with antennal length, but there is no obvious relationship with the numbers in other Orthopteroidea or even other Dictyoptera (Fig. 8).

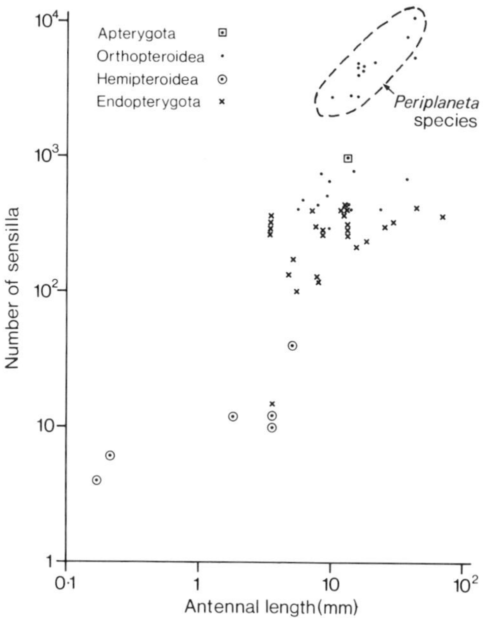

Fig. 8 Numbers of contact chemoreceptors on the antennae of various insects in relation to their antennal lengths. (Data from tables)

The only developmental data for an hemipteroid relates to *Acyrthosiphon pisum;* there is no regular change with size (Table 15). Adult Hemipteroidea have very few contact chemoreceptors on the antennae even allowing for their small size and the antennae of Endopterygota generally bear fewer contact chemoreceptors than those of Orthopteroidea of similar length (Fig. 8).

The situation with olfactory sensilla on the antennae is quite different; here there is a good correlation between antennal length and sensillum number in both Orthopteroidea and Endopterygota (Fig. 9). The more fragile

Hemipteroidea again have fewer than would be expected for their size, but the data for *Triatoma*, a much more robust species, fit the general correlation. There is also a correlation between antennal length and the total numbers of antennal chemosensory neurones, but the scatter is greater than for olfactory sensilla because of the wide variation in the numbers of neurones in each sensillum, ranging from two to over 50. So it may be concluded that Orthopteroidea do tend to have more antennal sensilla and chemosensitive neurones than other insects, but this is primarily a function of their greater size.

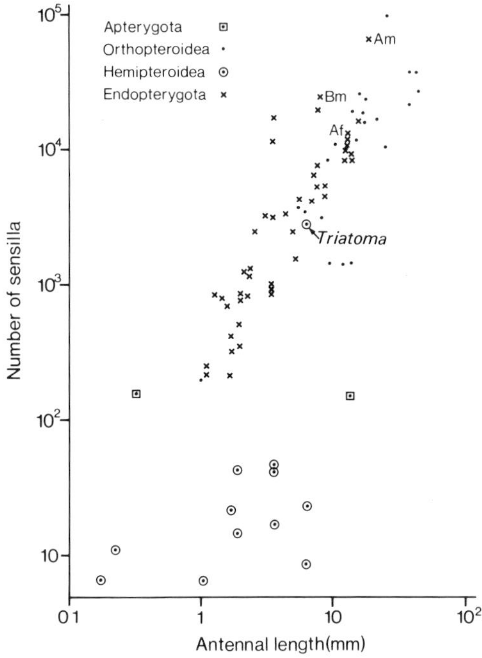

Fig. 9 Numbers of olfactory sensilla on the antennae of various insects in relation to their antennal lengths. Am and Af refer to male and female, respectively, of *Antheraea polyphemus*, Bm to the male of *Bombyx mori*. (Data from tables)

3.2 SEX

In general there are no differences between the sexes in the number of contact chemoreceptors on the mouthparts or antennae which cannot be related to size (see Tables 11, 14, 18 and 24; and Chapman and Thomas, 1978). There are some exceptions in species where the feeding habits of the sexes differ; blood-feeding female mosquitoes have more chemoreceptors on the labrum and maxillary palps than nectar-feeding males (Table 4).

Sometimes, however, there are big differences in the numbers of antennal olfactory sensilla in the two sexes. In many Dictyoptera (Table 11), some Acrididae (Table 14), alate aphids (Table 15), many moths (Table 18) and *Apis mellifera* (Table 20), males have many more sensilla than females. Most of these are insects in which the males are known to be attracted to the female by pheromones (and see Section 3.3). Where the numbers are similar it either reflects the fact that odour is not important in attraction, which depends primarily on auditory or visual responses as in Odonata and butterflies, or the fact that both sexes are attracted by the same odour, as with aggregation pheromones in the bark beetles (Table 24). In some species sexual dimorphism in the numbers of chemoreceptors is related to differences in behaviour not directly associated with sex; the relatively large number in female mosquitoes is presumably related to host-finding behaviour.

3.3 THE NEED FOR SENSITIVITY

In the males of some moths in which attraction by a female sex attractant is particularly important in reproductive behaviour, the antennae have a complex form which is associated with an increase in the numbers of sensilla present. This is illustrated in Fig. 9 by *Bombyx mori* and *Antheraea polyphemus* in which the number of sensilla on the male antenna (Bm, Am), which is pectinate, is well above the value expected from the overall regression of number on antennal length, while the number in the female *A. polyphemus* (Af) is within the expected range.

In Fig. 10 the number of olfactory sensilla on the antenna of the female of each species which has been studied is plotted against the difference in number between the male and female. In nearly all the species in which a female sex attractant pheromone is known or is suspected to occur, the male has at least 1000, and often 10 000, more sensilla on one antenna than the female. This is the case with many cockroaches (Jacobson, 1972; Persoons and Ritter, 1979), although olfactory attraction is not recorded for *Arenivaga* (No. 8 in Fig. 10). *Blaberus craniifer* (No. 6) and *Leucophaea maderae* (No. 7) have similar numbers of sensilla in the two sexes; in both species there are over 10 000 olfactory sensilla. The male mantis, *Tenodera angustipennis* (No. 9), probably responds to a female attractant as is known to occur in *Mantis religiosa* (Jacobson, 1972). All of the Lepidoptera shown in Fig. 10 are known to have female sex attractants (see e.g. Tamaki, 1977) except for the butterflies (Nos 19, 20), which probably depend on vision to bring the sexes together, and *Homodaula anisocentra* (No. 24).

At least some of these insects are known to respond behaviourally to very low concentrations of pheromone and the possession of a large number of sensilla probably confers a high degree of sensitivity to these odours. This

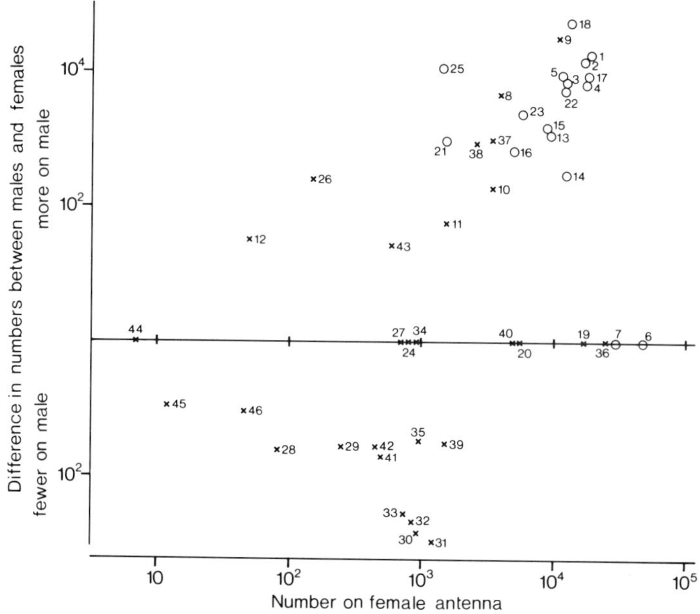

Fig. 10 Numbers of olfactory sensilla on the antenna of various female insects in relation to the difference in numbers between the sexes (number on male antenna–number on female antenna) of the same species. (○) indicates a species in which the female is known to produce a pheromone which attracts the male; (×) no sex attractant pheromone known to exist. Numbers refer to species:

Orthopteroidea
1 *Periplaneta americana*
2 *P. australasiae*
3 *P. brunnea*
4 *P. fuliginosa*
5 *P. japonica*
6 *Blaberus craniifer*
7 *Leucophaea maderae*
8 *Arenivaga* sp.
9 *Tenodera angustipennis*
10 *Melanoplus differentialis*
11 *M. bivittatus*

Hemipteroidea
12 *Aphis fabae*
44 *Trioza erytreae*
45 *Bagnalliella yuccae*

Endopterygota
13 *Trichoplusia ni*
14 *Heliothis zea*
15 *Prodenia ornithogalli*
16 *Spodoptera exigua*
17 *Bombyx mori*
18 *Antheraea polyphemus*
19 *Danaus gilippus*
20 *Colias eurytheme*
21 *Choristoneura fumiferana*
22 *Cydia nigricana*
23 *Ostrinia nubilalis*
24 *Homadaula anisocentra*
25 *Apis mellifera**
26 *Aphidius smithi**
27 *Itoplectis conquisitor**
28 *Nasonia vitripennis**
29 *Wyeomyia smithii*
30 *Aedes aegypti*
31 *Culex pipiens*
32 *Culex territans*
33 *Anopheles maculipennis*
34 *Nebria brevicollis*
35 *Aphaenops crypticola*
36 *Necrophorus vespilloides*
37 *Monochamus notatus*
38 *M. scutellatus*
39 *Curculio caryae*
40 *Hylobius abietis*
41 *Hypera meles*
42 *Ips confusus*
43 *Trypodendron lineatum*
46 *Cediopsylla simplex*

* Plate organs only

would be expected simply from the increased surface area available for the capture of odour molecules (Boeckh, 1980). In addition, where the increase in number is associated with an increase in the structural complexity of the antenna, the arrangement is such that a large volume of air is sampled with the greatest efficiency (Boeckh *et al.*, 1960).

But why do the females of these same species often have such large numbers of sensilla? There is no evidence for any comparable attraction of the females to the males. The cockroach species are non-specific feeders, and most of the Lepidoptera for which data exist are polyphagous so that there is no reason to expect the insects to be particularly sensitive to specific chemicals; the reason for the large number of olfactory receptors on the antenna of the female mantis which depends on vision to catch its food is equally obscure. The two butterflies, however, are relatively specific in their host-plant requirements and males of *Danaus gilippus* need to locate plants which provide a source of pyrrolizidine alkaloids for their own pheromones (Boppré, 1978). In a similar way, the beetle *Necrophorus vespilloides* is dependent on locating carrion for feeding and reproduction. Perhaps the need to locate relatively sparsely distributed habitats demands a high level of sensitivity and so a large number of sensilla in the insects.

In those species in which the female has relatively few sensilla the male also lacks large numbers and the difference between the sexes is small. Pheromones may still be important, as in the bark beetles (Nos 26–28 in Fig. 10), but these insects are associated with host plants producing characteristic odours (e.g. Rudinsky and Ryker, 1977). In these cases the plant serves as a relatively large and continuous source of odour, perception of which does not demand a high level of sensitivity. In the Culicidae (Nos 29–33), sound is important in bringing the sexes together; at least in the species studied food sources and oviposition sites are widely distributed and recognised by non-specific odours. Pheromones are not known to be produced by *Melanoplus* species (Nos 10 and 11), but the mechanism bringing the sexes together is not known. In none of the other species with small numbers of antennal sensilla shown in Fig. 10 are pheromones or specific odours known to be important in attracting the insects from a distance. Many of them are small insects, relatively feeble fliers in which attraction from a distance, necessitating upwind flight, is hardly feasible. In these cases a very high level of sensitivity might be disadvantageous to the insect, causing it to attempt to find a distant host which it is physically incapable of reaching. High levels of sensitivity are primarily of value to relatively strong fliers which can make a successful approach from some distance away.

Further, indirect, evidence of the importance of large numbers of sensilla in conferring a high degree of sensitivity to a stimulus is provided by a consideration of the receptors concerned with the registration of ubiquitous environmental parameters such as temperature and humidity. Table 26 lists the numbers of temperature/humidity-sensitive sensilla on the antennae of various insects. It includes only those where the function has been proved electrophysiologically or which include one dendrite with a lamellate form (Altner and Prillinger, 1980). The numbers are small even in *Cimex lectularius* and *Triatoma infestans* in which temperature plays an important role in feeding on warm-blooded vertebrates. A few additional temperature/humidity receptors are present on the palps of blood-feeding mosquitoes and Ceratopogonidae (McIver, 1972; Chu-Wang *et al.*, 1975), and *Periplaneta americana* has a temperature-sensitive neurone in some olfactory sensilla on the antennae (Sass, 1978). *Apis mellifera* has about 200 coeloconic and ampullaceous sensilla on each antenna; they have neurones responding to temperature, humidity and CO_2 (Lacher, 1964). Caterpillars have three neurones in antennal sensilla which respond to temperature change (Schoonhoven, 1967).

TABLE 26 Numbers of temperature/humidity receptors on the antennae of various insects

Insect	No. of sensilla	Ref.
Hypogastrura socialis	1	Altner and Thies (1973)
Periplaneta americana	approx. 80	Schaller (1978)
Aphids	4–6	Table 15
Triatoma infestans	approx. 30	Bernard (1974)
Cimex lectularius	3	Steinbrecht and Müller (1976)
Lepidoptera	17–90	Table 18
Mosquitoes	2–11	Table 23; McIver and Siemicki (1976)
Musca domestica larva	1	Chu-Wang and Axtell (1972a)
Coleoptera	0–8[a]	Corbière-Tichané (1971a); Corbière-Tichané and Bermond (1972)

[a] cave dwelling species have additional receptors in the "vesicles olfactives" (Corbière-Tichané, 1974; Peck, 1977)

Thus the total number of temperature sensitive neurones on the head appendages of these insects probably rarely exceeds 200 and is often less than 50; there are probably about twice as many humidity receptor cells. Yet it is well known that insects can respond to changes of 1°C and less, and it is inferred from this that only a few receptors are necessary to regulate behaviour where the stimulus is such that every cell is bound to be affected. Large numbers of sensilla are necessary where, because of its dispersion, the chances of the stimulating material making contact with the receptors is low.

3.4 SENSILLA WITH LARGE NUMBERS OF NEURONES

Most chemoreceptor sensilla are associated with small numbers of neurones. Uniporous, contact chemoreceptors usually have from three to six neurones, although some have as many as 10, while multiporous, olfactory sensilla have variable numbers, between one and six (Zacharuk, 1980). However, some multiporous sensilla on the antennae of a variety of different insects have much bigger numbers of neurones, generally more than 10 and in some cases over 100 (Table 27). The structure of these sensilla in the Homoptera and larval Endopterygota seems to indicate that they are compound organules derived from a number of sensilla each with fewer neurones. There is no evidence for such an origin, however, in the Orthoptera.

TABLE 27 Occurrence of multiporous sensilla with large numbers of neurones on the antennae of different insects

Insect	Form of sensillum	No. of neurones	Ref.
Orthoptera			
Melanoplus	basiconic peg	~40	Slifer *et al.* (1959)
Homoptera			
aphids	plate organ	3–14	Bromley *et al.* (1979)
Pyrops candelaria	plate organ	150–350	Lewis and Marshall (1970)
Heteroptera			
Lygaeus kalmii	thin-walled peg	50–60	Slifer and Sekhon (1963)
Hymenoptera			
Neodiprion sertifer	thin-walled trichoid	10–14	Hallberg (1979)
Lasius fuliginosus	thin-walled trichoid	30	Dumpert (1972a)
Camponotus vagus	basiconic peg	30	Masson *et al.* (1972)
Apis mellifera	plate organ	12–18	Slifer and Sekhon (1961)
Aphidius smithii	plate organ	37	Borden *et al.* (1978b)
Itoplectis conquisitor	plate organ	27	Borden *et al.* (1978a)
Coeloides brunneri	plate organ	13	Richerson *et al.* (1972)
Chalcidoidea	plate organ	~50	Barlin and Vinson (1981)
Coleoptera			
elaterid larvae	antennal cone	8–30	Zacharuk (1962)
Tribolium larva	plate organ	130	Behan and Ryan (1978)
Speophyes lucidulus larva	membranous lobe	24	Corbière-Tichané (1971a)
Diptera			
Musca domestica larva	dome	21	Chu and Axtell (1971)

Whatever the origin of these sensilla, the area of porous cuticle exposed per dendrite is less than if the same number of neurones was associated with a number of basiconic sensilla each with a few neurones. The effect of this is to lessen the chances of contact with odour molecules so that sensitivity is reduced. This must be true of all plate organs irrespective of the number of neurones associated with them; there are nearly always large numbers.

It is to be expected that such sensilla may occur where the possession of a large number of cells is at a greater premium than sensitivity. Assuming that these cells respond to different ranges of chemicals, as they do in the plate organs of *Apis mellifera* (Lacher, 1964), this arrangement permits great versatility of response by the organism. It can be argued that in larval Endopterygota, living in close association with their food, a slight loss of sensitivity is unimportant, while in weak flying insects, like the aphids and parasitic Hymenoptera, a low level of sensitivity to some odours could be an advantage (see p. 315). However the large numbers of plate organs on the antennae of drone bees and of Scarabaeoidea (Meinecke, 1975), where high sensitivity is to be expected, seems to conflict with the suggestion that plate organs occur in situations where high sensitivity is not of great importance.

3.5 CHEMICAL RECOGNITION AND NUMBERS OF RECEPTORS

Where sharply tuned neurones exist together with discrete pathways to the coordinating centres of the brain, single labelled lines can provide all the information required to initiate a particular behaviour pattern. This is the case, for instance, in the male of *Bombyx mori* where a few molecules of the female attractant pheromone can initiate reproductive behaviour (Kaissling, 1971). Cells sharply tuned to attractant pheromones are known to occur in a number of insects and there is good evidence for the existence of distinct separate neural pathways from the antennae to the corpora pedunculata in some cockroaches and moths (Boeckh and Boeckh, 1979; Chambille *et al.*, 1980; Hildebrand *et al.*, 1980). Sharply tuned cells also occur in some contact chemoreceptors, such as the glucosinolate-sensitive cells in *Pieris brassicae* (Ma and Schoonhoven, 1973) and *Mamestra brassicae* (Weiczorek, 1976), and the hypericin-sensitive cell of *Chrysolina brunsvicensis* (Rees, 1969).

However, there are many instances where the behaviour of insects shows that they can discriminate between different chemicals, but where the electrophysiological evidence indicates a lack of sharply tuned receptor cells. A simple example is the ability of *Locusta migratoria* to distinguish a solution of sodium chloride, from a solution of sodium chloride and fructose; the neurones which respond are not particularly sensitive to either compound and the electrophysiological responses of single sensilla to the two solutions do not provide a basis for discriminating between the two solutions by the observer (Blaney, 1975). *Leptinotarsa decemlineata* can distinguish the odour of potato from that of other plants, yet potato odour is not produced by a characteristic chemical, but by a particular combination of a number of volatile chemicals which occur in many plants. Ma and Visser (1978) have shown that the antennal olfactory neurones of *L. decemlineata* fall into five main categories, one with a wide spectrum of response to numerous compounds, the others with narrower and different response spectra. Distinguishing the odour of potato from the odours of other plants can only be achieved by integrating the responses of a number of receptors. Similar conclusions have been reached in several instances by other authors working on different insects (e.g. Dethier and Schoonhoven, 1969; Dethier, 1980a). Most of these authors, like Ma and Visser (1978), consider the individual neurones to be the basic units, but, in *Locusta migratoria*, Blaney (1975) presents evidence to suggest that the sensillum, commonly containing six neurones, is the basic unit.

Hence it is established that many odours and flavours can only be distinguished by an array of receptors with qualitatively different response characteristics. However, in attempting to relate sensory input to the observed behaviour of *Locusta migratoria* and *Schistocerca gregaria*, Blaney (1980) treats most of the inputs as qualitatively similar. A re-examination of his data suggests that this is not valid and that qualitatively different inputs must, and do, exist. Blaney postulates that for *S. gregaria* a high level of sensory input inhibits feeding, and his regression line indicates maximum food intake with zero sensory input. Yet it is known that feeding by acridids is not normally initiated or maintained in the absence of a phagostimulant (Bernays and Chapman, 1974). Clearly sensory inputs must exhibit sufficient qualitative difference to permit phagostimulants and deterrents to be distinguished. Azadirachtin totally inhibits feeding on glass fibre discs by *S. gregaria* irrespective of the sucrose concentration present; it is inhibitory even at 0.00001 M (Haskell and Mordue, 1969). This extreme sensitivity suggests the existence of a receptor which is particularly sensitive to azadirachtin perhaps associated with a labelled line to the brain, as Blaney himself suggests. Hence an alternative interpretation of Blaney's (1980) data would postulate the existence of a stimulatory input, of one or more unspecialised inhibitory inputs, and possibly of a specialised inhibitory input tuned to azadirachtin. Similar arguments can be applied in reinterpreting his data for *Locusta migratoria*. In this species there is no evidence for a tuned "azadirachtin receptor", although qualitatively different inhibitory, or deterrent, cells may exist.

So it may be that in these insects, as in others studied from an olfactory point of view, chemical discrimination in the absence of labelled lines necessitates an array of receptors with different response characteristics. Blaney and Chapman (1970) concluded that, during palpation, *Locusta migratoria* touched a surface with about 30 sensilla at the tip of each palp and Blaney (1975) showed that it was possible for the experimenter to distinguish different compounds by analysing the inputs from 15 sensilla. Even in the smallest first instar nymphs of acridids there are generally more than 20 sensilla in each of the sensillum groups, and it is probable that the sensillum group acts as a functional unit. Ma and Visser (1978) concluded that 25 cells, falling into five main response categories, were representative of the sensillum population of the whole antenna of *Leptinotarsa decemlineata* and the insect could distinguish between complex odours with this array of cells. Maes (1980) has shown that three types of sensillum, each with two cells responding, are necessary for *Calliphora vicina* to distinguish between a range of inorganic salts. Other authors have come to similar conclusions in other insects (Dethier and Schoonhoven, 1969; Dethier, 1973, 1980a) and van der Molen *et al.* (1978) suggest that the whole leg of *C. vicina* with its array of sensilla should be regarded as a single sense organ.

However, these insects can differentiate a simple acceptable from a simple unacceptable stimulus on the basis of the input from a single sensillum. For example, stimulation of one labellar hair of *Phormia regina* with a sucrose solution leads to proboscis extension and feeding; stimulation of a second hair with NaCl leads to the cessation of feeding (Dethier, 1976). Electrophysiological analysis indicates the presence of "sugar" and "salt" receptors.

There is no conflict in these results. Single compounds such as salt and sugar sometimes stimulate only one neurone but often cause several neurones to fire (van der Starre, 1972; Blaney, 1974). In this case they commonly evoke a greater response in one neurone, presumably the "salt" or "sugar" receptor. Rejection or acceptance results from the balance of inputs from the different cells. With mixtures of chemicals a range of units will be stimulated, some indicating acceptability, others unacceptability, and the behavioural response results from the integration of these different inputs in a way which enables the organism to "assess" the quality of the substance perceived. This is the concept of across-fibre patterning (see e.g. Dethier, 1976). It gives the insect the ability to distinguish a wide spectrum of chemicals and necessitates the possession of an array of receptors with different response characteristics.

3.6 FEEDING HABITS AND THE NUMBERS OF SENSILLA

In Tables 28 and 29 the feeding habits of the insects discussed previously have been roughly categorised and the numbers of sensilla on mouthparts and antennae are shown in relation to feeding habit and systematic position.

TABLE 28 Numbers of chemoreceptors on the mouthparts of various insect taxa in relation to feeding habits. (Data from previous tables, numbers for larvae include antennal sensilla)

Taxon	Food habit						
	scavenger	predator	herbivore	nectar feeder	blood sucker	parasitoid	filter-feeder
Diplura	700 (palp)						
Thysanura	80–700 (palp)						
Odonata larva		50					
adult		220					
Orthoptera							
Acridoidea			600–3300				
Grylloidea	7000						
Dictyoptera	6100						
Dermoptera	200						
Homoptera			20				
Heteroptera	190		40		40		
Lepidoptera larva			38				
adult				200			
Diptera							
Nematocera larva		40					50
adult				50–150	200–300		
Cyclorrhapha larva	20–60						
adult	40–250			40–250	>20		
Hymenoptera							
Apis mellifera				500			
Parasitic species						14	
Coleoptera larva	60–200		100				
adult			100				

TABLE 29 Numbers of chemoreceptors on each antenna of various insect taxa in relation to feeding habits. (Data from previous tables)

Taxon	Food habit						
	scavenger	predator	herbivore	nectar feeder	blood sucker	parasitoid	vertebrate ectoparasite
Collembola	150						
Thysanura	1130						
Odonata		50–100					
Orthoptera							
Acridoidea			400–5000				
Tettigonioidea			1000				
Dictyoptera	3500–50 000	14 000–38 000					
Embioptera		1700					
Zoraptera	200						
Homoptera			5–80				
Heteroptera			2000–8000		50–2900		
Thysanoptera			20				
Mallophaga							15
Siphunculata							13–14
Lepidoptera				1100–98 000			
Diptera							
Nematocera			300	220–400	170–1200		
Cyclorrhapha				500–5000	1600–5000		
Hymenoptera							
Apis mellifera				4700			
Parasitic species						50	
Coleoptera	23 500	900–1000	400–6000				
Neuroptera		5000					
Siphonaptera							50

3.6.1 *Scavenging insects*

Cockroaches have large numbers of sensilla on the antennae even in the nymphal stages where sex pheromones are not important (see Table 12). Sass (1980) suggests that in adult *Periplaneta americana* there are 2000 neurones in each of seven classes responding to different ranges of food odours. There are also several thousand sensilla on the mouthparts. Thysanura, with similar habits, also have relatively large numbers of sensilla. With insects which are virtually omnivorous a high degree of sensitivity requiring a large number of sensilla would not be expected, but possibly food selection requires a series of receptors with different ranges of responsiveness.

The large number of sensilla on the antennae of *Necrophorus vespilloides* may be related to a high degree of sensitivity needed for locating relatively widely dispersed carcases. Flies which also feed and breed in carrion have many fewer sensilla, but are probably less selective than the beetle in their choice of oviposition and feeding sites. They have relatively few sensilla on the mouthparts for monitoring the food. Saprophagous larval Cyclorrhapha and Coleoptera also have small numbers of chemoreceptors. Distance perception of food sources is not necessary for them because the eggs are laid by the parent on or close to appropriate sources of food. They respond to relatively high concentrations of non-specific odours, such as CO_2 and ammonia (Jones and Coaker, 1978), for which small numbers of sensilla are adequate.

3.6.2 *Predaceous insects*

Amongst predaceous insects, vision is certainly of primary importance in prey-catching by Odonata and Mantodea. In the former there are very few chemoreceptors on the antennae or mouthparts, but mantids have very large numbers of antennal chemoreceptors. In the male this can be attributed partly to the need to perceive a sex attractant pheromone, but the significance of a large number in female mantids is unknown. Relatively large numbers of sensilla are also present on the antennae of both sexes of predaceous beetles and of *Chrysops carnea* (Neuroptera). The predaceous larvae of the mosquito *Toxorhynchites brevipalpis* have similar numbers of sensilla to filter-feeding species.

3.6.3 *Phytophagous insects*

Nymphal Acridoidea have relatively large numbers of olfactory sensilla on the antennae even in species, such as *Melanoplus bivittatus*, where no social interaction occurs (Table 14). It is known that acridids have the capacity to orientate to plant odours from some distance away (Kennedy and Moorhouse, 1969), although most species are relatively unselective with respect to their food (see below) and a high degree of sensitivity is probably not necessary. The olfactory neurones have a range of sensitivities to different plant compounds (Kafka, 1970) and the large number may facilitate discrimination at a distance, although there is no evidence that this occurs. Once on the plant, acridids monitor it first with the terminal sensilla of the palps (Blaney and Chapman, 1970) and then, after biting has released the fluids from within the plant, with the sensilla of the cibarial cavity. This process is repeated as the insect moves from leaf to leaf or plant to plant. Large numbers of contact chemoreceptors are involved, presumably to facilitate discrimination by across-fibre patterning.

The phytophagous Heteroptera studied have comparable numbers of antennal sensilla to the Acridoidea, but far fewer sensilla on the mouthparts. Tree boring beetles have variable numbers of antennal sensilla. In *Hylobius abietis* some cells are specialised for the perception of host odours (Table 25) and this may also be the case in the cerambycid *Monochamus notatus* (Dyer and Seabrook, 1978). The former also has cells which are highly sensitive to an aggregation pheromone. Nevertheless most of the olfactory cells associated with antennal sensilla of *H. abietis* have a more generalised response, perhaps enabling the insect to discriminate between host trees of different qualities. The scolytid beetles have fewer sensilla and some species may lack cells specific for the host tree odour (Dickens and Payne, 1977), although they are attracted by the odours of their hosts (Rudinsky and Ryker, 1977). Host odours may also synergise the response to the aggregation pheromone. In these species, the host trees have characteristic odours which may provide the initial stimulus for host-finding. Closer to the hosts, aggregation pheromones may be perceived and plant quality determined. At no stage is a high level of sensitivity necessary and this may account for the relatively small number of olfactory receptors in the antennae of bark beetles (Table 24). These insects have relatively few mouthpart sensilla.

Apterous aphids have little capacity for moving to new hosts and have very few antennal sensilla. Winged morphs of the same species have more plate organs (Table 15) and it is known that odours may influence landing (Chapman *et al.*, 1981) although in many species vision is of primary importance in the initial stages of host selection (van Emden *et al.*, 1969). Acceptance or rejection occurs after alighting, perhaps involving odours close to the surface of the leaf perceived by the antennae, or gustatory testing by the cibarial receptors. In neither case is great sensitivity necessary and the numbers of sensilla are small, both on the antennae and mouthparts. Thysanoptera also have very few antennal sensilla and, at least in some flower-dwelling species, vision plays a part in host-finding (Ward, 1973). It has been argued above (p. 315) that a very sensitive olfactory system could be a disadvantage to these weak fliers, so a large number of sensilla is not to be expected.

Larval Lepidoptera are in a different category since the initial stages of host finding are normally carried out by the parent moth or butterfly. Where this is not the case, as with some *Spodoptera* species, the insects are polyphagous or graminivorous so that precise location of a host plant is necessary. The small number of olfactory sensilla possessed by caterpillars reflects this lack of a distance response. Nevertheless, caterpillars can discriminate host plant from non-host plants when close to them (Jermy, 1966), having a similar capacity to Acrididae in this respect although possessing only a few sensilla. In making their choice, the odour close to the leaf surface is important (Dethier, 1937), while ablation of the maxillae, which carry a "deterrent cell" in one contact chemoreceptor, results in the acceptance of a wider range of foods (Ma, 1972)

Among nectar-feeding forms, vision certainly plays a major part in locating the food in Lepidoptera and *Apis mellifera,* and this is probably also true of adult Nematocera, which commonly feed at white or yellow flowers, and adult Cyclorrhapha which often feed at the heads of umbels. However, in the latter case and with those insects which feed on the sugary flows from some plant wounds or decomposing fruits, attraction to widely occurring compounds, such as alcohols, is almost certainly involved in locating the food and a high level of sensitivity would not be expected. Adult Lepidoptera do often possess large numbers of antennal sensilla, but these are probably associated mainly with pheromone perception by males and oviposition site selection by females. However, if the associated neurones respond to a wide range of odours, these insects may as a consequence also be highly sensitive to the odours of sources of sugars.

3.6.4 *Blood sucking insects*

Among the species feeding on vertebrate blood the numbers of antennal sensilla range from relatively small to large, larger numbers being present in the bigger, more mobile insects. For example, amongst the Heteroptera, *Cimex lectularius* has fewer sensilla than *Triatoma infestans*, and, in the Diptera, *Culicoides furens* has few compared with *Stomoxys calcitrans.* Host-finding by most of these insects is known to involve visual and olfactory responses, the odours being such non-specific chemicals as carbon dioxide, lactic acid and acetone (see e.g. Clements, 1963; Vale, 1980). A high level of sensitivity to such widely occurring compounds would be disadvantageous unless coupled with other means of identifying the potential host and with the capacity to reach it from some distance away. The greater number of antennal sensilla in the more strongly flying insects is in keeping with this thesis. All these insects have only small numbers of chemoreceptors on the mouthparts and, in several, neurones with a particular sensitivity to ATP have been found

(Friend and Smith, 1977). Since host identification apparently occurs before the insect lands and starts to probe a small number of sensilla monitoring food quality is all that is required.

3.6.5 *Parasitoids*

Only one parasitoid has been studied in detail, but all the parasitic Hymenoptera studied have relatively few plate organs on the antennae although they are associated with large numbers of neurones (Table 22). There is no information on the ability of parasitoids to locate their hosts from a distance, but at least in some cases they are known to be attracted initially by odours emanating from the habitat of the host. For example, *Diaretiella rapae*, a parasite of the aphid *Brevicoryne brassicae*, is attracted to the host plants of the aphids by the allylisothiocyanate produced by the plant (see van Emden, 1978). Attraction from a distance is not likely to occur in weak fliers so that large numbers of antennal sensilla are not to be expected. Subsequently, within the habitat of the host, more specific odours serve to identify and locate it (Matthews, 1974; W. J. Lewis *et al.*, 1977). The active searching behaviour of the insects at this time, commonly coupled with vibratory movements of the antennae, permits a relatively large volume of air to be sampled, perhaps offsetting the relative lack of sensitivity of the plate organs.

3.6.6 *General conclusions*

The data are still too few to draw more than the most tentative conclusions but it appears that in those insects where odour perception at some distance from the source is important a high degree of sensitivity is an asset and large numbers of sensilla are present on the antennae. Such distance perception is only of relevance if the insects are sufficiently active and robust to reach the source; in relatively weak fliers, such as aphids and mosquitoes, large numbers of sensilla are not present. This is also true in insects which are attracted by widely occurring odours. In this case the concentration of the odour is important; responding to low concentrations would often be misleading and hence numbers of sensilla and sensitivity are relatively low. There are no obvious differences in numbers of antennal sensilla associated with the taxonomic positions of the insects.

With the contact chemoreceptors of the mouthparts, the position is different. There is no obvious relationship between numbers of sensilla and feeding habits, but it is generally true that orthopteroid, and possibly apterygote, insects have large numbers of chemoreceptors irrespective of their feeding habits, while Hemipteroidea and Endopterygota have only a few.

3.7 FOOD SPECIFICITY OF PHYTOPHAGOUS INSECTS

Differences in the numbers of sensilla on the mouthparts reflect broad differences in the degree of food specificity of phytophagous insects belonging to the major groups. Table 30 gives an analysis of the degree of specificity in British phytophagous insects based on data supplied by Dr L. K. Ward. The insects are classified as feeding only on one genus of plants, on one family (excluding Gramineae and Cyperaceae), on Gramineae and/or Cyperaceae only, or on representatives of more than one family. The grasses are treated separately because the insects feeding on them are considered to be unspecialised in their chemical responses (Bernays and Chapman, 1977), whereas insects specialising on other families of plants are often, perhaps always, specialised in their responses to chemicals characteristic of the plants concerned (e.g. Thorsteinson, 1958; Hsiao and Fraenkel, 1968; Nielsen, 1978). Polyphagous insects are also unspecialised in being relatively tolerant of a wide range of plant secondary compounds.

Table 30 shows that none of the British phytophagous orthopteroids is specialised with respect to the food eaten, whereas the majority of Hemipteroidea and Endopterygota are limited to one family and, in many cases, to one genus of plants. Orthopteroidea are poorly represented in the British fauna, and a further analysis of the food eaten by Acridoidea has been made of data from elsewhere (Table 31). This is based on studies where faecal analysis has established unequivocally the plants eaten. Relatively few species are restricted to one family or genus of plants and, although the breadth of feeding of a number of species has not been determined, it remains true that over 50% (in most cases over 75%) of the Acridoidea are unspecialised in the food they eat. The converse is true of all the other groups of insects for which evidence has been obtained from the literature (Table 31), although the apparent specificity of many Aleyrodidae may simply reflect a lack of records (Mound and Halsey, 1978). So the generalisation holds that phytophagous Orthopteroidea with many chemoreceptors on the mouthparts, are unspecialised in their food selection compared with Hemipteroidea and Endopterygota which have only a few contact chemoreceptors on the mouthparts. The lack of specificity in Orthopteroidea is in keeping with the view that the primitive phytophagous insects were polyphagous as a consequence of the omnivorous feeding habits of their ancestors (Takahashi, 1971).

Is the association between numbers of sensilla and breadth of diet more than a coincidence? In *Locusta migratoria* and *Schistocerca gregaria*, Blaney (1974, 1975, 1980) found no evidence for sharply tuned receptors, with the possible exception of an "azadirachtin receptor" (see above). Nor is there any behavioural evidence that any particular chemicals, apart from azadirachtin, are particularly significant to these insects (Bernays and Chapman, 1978). One would, therefore, expect them to require an array of chemoreceptors

TABLE 30 Range of feeding habits of different orders of British phytophagous insects. (By courtesy of Dr L. K. Ward, Institute of Terrestrial Ecology, Furzebrook Experimental Station, Dorset, UK)

Order	No. of species	% feeding on one genus	% feeding on one family (not including grasses)	% feeding on Gramineae and/or Cyperaceae	% feeding on more than one family
Orthopteroidea					
Orthoptera	27	0	0	41	59
Hemipteroidea					
Heteroptera	286	41	14	8	37
Homoptera	1010	60	10	16	13
Thysanoptera	114	46	10	17	27
Endopterygota					
Lepidoptera	1344	53	24	6	17
Coleoptera	997	38	30	2	30
Hymenoptera	632	67	14	6	12
Diptera	485	52	20	22	6

TABLE 31 Range of feeding habits of different taxa from various regions

Taxon	Area	No. of insects in sample	% feeding on one genus	% feeding on one family (not including grasses)	Unknown	% feeding on grass only	% feeding on more than one family	Ref.
Orthopteroidea								
Acridoidea	Spain	39	0	0	23	33	44	5
	Ghana	71	0	0	24	51	25	3
	United States	60	3	2	2	45	48	2,4
	Argentina	54	0	11	33	33	23	1,2
Hemipteroidea								
Psylloidea	World	1250	100 (genus and family combined)		0	0	0	6
Aphidoidea	World	3600	>50	–	–	–	–	7
Thysanoptera	World	~ 3000	>50	–	–	–	–	7
Alleyrodidae	World	994	61	6	0	4	29	11
Endopterygota								
Lepidoptera								
(Papilionoidea)	United States	86	34	22	0	23	21	8
(micro-Lepidoptera)	World	4151	?	91	0	4	5	9
(Papilionoidea)	Palaearctic	1368	12	24	19	28	17	10

[1] Gangwere and Ronderos (1975)
[2] Otte and Joern (1977)
[3] Chapman (1962, 1964)
[4] Mulkern *et al.* (1962, 1964); Banfill and Brusven (1973); Campbell *et al.* (1974)
[5] Gangwere and Agacino (1973)
[6] Eastop (1979)
[7] Eastop (1973)
[8] Slansky (1974)
[9] Powell (1980)
[10] Kostrowicki (1969)
[11] Mound and Halsey (1978)

with rather generalised, but differing, response characteristics enabling them to discriminate between different foods by across-fibre patterning. It may be significant that the few acridids with a restricted host plant range which have been examined have fewer sensilla than other, closely related, species with broader diets (see Fig. 3). Possibly these insects do respond to marker chemicals in their food plants.

By contrast, many of the Endopterygota and Hemipteroidea with a restricted host-plant range are known to be stimulated to feed by specific host-associated chemicals (Table 32) and some species are known to have receptor cells which are very sensitive to these chemicals (Table 7). Single cells which respond to a range of deterrent compounds are also known to occur in larval Lepidoptera (Schoonhoven, 1973).

TABLE 32 Insects which exhibit increased feeding activity in response to compounds which are characteristic of the host plant

Species	Host plant	Compounds	Ref.
Homoptera			
Brevicoryne brassicae	Cruciferae	Sinigrin	van Emden (1972)
Hydaphis erysimi	Cruciferae	Sinigrin	Nault and Styer (1972)
Lepidoptera			
Pieris brassicae	Cruciferae	Sinigrin	Blom (1978)
Plutella maculipennis	Cruciferae	Sinigrin	Thorsteinson (1958)
Serrodes parlita	Sapindaceae	Quebrachitol	Hewitt *et al.* (1969)
Ceratomia catalpae	Bignoniaceae	Catalposide	Nayer and Fraenkel (1963)
Laothoe populi	Salicaceae	Salicin	Schoonhoven (1972)
Bombyx mori	Moraceae	Morin	Hamamura *et al.* (1962)
Oidaematophorus monodactylus	Convolvulaceae	Specific alkaloid	Mohyuddin (1973)
Coleoptera			
Phyllotreta armoraciae	Cruciferae	Kaempferol 3-*O*-xylosylgalactoside	Nielsen *et al.* (1979)
Phyllotreta spp. } *Phaedon cochleariae* }	Cruciferae	Glucosinolates	Nielsen (1978) Hicks (1974)
Aulacophora foveicollis	Cucurbitaceae	Cucurbitacin	Sinha and Krishna (1970)
Deloyala guttata	Convolvulaceae	Specific alkaloid	Mohyuddin (1973)

It is concluded that the general association of large numbers of contact chemoreceptors with an unspecialised diet in phytophagous Orthopteroidea and of small numbers of sensilla with a specialised diet in other groups is not a coincidence. Lack of specialisation requires large numbers of sensilla and across-fibre patterning to give versatility to the system. Specialisation can readily be achieved by having tuned receptors and labelled lines. Since, in the case of contact chemoreception, sensitivity is not at a premium, large numbers of receptors are not essential once labelled lines exist and a considerable reduction in number and a simplification of the central processing of information can be achieved. This appears to be the situation in Hemipteroidea and Endopterygota.

4 Chemoreceptor populations and the evolution of insects

As long as insects possessed large numbers of sensilla on the mouthparts, acceptance or rejection of food will have followed from the integration of a large number of sensory inputs. Small changes in the spectrum of sensitivity of individual receptors or classes of receptor, are unlikely to have been selected for and sustained because their inputs will have been swamped by a multitude of other inputs. Hence it can be envisaged that feeding specialisation based on the chemical characteristics of the food will not have occurred readily and the insects will have remained catholic in their choice of food. This is the situation in the Apterygota and Orthopteroidea. Adaptations to new ecological situations did, of course, occur, but it is argued that these adaptations were not primarily associated with the chemical qualities of the food (see for example, Lawton, 1978). Bernays and Chapman (1978) concluded from a consideration of the chemical interactions between Acridoidea and their host-plants that there will have been little tendency for chemical co-evolution to have occurred. Some food specialisation in Acridoidea has occurred in habitats where extreme conditions have enforced an almost total dependance on certain plants. This is true of the desert grasshoppers of the United States and Argentina (Otte and Joern, 1977), of *Xenocheila zarudnyi* from the deserts of Iran (Chapman, 1966) and of *Paulinea acuminata* and *Cornops longicorne* living in certain aquatic habitats (Bennett, 1966, 1970). In arid regions, *Calotropis procera* provides a habitat for *Poekilocerus* species (Fishelson, 1960) when most other vegetation is dry and dead. But even in these species chemical adaptation to the host plant is not complete and, in the absence of the host, some of them are known to accept other plants (see e.g. Uvarov, 1977).

Food specialisation has also occurred in some Acridoidea in neotropical rainforest (Rowell, 1978). In this case, as with grass-feeding Acrididae (Bernays and Chapman, 1977), the specialisation may have arisen from an avoidance of the unacceptable plants due to the diversity of their secondary chemicals rather than as a positive adaptation to the chemical properties of the favoured host plant.

In the examples given, chemosensory specialisation facilitating recognition of the host by its characteristic chemicals may have developed secondarily and in several acridoids with a restricted diet there are fewer sensilla on the mouthparts than in related more catholic species (Fig. 3). In no case is it known whether or not tuned receptors are present.

With small numbers of receptors on the mouthparts, as in Hemipteroidea and Endopterygota, a change in the sensitivity of one receptor is of much greater significance in the total picture perceived within the central nervous system. As a consequence, tuned receptors and labelled lines are likely to develop and adaptations to chemically new food can occur readily. It can be envisaged that a chemical might change from being deterrent to being phagostimulatory by a relatively simple modification of dendritic connections within the brain. As a result, existing chemical differences between plants could serve as primary isolating mechanisms in the development of insect species. The close association with particular plant species which thus results will have favoured chemical adaptations by the plant and step-wise chemical co-evolution of plants and insects could occur (Gilbert and Raven, 1975; Harborne, 1978).

In the Endopterygota the sensory system and associated parts of the central nervous system is extensively rebuilt at metamorphosis. It is arguable that the possession of only small numbers of receptors in holometabolous larval forms is advantageous in such a situation, minimising the expenditure of resources when the adult systems develop.

It is suggested that the small numbers of chemoreceptors on the mouthparts of Hemipteroidea and Endopterygota are an essential prerequisite to the diversification of the phytophagous species within these groups and may have facilitated the development of holometabolous insects. The reduction in numbers of sensilla from the earlier Orthopteroidea is thus seen as a major step in the evolution of the insects.

5 Conclusions

Large numbers of chemoreceptors are significant principally in two ways: they confer a high degree of sensitivity on the system by exposing a large surface area for the collection of molecules, and, if they respond differentially to different chemical compounds, they provide for a high level of versatility in discriminating between chemicals. With small numbers of receptors much of this versatility is lost, but presumably the process of integration within the central nervous system is simplified considerably, both anatomically and physiologically. At the same time adaptation to new chemical environments is facilitated. A high level of sensitivity is not always an advantage since it may lead to a response in a situation where the insect cannot complete its behaviour sequence and is unable to reach the source of the odour.

It is apparent that observed behaviour will often be the outcome of interactions between several, even many, chemosensory inputs. It is not, therefore, to be expected that the input from a single receptor will be directly related to behaviour except in very simple situations. This may be true even where a receptor is tuned to a key stimulus. For example, bark beetles have very few olfactory sensilla compared with many moths and, as a consequence, they are probably less sensitive to their attractant pheromone. However, bark beetle pheromones are produced at host trees which have a characteristic odour. This is also attractant to the beetles and occurs in much greater concentrations than the pheromone so that attraction from a distance can occur without a very sensitive sensory system.

Clearly there are good reasons why insects have different numbers of chemoreceptors; equally clearly their significance to the insect is unlikely to be understood from the study of single neurones and sensilla. Only by studying populations of receptors will a proper understanding of the chemosensory regulation of insect behaviour be gained. Who would think of trying to interpret the visual responses of an insect from the study of one ommatidium?

Acknowledgements

I am indebted to many colleagues within and outside the Centre for Overseas Pest Research for contributing to this review in many ways. In particular I should like to express my appreciation to Dr L. K. Ward for allowing me to use her extensive unpublished information of the food of British phytophagous insects and to Professor G. B. Mulkern for providing material of North American grasshoppers on which much of the work on Acrididae was based. Finally, I must thank Dr E. A. Bernays who has encouraged and discussed the work through all its stages. She has contributed much of value.

References

Abushama, F. T. (1968). Food-plant selection by *Poecilocerus hieroglyphicus* (Klug) (Acrididae: Pyrgomorphinae) and some of the receptors involved. *Proc. Roy. Entomol. Soc. London A* **43,** 96–104

Adler, V. E., Doolittle, R. E., Shimanuki, H. and Jacobson, M. (1973). Electrophysiological screening of queen substance and analogues for attraction to drone, queen, and worker honey bees. *J. econ. Ent.* **66,** 33–36

Akhtar, M. (1971). Laboratory feeding tests with *Chrotogonus trachypterus* (Blanchard) (Orthoptera, Acrididae). *Pakist. J. Zool.* **3,** 163–167

Albert, P. J. (1980). Morphology and innervation of mouthpart sensilla in larvae of the spruce budworm, *Choristoneura fumiferana* (Clem.) (Lepidoptera: Tortricidae). *Can. J. Zool.* **58,** 842–851

Albert, P. J. and Seabrook, W. D. (1973). Morphology and histology of the antenna of the male eastern spruce budworm, *Choristoneura fumiferana* (Clem.) (Lepidoptera: Tortricidae). *Can. J. Zool.* **51,** 443–448

Albert, P. J., Seabrook, W. D. and Paim, U. (1974). Isolation of a sex pheromone receptor in males of the eastern spruce budworm, *Choristoneura fumiferana* (Clem.), (Lepidoptera: Tortricidae). *J. comp. Physiol.* **91,** 79–89

Altner, H. and Ernst, K.-D. (1974). Struktureigentumlichkeiten antennaler Sensillen bodenlebender Collembolen. *Pedobiologia* **14,** 118–122

Altner, H. and Prillinger, L. (1980). Ultrastructure of invertebrate chemo-, thermo-, and hygroreceptors and its functional significance. *Int. Rev. Cytol.* **67,** 69–139

Altner, H. and Stetter, H. (1980). Olfactory input from the maxillary palps in the cockroach as compared with the antennal input. In: *Olfaction and Taste VII* (H. van der Starre, ed.) Information Retrieval Ltd., London

Altner, H. and Thies, G. (1972). Reizleitende Strukturen und Ablauf der Häutung an Sensillen einer euedaphischen Collembolenart. *Z. Zellforsch.* **129,** 196–216

Altner, H. and Thies, G. (1973). A functional unit consisting of an eversible gland with neurosecretory innervation and a proprioceptor derived from a complex sensillum in an insect. *Z. Zellforsch.* **145,** 503-519

Altner, H. and Thies, G. (1976). The postantennal organ: a specialized unicellular sensory input to the protocerebrum in Apterygotan insects (Collembola). *Cell Tiss. Res.* **167,** 97–110

Altner, H. and Thies, G. (1978). The multifunctional sensory complex in the antennae of *Allacma fusca* (Insecta). *Zoomorphologie* **91,** 119–131

Altner, H., Ernst, K.-D. and Karuhize, G. (1970). Untersuchungen am Postantennalorgan der Collembolen (Apterygota) I. Die Feinstruktur der postantennalen Sinnesborste von *Sminthurus fuscus* (L.). *Z. Zellforsch.* **111,** 263–285

Altner, H., Karuhize, G. and Ernst, K.-D. (1971). Untersuchungen am Postantennalorgan der Collembolen II. Cuticulärer Apparat und Dendritenendigung bei *Onychiurus* spec. *Rev. Ecol. Biol. Sol.* **8,** 31–35

Altner, H., Sass, H. and Altner, I. (1977). Relationship between structure and function of antennal chemo-, hygro-, and thermoreceptive sensilla in *Periplaneta americana*. *Cell Tiss. Res.* **176,** 389–405

Amrine, J. W. and Lewis, R. E. (1978). The topography of the exoskeleton of *Cediopsylla simplex* (Baker 1895) (Siphonaptera: Pulicidae). 1. The head and its appendages. *J. Parasit.* **64,** 343–358

Avé, D. A. (1980). Morphology of mouthpart sensory structures on larvae of *Heliothis zea* (Boddie) (Lepidoptera, Noctuidae). In: *Olfaction and Taste VII* (H. van der Starre, ed). Information Retrieval Ltd., London

Avé, D., Frazier, J. L. and Hatfield, L. D. (1978). Contact chemoreception in the tarnished plant bug *Lygus lineolaris*. *Ent. exp. Appl.* **24,** 217–227

Aziz, S. A. (1958). Probable hygroreceptors in the desert locust, *Schistocerca gregaria* Forsk. (Orthoptera: Acrididae). *Indian J. Ent.* **19,** 164–170

Banfill, J. C. and Brusven, M. A. (1973). Food habits and ecology of grasshoppers in the Seven Devils Mountains and Salmon River breaks of Idaho. *Melanderia* no. 12, 33 pp.

Bangying, X. (1979). Structure and function of the mouthparts of *Trichogramma dendrolimi*. *Scientia Sin.* **22,** 975–977

Barbier, R. (1961). Contribution a l'étude da l'anatomie sensori-nerveuse des insectes trichoptères. *Annls Sci. nat. Zool.* **3,** 171–183

Bareth, C. and Juberthie-Jupeau, L. (1977). Ultrastructure des soies sensorielles des palpes labiaux de *Campodea sensillifera* (Conde et Mathien) (Insecta: Diplura). *Int. J. Insect Morphol. & Embryol.* **6,** 191–200

Barlin, M. R. and Vinson, S. B. (1981). Multiporous plate sensilla in antennae of the Chalcidoidea (Hymenoptera). *Int. J. Insect Morphol. & Embryol.* **10,** 29–42

Bassemir, U. and Hansen, K. (1980). Single-pore sensilla of damselfly-larvae: representatives of phylogenetically old contact chemoreceptors? *Cell Tiss. Res.* **207,** 307–320

Bay, D. E. and Pitts, C. W. (1976). Antennal olfactory sensilla of the face fly, *Musca autumnalis* Degeer (Diptera: Muscidae). *Int. J. Insect Morphol. & Embryol.* **5,** 1–16

Bedini, C. and Tongiorgi, P. (1971). The fine structure of the pseudoculus of Acerentomids Protura (Insecta Apterygota). *Monitore Zool. ital.* (n.s.) **5,** 25–38

Behan, M. and Ryan, M. F. (1977). Sensory receptors on the ovipositor of the carrot fly (*Psila rosae* (F.)) (Diptera: Psilidae) and the cabbage root fly (*Delia brassicae* Wiedemann) (Diptera: Anthomyiidae). *Bull. ent. Res.* **67,** 383–389

Behan, M. and Ryan, M. F. (1978). Ultrastructure of antennal sensory receptors of *Tribolium* larvae (Coleoptera: Tenebrionidae). *Int. J. Insect. Morphol & Embryol.* **7,** 221–236

Behan, M. and Schoonhoven, L. M. (1978). Chemoreception of an oviposition deterrent associated with eggs in *Pieris brassicae. Ent. exp. Appl.* **24,** 163–179

Behrend, K. (1971). Riechen in Wasser und in Luft bei *Dytiscus marginalis* L. *Z. vergl. Physiol.* **75,** 108–122

Bennett, F. D. (1966). Investigations on the insects attacking the aquatic ferns, *Salvinia spp.* in Trinidad and Northern South America. *Proc. Sth. Weed Control Conf.* **19,** 497–504

Bennett, F. D. (1970). Insects attacking water hyacinth in the West Indies, British Honduras and the USA. *Hyacinth Control J.* **8,** 10–13

Bernard, J. (1974). *Étude électrophysiologique de récepteurs impliqués dans l'orientation vers l'hôte et dans l'acte hématophage chez un Hémiptère: Triatoma infestans.* Thesis, University of Rennes

Bernays, E. A. and Chapman, R. F. (1974). The regulation of food intake by acridids. In: *Experimental Analysis of Insect Behaviour* (L. Barton Browne, ed.). Springer Verlag, Berlin

Bernays, E. A. and Chapman, R. F. (1977). Deterrent chemicals as a basis of oligophagy in *Locusta migratoria* (L.). *Ecol. Ent.* **2,** 1–18

Bernays, E. A. and Chapman, R. F. (1978). Plant chemistry and acridoid feeding behaviour. In: *Biochemical Aspects of Plant and Animal Co-evolution* (J. A. Harborne, ed.). Academic Press, London

Bernays, E. A., Chapman, R. F., Cook, A. G., McVeigh, L. J. and Page, W. W. (1975). Food plants in the survival and development of *Zonocerus variegatus* (L.). *Acrida* **4,** 33–46

Blaney, W. M. (1974). Electrophysiological responses of the terminal sensilla on the maxillary palps of *Locusta migratoria* (L.) to some electrolytes and non-electrolytes. *J. exp. Biol.* **60,** 275–293

Blaney, W. M. (1975). Behavioural and electrophysiological studies of taste discrimination by the maxillary palps of larvae of *Locusta migratoria* (L.). *J. exp. Biol.* **62,** 555–569

Blaney, W. M. (1980). Chemoreception and food selection by locusts. In: *Olfaction and Taste VII* (H. van der Starre, ed.). Information Retrieval Ltd., London

Blaney, W. M. and Chapman, R. F. (1969). The anatomy and histology of the maxillary palp of *Schistocerca gregaria. J. Zool., Lond.* **157,** 509–535

Blaney, W. M. and Chapman, R. F. (1970). The functions of the maxillary palps of Acrididae (Orthoptera). *Ent. exp. Appl.* **13,** 363–376

Blaney, W. M., Chapman, R. F. and Cook, A. G. (1971). The structure of the terminal sensilla of the maxillary palps of *Locusta migratoria* (L.), and changes associated with moulting. *Z. Zellforsch.* **121,** 48–68

Block, E. F. and Bell, W. J. (1974). Ethometric analysis of pheromone receptor function in cockroaches. *J. Ins. Physiol.* **20,** 993–1003

Blom, F. (1978). Sensory input behavioural output relationships in the feeding activity of some lepidopterous larvae. *Ent. exp. Appl.* **24,** 258–263

Boeckh, J. (1962). Elektrophysiologische Untersuchungen an einzelnen Geruchsrezeptoren auf den Antennen des Totengräbers (*Necrophorus*, Coleoptera). *Z. vergl, Physiol.* **46,** 212–248

Boeckh, J. (1973). Die Reaktionen von Neuronen in Deutocerebrum der Wanderheuschrecke bei Duftreizung der Antennen. *Verh. dtsch. zool. ges.* **66,** 189–193

Boeckh, J. (1980). Neural mechanisms of odour detection and odour discrimination in the insect olfactory pathway. In: *Insect Neurobiology and Pesticide Action.* Society of Chemical Industry, London

Boeckh, J. and Boeckh, V. (1979). Threshold and odor specificity of pheromone-sensitive neurons in the deutocerebrum of *Antheraea pernyi* and *A. polyphemus* (Saturnidae). *J. comp. Physiol.* **132,** 235–242

Boeckh, J., Kaissling, K.-E. and Schneider, D. (1960). Sensillum und Bau der Antennengeissel von *Telea polyphemus* (Vergleiche mit weiteren Saturniden: *Antheraea, Platysamia* und *Philosamia*). *Zool. Jahrb. Abt. Anat. Ontog. Tiere* **78,** 559–584

Boeckh, J., Kaissling, K.-E. and Schneider, D. (1965). Insect olfactory receptors. *Cold Spring Harb. Symp. quant. Biol.* **30,** 263–280

Boeckh, J., Sandri, C. and Akert, K. (1970a). Sensorische Eingänge und synaptische Verbindungen im Zentralnervensystem von Insekten. *Z. Zellforsch.* **103,** 429–446

Boeckh, J., Sass, H. and Wharton, D. R. A. (1970b). Antennal receptors: reactions to female sex attractant in *Periplaneta americana. Science, N.Y.* **168,** 589

Boeckh, J., Boeckh, V. and Kühn, A. (1977). Further data on the topography and physiology of central olfactory neurons in insects. In: *Olfaction and Taste VI* (J. LeMagnen and P. MacLeod eds). Information Retrieval Ltd., London

Boer, G. de, Dethier, V. G. and Schoonhoven, L. M. (1977). Chemoreceptors in the preoral cavity of the tobacco hornworm, *Manduca sexta*, and their possible function in feeding behaviour. *Ent. exp. Appl.* **21,** 287–298

Bolwig, N. (1946). Senses and sense organs of the anterior end of the house fly larva. *Vidensk. Meddr dansk naturh. Foren.* **109,** 81–217

Boo, K. S. and McIver, S. B. (1975). Fine structure of sunken thick-walled pegs (sensilla ampullacea and coeloconica) on the antennae of mosquitoes. *Can. J. Zool.* **53,** 262–266

Boppré, M. (1978). Chemical communication, plant relationships, and mimicry in the evolution of danaid butterflies. *Ent. exp. Appl.* **24,** 264–267

Borden, J. H. (1968). Antennal morphology of *Ips confusus* (Coleoptera: Scolytidae). *Ann. ent. Soc. Am.* **61,** 10–13

Borden, J. H. and Wood, D. L. (1966). The antennal receptors and olfactory response of *Ips confusus* (Coleoptera: Scolytidae) to male sex attractant in the laboratory. *Ann. ent. Soc. Am.* **59,** 253–261

Borden, J. H., Chong, L. and Rose, A. (1978a). Morphology of the elongate placoid sensillum on the antennae of *Itoplectis conquisitor. Ann. ent. Soc. Am.* **71,** 223–227

Borden, J. H., Rose, A. and Chorney, R. J. (1978b). Morphology of the elongate sensillum placodeum on the antennae of *Aphidius smithi* (Hymenoptera: Aphidiidae). *Can. J. Zool.* **56,** 519–525

Borg, T. K. and Norris, D. M. (1971). Ultrastructure of sensory receptors on the antennae of *Scolytus multistriatus* (Marsh.). *Z. Zellforsch.* **113** 13–28

Braverman, Y. and Hulley, P. E. (1979). The relationship between the numbers and distribution of some antennal and palpal sense organs and host preference in some *Culicoides* (Diptera: Ceratopogonidae) from southern Africa. *J. med. Ent.* **15,** 419–424

Bromley, A. K., Dunn, J. A. and Anderson, M. (1979). Ultrastructure of the antennal sensilla of aphids I. Coeloconic and placoid sensilla. *Cell Tiss. Res.* **203,** 427–442

Buerger, G. (1967). Sense organs on the labra of some blood-feeding Diptera. *Quaest. Ent.* **3,** 283–290

Callahan, P. S. (1969). The exoskeleton of the corn earworm moth, *Heliothis zea* Lepidoptera: Noctuidae with special reference to the sensilla as polytubular dielectric arrays. *Univ. Georgia Coll. of Agric. Exp. Stations Res. Bull.* no. 54

Campbell, J. B., Arnett, W. H., Lambley, J. D., Jantz, O. K. and Knutson, H. (1974). Grasshoppers (Acrididae) of the Flint Hills native tallgrass prairie in Kansas. *Agric. exp. Sta., Kansas State Univ., Research paper* no. 19, 147 pp.

Chadha, G. K. and Roome, R. E. (1980). Oviposition behaviour and the sensilla of the ovipositor of *Chilo partellus* and *Spodoptera littoralis* (Lepidoptera Noctuidae). *J. Zool. London* **192,** 169–178

Chambille, I., Masson, C. and Rospars, J. R. (1980). The deutocerebrum of the cockroach *Blaberus craniifer* Burm. Spatial organization of the sensory glomeruli. *J. Neurobiol.* **11,** 135–157

Chambille, I. and Rospars, J. P. (1981). Deutocerebron de la blatte *Blaberus craniifer* Burm. (Dictyoptera: Blaberidae): étude qualitative et identification morphologique des glomerules. *Int. J. Insect Morphol. & Embryol.* **10,** 141–165

Chapman, R. F. (1962). The ecology and distribution of grasshoppers in Ghana. *Proc. zool. Soc. London* **139,** 1–66

Chapman, R. F. (1964). The structure and wear of the mandibles in some African grasshoppers. *Proc. zool. Soc. London* **142,** 107–121

Chapman, R. F. (1966). The mouthparts of *Xenocheila zarudnyi* (Orthoptera, Acrididae). *J. Zool. London* **148,** 277–288

Chapman, R. F. and Thomas, J. G. (1978). The numbers and distribution of sensilla on the mouthparts of Acridoidea. *Acrida* **7,** 115–148

Chapman, R. F., Bernays, E. A. and Simpson, S. (1981). Attraction and repulsion of the aphid, *Cavariella aegopodii*, by plant odours. *J. chem. Ecol.* **7,** 881–888

Chin, Chun-Teh (1950). Studies on the physiological relations between the larvae of *Leptinotarsa decemlineata* Say and some solanaceous plants. *Tijdschr. PlZiekt.* **56,** 1–88

Chu, L.-W. and Axtell, R. C. (1971). Fine structure of the dorsal organ of the house fly larva, *Musca domestica* L. *Z. Zellforsch.* **117,** 17–34

Chu-Wang, I.-W., and Axtell, R. C. (1972a). Fine structure of the terminal organ of the house fly larva, *Musca domestica* L. *Z. Zellforsch.* **127,** 287–305

Chu-Wang, I.-W., and Axtell, R. C. (1972b). Fine structure of the ventral organ of the house fly larva, *Musca domestica* L. *Z. Zellforsch.* **130,** 489–495

Chu-Wang, I.-W., Axtell, R. C. and Kline, D. L. (1975). Antennal and palpal sensilla of the sand fly *Culicoides furens* (Poey) (Diptera: Ceratopogonidae). *Int. J. Insect Morphol. & Embryol.* **4,** 131–149

Clements, A. N. (1963). *The Physiology of Mosquitoes.* Pergamon Press, Oxford

Cobben, R. H. (1978). Evolutionary trends in Heteroptera Part II. Mouthpart-structures and feeding strategies. *Meded. LandbHoogesch. Wageningen* 78–5

Colbo, M. H., Lee, R. M. K. W., Davies, D. M. and Yang, Y. J. (1979). Labro-cibarial sensilla and armature in adult black flies (Diptera: Simuliidae). *J. med. Ent.* **15,** 166–175

Cook, A. G. (1972). The ultrastructure of the A1 sensilla on the posterior surface of the clypeo-labrum of *Locusta migratoria migratorioides* (R & F). *Z. Zellforsch.* **134,** 539–544
Cook, A. G. (1977). The anatomy of the clypeo-labrum of *Locusta migratoria* (L.) (Orthoptera: Acrididae). *Acrida* **6,** 287–306
Cook, A. G. (1979). The ultrastructure of some sensilla on the clypeo-labrum of *Locusta migratoria* (L.) (Orthoptera: Acrididae). *Acrida* **8,** 47–62
Corbière-Tichané, G. (1969). Ultrastructure du labium de la larve du *Speophyes lucidulus* Delarouzeei (Coleoptera, Catopidae). *Z. Morph. Tiere* **66,** 73–86
Corbière-Tichané, G. (1970). Ultrastructure du labre des larves du *Speophyes lucidulus* (Delarouzeei), Coléoptère cavernicole de la sous-famille des Bathysciinae (Coleoptera, Catopidae). *Z. Morph. Tiere* **67,** 86–96
Corbière–Tichané, G. (1971). Ultrastructure de l'équipement sensoriel de la mandibule chez la larve due *Speophyes lucidulus* Delar. (Coléoptère cavernicole de la sous-famille des Bathysciinae). *Z. Zellforsch.* **112,** 129–138
Corbière-Tichané, G. (1971a). *Recherches sur l'équipement sensoriel du coléoptère cavernicole.* Thesis, Faculté de Sciences (Aix-Marseille), Université de Provence
Corbière-Tichané, G. (1974). Fine structure of an antennal sensory organ ('vésicule olfactive') of *Speophyes lucidulus* Delar. (Cave Coleoptera of the Bathysciinae sub-family). *Tissue & Cell* **3,** 535–550
Corbière-Tichané, G. and Bermond, N. (1971). Ultrastructure et électrophysiologie des styles antennaires de la larve de *Speophyes lucidulus* Delar. (Coléoptère Bathysciinae). *Annls Sci. nat. Zool.* **13,** 505–542
Corbière-Tichané, G. and Bermond, N. (1972). Sensilles énigmatiques de l'antenne de certains Coléoptères. Étude comparative au microscope électronique. *Z. Zellforsch.* **127,** 9–33
Cornet, M., Neville, E. M. and Walker, A. R. (1974). Note sur les *Culicoides* (Diptera, Ceratopogonidae) du groupe de *C. milnei* Austen, 1909, en Afrique orientale et australe. *Cah. O.R.S.T.O.M., sér. Ent. méd. et parasitol.* **12,** 231–243
Cornford, M. E., Rowley, W. A. and Klun, J. A. (1973). Scanning electron microscopy of antennal sensilla of the European corn borer, *Ostrinia nubilalis. Ann. ent. Soc. Am.* **66,** 1079–1088
Craig, D. A. and Borkent, A. (1980). Intra- and inter-familial homologies of maxillary palpal sensilla of larval Simuliidae (Diptera: Culicomorpha). *Can. J. Zool.* **58,** 2264–2279
Crnjar, R. and Prokopy, R. J. (in prep.). Morphological and electrophysiological mapping of tarsal chemoreceptors of oviposition-deterring pheromone in *Rhagoletis pomonella* (Walsh)
Dallai, R. (1971). First data on the ultrastructure of the postantennal organ of Collembola. *Rev. Ecol. Biol. Sol.* **8,** 11–29
Daly, P. F. and Ryan, M. F. (1979). Ultrastructure of antennal sensilla of *Nebria brevicollis* (Fab.) (Coleoptera: Carabidae). *Int. J. Insect Morphol. & Embryol.* **8,** 169–181
Davis, E. E. (1976). A receptor sensitive to oviposition site attractants on the antennae of the mosquito, *Aedes aegypti. J. Ins. Physiol.* **22,** 1371–1376
Davis, E. E. and Sokolove, P. G. (1975). Temperature responses of antennal receptors of the mosquito, *Aedes aegypti. J. comp. Physiol.* **96,** 223–236
Davis, E. E. and Sokolove, P. G. (1976). Lactic acid-sensitive receptors on the antennae of the mosquito, *Aedes aegypti. J. comp. Physiol.* **105,** 43–54.
Dethier, V. G. (1937). Gustation and olfaction in lepidopterous larvae. *Biol. Bull.* **72,** 7–23

Dethier, V. G. (1955). The tarsal chemoreceptors of the house fly. *Proc. Roy. Entomol. Soc. London A* **30,** 87–90

Dethier, V. G. (1973). Electrophysiological studies of gustation in lepidopterous larvae. II. Taste spectra in relation to food-plant discrimination. *J. comp. Physiol.* **82,** 103–134

Dethier, V. G. (1974). The specificity of the labellar chemoreceptors of the blowfly and the response to natural foods. *J. Ins. Physiol.* **20,** 1859–1869

Dethier, V. G. (1976). *The Hungry Fly. A Physiological Study of the Behaviour Associated with Feeding.* Harvard University Press

Dethier, V. G. (1980). Evolution of receptor sensitivity to secondary plant substances with special reference to deterrents. *Am. Nat.* **115,** 45–66

Dethier, V. G. (1980a). Responses of some olfactory receptors of the eastern tent caterpillar (*Malacosoma americanum*) to leaves. *J. chem. Ecol.* **6,** 213–220

Dethier, V. G. and Kuch, J. H. (1971). Electrophysiological studies of gustation in lepidopterous larvae. I. Comparative sensitivity to sugars, amino acids, and glycosides. *Z. vergl. Physiol.* **72,** 343–363

Dethier, V. G. and Schoonhoven, L. M. (1969). Olfactory coding by lepidopterous larvae. *Ent. exp. Appl.* **12,** 535–543

Dickens, J. C. and Payne, T. L. (1977). Bark beetle olfaction: pheromone receptor system in *Dendroctonus frontalis. J. Ins. Physiol.* **23,** 481–489

Dickens, J. C. and Payne, T. L. (1978). Structure and function of the sensilla on the antennal club of the southern pine beetle, *Dendroctonus frontalis* (Zimmerman) (Coleoptera: Scolytidae). *Int. J. Insect Morphol. & Embryol.* **7,** 251–265

Drongelen, W. van (1979). Contact chemoreception of host plant specific chemicals in larvae of various *Yponomeuta* species (Lepidoptera). *J. comp. Physiol.* **134,** 265–279

Dubose, W. P. and Axtell, R. C. (1968). Sensilla on the antennal flagella of *Hippelates* eye gnats. *Ann. ent. Soc. Am.* **61,** 1547–1561

Dumpert, K. (1972a). Bau und Verteilung der Sensillen auf der Antennengeissel von *Lasius fuliginosus* (Latr.) (Hymenoptera, Formicidae). *Z. Morph. Tiere* **73,** 95–116

Dumpert, K. (1972b). Alarmstoffrezeptoren auf der Antenne von *Lasius fuliginosus* (Latr.) (Hymenoptera, Formicidae). *Z. vergl. Physiol.* **76,** 403–425

Dunn, J. A. (1978). Antennal sensilla of vegetable aphids. *Ent. exp. Appl.* **24,** 348–349

Dyer, L. J. and Seabrook, W. D. (1975). Sensilla on the antennal flagellum of the sawyer beetles *Monochamus notatas* (Drury) and *Monochamus scutellatus* (Say) (Coleoptera: Cerambycidae). *J. Morph.* **146,** 513–532

Dyer, L. J. and Seabrook, W. D. (1978). Evidence for the presence of acceptor sites for different terpenes on one receptor cell in male *Monochamus notatus* (Drury) (Coleoptera: Cerambycidae). *J. chem. Ecol.* **4,** 523–529

Eastop, V. F. (1973). Deductions from the present day host plants of aphids and related insects. *Symp. Roy. Entomol. Soc. London* **6,** 157–178

Eastop, V. (1979). Sternorrhyncha as Angiosperm taxonomists. *Symb. bot. upsal.* **22,** 120–134

Eaton, J. L. (1979). Chemoreceptors in the cibario-pharyngeal pump of the cabbage looper moth, *Trichoplusia ni* (Lepidoptera: Noctuidae). *J. Morph.* **160,** 7–16

Ellis, P. E. and Brimacombe, L. C. (1980). The mating behaviour of the Egyptian cotton leafworm moth, *Spodoptera littoralis* (Boisd.). *Anim. Behav.* **28,** 1239–1248

Emden, H. F. van (1972). Aphids as phytochemists. In: *Phytochemical Ecology* (J. B. Harborne, ed.). Academic Press, London

Emden, H. F. van (1978). Insects and secondary plant substances – an alternative viewpoint with special reference to aphids. In: *Biochemical Aspects of Plant and Animal Co-evolution* (J. B. Harborne, ed.). Academic Press, London

Emden, H. F. van, Eastop, V. F., Hughes, R. D. and Way, M. J. (1969). The ecology of *Myzus persicae*. *Ann. Rev. Ent.* **14,** 197–270

Ernst, K.-D., Boeckh, J. and Boeckh, V. (1977). A neuroanatomical study on the organization of the central antennal pathways in insects II. Deutocerebral connections in *Locusta migratoria* and *Periplaneta americana*. *Cell Tiss. Res.* **176,** 285–308

Esslen, J. and Kaissling, K.-E. (1976). Zahl und Verteilung antennaler Sensillen bei der Honigbiene (*Apis mellifera* L.). *Zoomorphologie* **83,** 227–251

Exner, K. and Craig, D. A. (1976). Larvae of Alberta Tanyderidae (Diptera: Nematocera). *Quaest. Ent.* **12,** 219–237

Falk, R., Bleiser-Avivi, N. and Atidia, J. (1976). Labellar taste organs of *Drosophila melanogaster*. *J. Morph.* **150,** 327–341

Felt, B. T. and Vande Berg, J. S. (1976). Ultrastructure of the blowfly chemoreceptor sensillum (*Phormia regina*). *J. Morph.* **150,** 763–784

Ferguson, A. W. and Free, J. B. (1981). Factors determining the release of Nasonov pheromone by honeybees at the hive entrance. *Physiol. Ent.* **6,** 15–19

Fishelson, L. (1960). The biology and behaviour of *Poekilocerus bufonius* Klug with special reference to the repellent gland (Orth. Acrididae). *Eos, Madr.* **36,** 41–62

Friend, W. G. and Smith, J. J. B. (1977). Factors affecting feeding by bloodsucking insects. *Ann. Rev. Ent.* **22,** 309–331

Gaffal, K. P. (1981). Terminal sensilla on the labium of *Dysdercus intermedius* Distant (Heteroptera: Pyrrhocoridae). *Int. J. Insect Morphol. & Embryol.* **10,** 1–6

Galic, M. (1971). Die Sinnesorgane an der Glossa, dem Epipharynx und dem Hypopharynx der Arbeiterin von *Apis mellifica* L. (Insecta, Hymenoptera). *Z. Morph. Tiere* **70,** 201–228

Gangwere, S. K. and Agacino, E. M. (1973). Food selection and feeding behaviour in Iberian Orthopteroidea. *An. Inst. nac. Invest. agrarias.* (*Prot. Veg.*) **3,** 251–337

Gangwere, S. K. and Ronderos, R. A. (1975). A synopsis of food selection in Argentine Acridoidea. *Acrida* **4,** 173–194

Gardner, C. F., Nielsen, L. T., and Knight, K. L. (1973). Morphology of the mouthparts of larval *Aedes communis* (DeGeer) (Diptera: Culicidae). *Mosquito Syst.* **5,** 163–182

Gernet, G. von and Buerger, G. (1966). Labral and cibarial sense organs of some mosquitoes. *Quaest. Ent.* **2,** 259–270

Gilbert, L. E. and Raven, P. H. (1975). *Coevolution of Animals and Plants*. University of Texas Press, Austin

Gillett, S. D. (1975). The action of the gregarisation pheromone on five non-behavioural characters of phase polymorphism of the desert locust, *Schistocerca gregaria* (Forskål). *Acrida* **4,** 137–149

Gillon, Y. (1972). Caractéristiques quantitatives du développement et de l'alimentation d'*Anablepia granulata* (Ramme 1929) (Orthoptera: Gomphocerinae). *Annls Univ. Abidjan* (E) **5,** 373–393

Grabowski, C. T. and Dethier, V. G. (1954). The structure of the tarsal chemoreceptors of the blowfly, *Phormia regina* Meigen. *J. Morph.* **94,** 1–20

Grant, G. G. (1971a). Electroantennogram responses to the scent brush secretions of several male moths. *Ann. ent. Soc. Am.* **64,** 1428–1431

Grant, G. G. (1971b). Feeding activity of adult cabbage loopers on flowers with strong olfactory stimuli. *J. econ. Ent.* **64,** 315–316

Grant, G. G., Brady, U. E. and Brand, J. M. (1972). Male armyworm scent brush secretion: identification and electroantennogram study of major components. *Ann. ent. Soc. Am.* **65,** 1224–1227

Greenberg, B. and Ash, N. (1972). Setiferous plaques on antennal pedicels of muscoid Diptera: appearance in various species and tests of function. *Ann. ent. Soc. Am.* **65,** 1340–1346

Grula, J. W. and Taylor, O. R. (1980). A micromorphological and experimental study of the antennae of the sulfur butterflies, *Colias eurytheme* and *C. philodice* (Lepidoptera: Pieridae). *J. Kansas ent. Soc.* **53,** 476–484

Hallberg, E. (1979). The fine structure of the antennal sensilla of the pine saw fly *Neodiprion sertifer* (Insecta: Hymenoptera). *Protoplasma* **101,** 111–126

Hamamura, Y., Hayashiya, K., Naito, K., Matsuura, K. and Nishida, J. (1962). Food selection by silkworm larvae. *Nature, London* **194,** 754–755

Hamon, M. (1961). Contribution a l'étude de la morphogenèse sensori-nerveuse des Dytiscidae (Insectes Coléoptères). *Annls Sci. nat. Zool.* **12,** 153–171

Harbach, R. E. and Knight, K. L. (1977). A mosquito taxonomic glossary XI. The larval maxilla. *Mosquito Syst.* **9,** 128–175

Harbach, R. E. and Larsen, J. R. (1976). Ultrastructure of sensilla on the distal antennal segment of adult *Oncopeltus fasciatus* (Dallas) (Hemiptera: Lygaeidae). *Int. J. Insect Morphol. & Embryol.* **5,** 23–33

Harbach, R. E. and Larsen, J. R. (1977). Fine structure of antennal sensilla of the adult mealworm beetle, *Tenebrio molitor* L. (Coleoptera: Tenebrionidae). *Int. J. Insect. Morphol. & Embryol.* **6,** 41–60

Harborne, J. B. (1978). *Biochemical Aspects of Plant and Animal Co-evolution.* Academic Press, London

Hardie, J. (1980). Reproductive, morphological and behavioural affinities between the alate gynopara and virginopara of the aphid, *Aphis fabae. Physiol. Ent.* **5,** 385–396

Haskell, P. T. and Mordue, A. J. (1969). The role of mouthpart receptors in the feeding behaviour of *Schistocerca gregaria. Ent. exp. Appl.* **12,** 591–610

Haskell, P. T. and Schoonhoven, L. M. (1969). The function of certain mouthpart receptors in relation to feeding in *Schistocerca gregaria* and *Locusta migratoria migratorioides. Ent. exp. Appl.* **12,** 423–440

Hatfield, L. D. and Frazier, J. L. (1980). Ultrastructure of the labial tip sensilla of the tarnished plant bug, *Lygus lineolaris* (P. de Beauvois) (Hemiptera: Miridae). *Int. J. Insect Morphol & Embryol.* **9,** 59–66

Hatfield, L. D., Frazier, J. L. and Coons, L. B. (1976). Antennal sensilla of the pecan weevil, *Curculio caryae* (Horn) (Coleoptera: Curculionidae). *Int. J. Insect Morphol. & Embryol.* **5,** 279–287

Haupt, J. (1972). Ultrastruktur des Pseudoculus von *Eosentomon* (Protura, Insecta). *Z. Zellforsch.* **135,** 539–551

Hawke, S. D. and Farley, R. D. (1971). Antennal chemoreceptors of the dwarf burrowing cockroach, *Arenivaga* sp. *Tissue & Cell* **3,** 649–664

Hawke, S. D., Farley, R. D. and Greany, P. D. (1973). The fine structure of sense organs in the ovipositor of the parasitic wasp, *Orgilus lepidus* Muesebeck. *Tissue & Cell* **5,** 171–184

Henig, B. (1930). O unerwienin tak zwanych niz szych organów zmyslowych gąsienic motyli. *Prace Wydz. Mat.-Przyrod. Tow. Przyj. Nauk w Wilnie* **6,** 41–81

Henning, B. (1974). Morphologie und Histologie des Tarsen von *Tettigonia viridissima* L. (Orthoptera, Ensifera). *Z. Morph. Tiere* **79,** 323–342

Hewitt, P. H., Whitehead, V. B. and Read, J. S. (1969). Quebrachitol: a phagostimulant for the larvae of the moth, *Serrodes partita*. *J. Ins. Physiol.* **15,** 1929–1933

Hicks, K. L. (1974). Mustard oil glucosides: feeding stimulants for adult cabbage flea beetles, *Phyllotreta cruciferae* (Coleoptera: Chrysomelidae). *Ann. ent. Soc. Am.* **67,** 261–264

Hildebrand, J. G., Matsumoto, S. G., Camazine, S. M., Tolbert, L. P., Blank, S., Ferguson, H. and Elker, V. (1980). Organisation and physiology of antennal centres in the brain of the moth *Manduca sexta*. In: *Insect Neurobiology and Pesticide Action.* Society of Chemical Industry, London

Hintze-Podufal, Ch. and Otto, B. (1975). External morphology of antennae and their sense organs in the roach *Gromphadorhina brunneri* (Blattoidea: Dictyoptera). *J. Morph.* **146,** 251–264

Hooper, R. L., Pitts, C. W. and Westfall, J. A. (1972). Sense organs on the ovipositor of the face fly, *Musca autumnalis*. *Ann. ent. Soc. Am.* **65,** 577–586

Hsiao, T. H. and Fraenkel, G. (1968). The role of secondary plant substances in the food specificity of the Colorado potato beetle, *Leptinotarsa decemlineata* (Coleoptera: Chrysomelidae). *Ann. ent. Soc. Am.* **61,** 44–54

Hummelen, P. and Gillon, Y. (1968). Étude de la nourriture des Acridiens de la savane de Lamto en Côte D'Ivoire. *Annls Univ. Abidjan* (E) **1,** 199–206

Ismail, I. A. H. (1962). Sense organs in the antennae of *Anopheles maculipennis atroparvus* (v. Thiel), and their possible function in relation to the attraction of female mosquito to man. *Acta. trop.* **19,** 1–58

Ismail, I. A. H. (1964). Comparative study of sense organs in the antennae of culicine and anopheline female mosquitoes. *Acta trop.* **21,** 155–168

Jacobson, M. (1972). *Insect Sex Pheromones*. Academic Press, New York

Jaisson, P. (1969). Études de la distribution des organes sensoriels de l'antenne et de leurs relations possible avec le comportement chez deux fourmis myrmicines: *Myrmica laevinodis* Nyl. et *Aphaenogaster gibbosa* Latr. récoltées dans la region des Eyzius. *Insectes Soc.* **16,** 279–312

Jefferson, R. N., Rubin, R. E., McFarland, S. U. and Shorey, H. H. (1970). Sex pheromones of noctuid moths. XXII. The external morphology of the antennae of *Trichoplusia ni*, *Heliothis zea*, *Prodenia ornithogalli*, and *Spodoptera exigua*. *Ann. ent. Soc. Am.* **63,** 1227–1238

Jermy, T. (1966). Feeding inhibitors and food preference in chewing phytophagous insects. *Ent. exp. Appl.* **9,** 1–12

Jez, D. H. and McIver, S. B. (1980). Fine structure of antennal sensilla of larval *Toxorhynchites brevipalpis* Theobald (Diptera: Culicidae). *Int. J. Insect Morphol. & Embryol.* **9,** 147–159

Jones, M. G. (1944). The structure of the antenna of *Aphis* (*Dorsalis*) *fabae* Scopoli, and of *Melanoxantherium salicis* L. (Hemiptera), and some experiments on olfactory responses. *Proc. Roy. Entomol. Soc. London A* **19,** 13–22

Jones, O. T. and Coaker, T. H. (1978). A basis for host plant finding in phytophagous larvae. *Ent. exp. Appl.* **24,** 472–484

Joyce, R. J. V. (1952). The ecology of grasshoppers in east central Sudan. *Anti-Locust Bull.* no. 11, 99 pp.

Juberthie, C. and Massoud, Z. (1977). L'equipment sensoriel de l'antenne d'un coléoptère troglobie, *Aphaenops cryticola* Linder (Coleoptera: Trechinae) *Int. J. Insect. Morphol. & Embryol.* **6,** 147–160

Kafka, W. A. (1970). Molekulare Wechselwirkungen bei der Erregung einzelner Riechzellen. *Z. vergl. Physiol.* **70,** 105–143

Kafka, W. A. (1971). Specificity of odor-molecule interaction in single cells. In: *Gustation and Olfaction* (G. Okloff and A. F. Thomas, eds). Academic Press, London

Kaib, N. (1974). Die Fleisch- und Blumenduftrezeptoren auf der Antenne der Schmeissfliege *Calliphora vicina*. *J. comp. Physiol.* **95,** 105–121

Kaissling, K. E. (1971). Insect olfaction. In: *Handbook of Sensory Physiology* (L. M. Beidler, ed.) Vol. 4. Springer Verlag, Berlin

Kaissling, K. E. (1974). Sensory transduction in insect olfactory receptors. In: *Biochemistry of Sensory Functions* (L. Jaenicke, ed.). Springer Verlag, Berlin

Kaissling, K. E. (1979). Recognition of pheromone by moths, especially in saturniids and *Bombyx mori*. In: *Odour Communication in Animals* (F. J. Ritter, ed.). Elsevier/North-Holland Biomedical Press, Amsterdam

Kaissling, K. E. and Renner, M. (1968). Antennale Rezeptoren für Queen Substance und Sterzelduft bei der Honigbiene. *Z. vergl. Physiol.* **59,** 357–361

Kaissling, K. E., Kasang, G., Bestmann, H. I., Stransky, W. and Vostrowsky, O. (1978). A new pheromone of the silkworm moth *Bombyx mori*. Sensory pathway and behaviour effect. *Naturwissenchaften* **65,** 382–384

Karuhize, G. R. (1971). The structure of the postantennal organ in *Onychiurus* sp. (Insecta: Collembola) and its connection to the central nervous system. *Z. Zellforsch.* **118,** 263–282

Kay, R. E. (1971). Olfactory unit potentials and receptor potential responses of *Lucilia sericata*. *Am. J. Physiol.* **220,** 1481–1487

Kellogg, F. E. (1970). Water vapour and carbon dioxide receptors in *Aedes aegypti*. *J. Ins. Physiol.* **16,** 99–108.

Kendall, M. D. (1970). The anatomy of the tarsi of *Schistocerca gregaria* Forskål. *Z. Zellforsch.* **109,** 112–137

Kennedy, J. S. and Moorhouse, J. E. (1969). Laboratory observations on locust responses to wind-borne grass odour. *Ent. exp. Appl.* **12,** 487–503

Kevan, D. K. McE. (1954). A study of the genus *Chrotogonus* Audinet-Serville, 1839 (Orthoptera-Acrididae). III, A review of available information on its economic importance, biology, etc. *Indian J. Ent.* **16,** 145–172

Kijima, H., Amakawa, T., Nakashima, M. and Morita, H. (1977). Properties of membrane-bound α-glucosidases: possible sugar receptor protein of the blowfly *Phormia regina*. *J. Ins. Physiol.* **23,** 469–479

Klein, U. (1979). *Zur Funktion und Struktur der Palpen und ihres Sensillenfeldes bei Gryllus bimaculatus De Geer*. Inaugural-Dissertation zur Erlangung des Doktorgrades des Fachbereichs Biologie der Luding-Maximilians-Universität, München

Klein, U. (1981). Sensilla on the cricket palp. Fine structure and spatial organization. *Cell Tiss. Res.* **219,** 229–252

Kostrowicki, A. S. (1969). Geography of the palaearctic Papilionoidea (Lepidoptera). *Zaklad Zoologii Systematycznej, Polskiej Akademii Nauk. Panstwowe Wydawnictwo Naukowe* 1–314

Kraus, C. (1957). Versuch einer morphologischen und neurophysiologischen Analyse des Stechaktes von *Rhodnius prolixus* Stal 1858. *Acta trop.* **14,** 36–87

Kryzywiec, D. (1968). Neue, unbekannte Sinnesorgane der Blattläuse (Homoptera, Aphidoidea). *Bull. Acad. pol. Sci. Cl. II Sér. Sci. biol.* **16,** 765–771

Kusano, T. and Sato, H. (1980). The sensitivity of tarsal chemoreceptors for sugars in the cabbage butterfly, *Pieris rapae crucivora* Boisduval. *Appl. Ent. Zool.* **15,** 385–391

Kuwabara, M. and Takeda, K. (1956). On the hygroreceptor of the honeybee, *Apis mellifica*. *Seiro-Seitai* **7,** 1–6

Lacher, V. (1964). Elektrophysiologische Untersuchungen an einzelnen Rezeptoren für Geruch, Kohlendioxyd, Luftfeuchtigkeit und Temperatur auf den Antennen der Arbeitsbiene und der Drohne (*Apis mellifica* L.). *Z. vergl. Physiol.* **48,** 587–623

Lacher, V. (1967). Electrophysiologische Untersuchungen an einzelnen Geruchsrezeptoren auf den Antennen weiblicher Moskitos (*Aedes aegypti* L.) *J. Ins. Physiol.* **13,** 1461–1470

Lacher, V. (1971). Arbeitsbereiche von Geruchsrezeptoren auf der Moskitoantenne (*Aedes aegypti*). *J. Ins. Physiol.* **17,** 507–517

Lambin, M. (1973). Les sensilles de l'antenne chez quelques blattes et en particulies chez *Blaberus* craniifer (Burm.). *Z. Zellforsch.* **143,** 183–206

Larink, O. (1978). Sensillenmuster auf den labium von Lepismatiden (Insecta: Zygentoma). *Zool. Anz.* **201,** 341–352

Larink, O. (1979). Neue taxonomische Merkmale bei Lepismatiden (Insecta, Zygentome). *Abh. naturw. Ver. Bremen* **39,** 77–82

Larsen, J. R. and Owen, W. B. (1971). Structure and function of the ligula of the mosquito *Culiseta inornata* (Williston). *Trans. Am. microsc. Soc.* **90,** 294–308

Lawton, J. H. (1978). Host-plant influences on insect diversity: the effects of space and time. *Symp. Roy. Entomol. Soc. London* **9,** 105–125

Lee, R. (1974). Structure and function of the fascicular stylets, and the labral and cibarial sense organs of male and female *Aedes aegypti* (L.) (Diptera, Culicidae). *Quaest. ent.* **10,** 187–215

Lee, R. M. K. W. and Davies, D. M. (1978). Cibarial sensilla of *Toxorhynchites* mosquitoes (Diptera: Culicidae). *Int. J. Insect Morphol. & Embryol.* **7,** 189–194

Lees, A. D. (1966). The control of polymorphism in aphids. *Adv. Insect Physiol.* **3,** 207–277

Levinson, A. R., Levinson, H. Z., Schwaiger, H., Cassidy, R. F. and Silverstein, R. M. (1978). Olfactory behaviour and receptor potentials of the Khapra beetle *Trogoderma granarium* (Coleoptera: Dermestidae) induced by the major components of its sex pheromone, certain analogues, and fatty acid esters. *J. chem. Ecol.* **4,** 95–108

Levinson, H. Z., Levinson, A. R., Müller, B. and Steinbrecht, R. A. (1974). Structure of sensilla, olfactory perception, and behaviour of the bedbug, *Cimex lectularis*, in response to its alarm pheromone. *J. Ins. Physiol.* **20,** 1231–1248

Lewis, C. T. (1954). Studies concerning the uptake of contact insecticides, I. The anatomy of the tarsi of certain Diptera of medical importance. *Bull. ent. Res.* **45,** 711–721

Lewis, C. T. (1970). Structure and function in some external receptors. *Symp. Roy. Entomol. Soc. London* **5,** 59–76

Lewis, C. T. (1971). Superficial sense organs of the antennae of the fly, *Stomoxys calcitrans*. *J. Ins. Physiol.* **17,** 449–461

Lewis, C. T. and Marshall, A. T. (1970). The ultrastructure of the sensory plaque organs of the antennae of the Chinese lantern fly, *Pyrops candelaria* L., (Homoptera, Fulgoridae). *Tissue & Cell* **2,** 375–385

Lewis, W. J., Jones, R. L., Hordlund, D. A. and Gross, H. R. (1977). Kairomones and their use for management of entomophagous insects. *Colloques int. Cent. natn. Rech. scient.* **265,** 455–469

Lhoste, J. (1957). Données anatomiques et histophysiologiques sur *Forficula auricularia* L. *Archs. Zool. exp. gén.* **95,** 75–252

Liebermann, A. (1925). Correlation zwischen den antennalen Geruchsorganen und der Biologie der Musciden. *Z. Morph. Ökol. Tiere.* **5,** 1–97

Lo, S. E. and Acton, A. B. (1969). The ultrastructure of the rostral sensory organs of the water bug, *Cenocorixa bifida* (Hungerford) (Hemiptera). *Can. J. Zool.* **47,** 717–722

Loftus, R. (1976). Temperature-dependent dry receptor on antenna of *Periplaneta.* Tonic response. *J. comp. Physiol.* **111,** 153–170

Louveaux, A. (1972). Equipment sensoriel et système nerveux périphérique des pièces buccales de *Locusta migratoria* L. *Insectes Soc.* **19,** 359–368

Louveaux, A. (1973). Étude d'une plage d'organes sensoriels sur les galea de *Locusta migratoria* L. *C.R. hebd. Séanc. Acad. Sci., Paris.* **277,** 1353–1356

Louveaux, A. (1975). Étude de l'innervation sensorielle de l'hypopharynx de larves de *Locusta migratoria migratorioides* R. et F. (Orthoptère, Acrididae). *Insectes Soc.* **22,** 3–12

Ma, W. C. (1972). Dynamics of feeding responses in *Pieris brassicae* Linn. as a function of chemosensory input: a behavioural, ultrastructural and electrophysiological study. *Meded. LandbHoogesch. Wageningen* **72,** 162 pp.

Ma, W. C. (1976). Mouthparts and receptors involved in feeding behaviour and sugar perception in the African armyworm, *Spodoptera exempta* (Lepidoptera, Noctuidae). *Symp. Biol. Hung.* **16,** 139–151

Ma, W. C. and Schoonhoven, L. M. (1973). Tarsal contact chemosensory hairs of the large white butterfly *Pieris brassicae* and their possible role in oviposition behaviour. *Ent. exp. Appl.* **16,** 343–357

Ma, W. C. and Visser, J. H. (1978). Single unit analysis of odour quality coding by the olfactory antennal receptor system of the Colorado beetle. *Ent. exp. Appl.* **24,** 520–533

Madge, D. S. (1965). The responses of cotton stainers (*Dysdercus fasciatus* Sign.) to relative humidity and temperature, and the location of their hygroreceptors. *Ent. exp. Appl.* **8,** 135–152

Maes, F. W. (1980). Neural coding of salt taste quality in blowfly labellar taste organ. In: *Olfaction and Taste VII* (H. van der Starre, ed.). Information Retrieval Ltd., London

Maes, F. W. and Vedder, C. G. (1978). A morphological and electrophysiological inventory of labellar taste hairs of the blowfly *Calliphora vicina. J. Ins. Physiol.* **24,** 667–672

Malz, D. and Hintze-Podufal, C. (1979a). Bau und Verteilung von Sinnesorganen auf dens clypeolabrum, Hypopharynx und der Mandibel der Schabe, *Gromphadorhina brunneri* Butler (Blaberoidea, Oxyhaloidae). *Zool. Jb. Anat.* **102,** 249–266

Malz, D. and Hintze-Podufal, C. (1979b). Bau und Verteilung von Sinnesorganen auf dens Labium und der Maxille der Schabe *Gromphadorhina brunneri* Butler (Blaberoidea, Oxyhaloidae). *Zool. Beitr.* **25,** 81–100

Marsh, D. (1975). Responses of male aphids to the female sex pheromone in *Megoura viciae* Buckton. *J. Ent. A* **50,** 43–64

Marshall, A. T. and Lewis, C. T. (1971). Structural variation in the antennal sense organs of fulgoroid Homoptera (Insecta). *Zool. J. Linn. Soc.* **50,** 181–184

Masson, C. (1972). Le système antennaire chez les fourmis I. Histologie et ultrastructure du deutocerebron. Étude comparée chez *Camponotus vagus* (Formicinae) et *Mesoponera caffraria* (Ponerinae). *Z. Zellforsch.* **134,** 31–64

Masson, C. and Friggi, A. (1971a). Étude morphologique et topographique des sensilla ampullacea de l'antenne de l'ouvrière de *Camponotus vagus* Scop. (Hym. Formicidae). *C.r. hebd. Séanc. Acad. Sci., Paris* **272,** 618–621

Masson, C. and Friggi, A. (1971b). Responses périphériques obtenues au niveau des récepteurs olfactifs de l'antenne de l'ouvrière de *Camponotus vagus* Scop. (Formicinae). *C.r. hebd. Séanc. Acad. Sci., Paris* **272,** 2346–2349
Masson, C. and Strambi, C. (1977). Sensory antennal organization in an ant and a wasp. *J. Neurobiol.* **8,** 537–548
Masson, C., Gabouriaut, D. and Friggi, A. (1972). Ultrastructure d'un nouveau type de récepteur olfactif de l'antenne d'insecte trouve chez la fourmi *Componotus vagus* Scop. (Hymenoptera, Formicinae). *Z. Morph. Tiere* **72,** 349–360
Massoud, Z. and Deboutteville, C. D. (1969). Étude de l'organisation sensorielle de l'antenne des Neelidae (Collemboles) au microscope electronique à balayage. *C.r. hebd. Séanc. Acad. Sci., Paris.* **269,** 2554–2556
Matthews, R. W. (1974). Biology of Braconidae. *A. Rev. Ent.* **19,** 15–32
McCutchan, M. C. (1969). Responses of tarsal chemoreceptive hairs of the blowfly, *Phormia regina. J. Ins. Physiol.* **15,** 2059–2068
McIver, S. B. (1969). Antennal sense organs of female *Culex tarsalis* (Diptera: Culicidae). *Ann. ent. Soc. Am.* **62,** 1455–1461
McIver, S. B. (1970). Comparative study of antennal sense organs of female culicine mosquitoes. *Can. Ent.* **102,** 1258–1267
McIver, S. (1971). Comparative studies on the sense organs on the antennae and maxillary palps of selected male culicine mosquitoes. *Can. J. Zool.* **49,** 235–239
McIver, S. B. (1972). Fine structure of pegs on the palps of female culicine mosquitoes. *Can. J. Zool.* **50,** 571–576
McIver, S. B. (1974). Fine structure of antennal grooved pegs of the mosquito, *Aedes aegypti. Cell Tiss. Res.* **153,** 327–337
McIver, S. B. (1975). Structure of cuticular mechanoreceptors of arthropods. *Ann. Rev. Ent.* **20,** 381–397
McIver, S. and Charlton, C. (1970). Studies on the sense organs of selected culicine mosquitoes. *Can. J. Zool.* **48,** 293–295
McIver, S. and Hudson, A. (1972). Sensilla on the antennae and palps of selected *Wyeomyia* mosquitoes. *J. med. Ent.* **9,** 337–345
McIver, S. and Siemicki, R. (1975). Palpal sensilla of selected anopheline mosquitoes. *J. Parasit.* **61,** 535–538
McIver, S. and Siemicki, R. (1976). Fine structure of the antennal tip of the crabhole mosquito, *Deinocerites cancer* Theobald (Diptera: Culicidae). *Int. J. Insect Morphol. & Embryol.* **5,** 319–334
McIver, S. and Siemicki, R. (1978). Fine structure of tarsal sensilla of *Aedes aegypti* (L.) (Diptera: Culicidae). *J. Morph.* **155,** 137–156
McIver, S. and Siemicki, R. (1981). Innervation of cibarial sensilla of *Aedes aegypti* (L.) (Diptera: Culicidae). *Int. J. Insect Morphol. & Embryol.* **10,** 355–357
Meinecke, C. C. (1975). Riechsensillen und Systematik der Lamellicornia (Insecta, Coleoptera). *Zoomorphologie* **82,** 1–42
Mercer, K. L. and McIver, S. B. (1973). Sensilla on the palps of selected blackflies (Diptera: Simuliidae). *J. med. Ent.* **10,** 236–239
Miller, F. H. (1970). Scanning electron microscopy of antennal structure of *Polyplax serrata* (Burmeister) (Anoplura: Hoplopleuridae). *J.N.Y. ent. Soc.* **78,** 33–37
Mitchell, B. K. (1974). Behavioural and electrophysiological investigations on the responses of larvae of the Colorado potato beetle (*Leptinotarsa decemlineata*) to amino acids. *Ent. exp. Appl.* **17,** 255–264
Mitchell, B. K. (1976). Physiology of an ATP receptor in labella sensilla of the tsetse fly *Glossina morsitans* Westw. (Diptera: Glossinidae). *J. exp. Biol.* **65,** 259–271

Mitchell, B. K. (1978). Some aspects of gustation in the larval red turnip beetle, *Entomoscelis americana*, related to feeding and host plant selection. *Ent. exp. Appl.* **24,** 540–549

Mitchell, B. K. and Gregory, P. (1979). Physiology of the maxillary sugar sensitive cell in the red turnip beetle, *Entomoscelis americana. J. comp. Physiol.* **132,** 167–178

Mitchell, B. K. and Schoonhoven, L. M. (1974). Taste receptors in Colorado beetle larvae. *J. Ins. Physiol.* **20,** 1787–1793

Moeck, H. A. (1968). Electron microscopic studies of antennal sensilla in the ambrosia beetle *Trypodendron lineatum* (Olivier) (Scolytidae). *Can. J. Zool.* **46,** 521–556

Mohyuddin, A. I. (1973). Chemical basis of host selection in four stenophagous species that feed on *Convolvulus* and *Calystegia* at Belleville, Ontario. *Ent. exp. Appl.* **16,** 201–212

Molen, J. N. v.d., Meulen, J. W. v.d., Kramer, J. J. de and Pasveer, F. J. (1978). Computerised classification of taste cell responses. *J. comp. Physiol.* **128,** 1–11

Moran, V. C. and Brown, R. P. (1973). The antennae, host plant chemoreception and probing activity of the citrus psylla, *Trioza erytreae* (Del Guercio) (Homoptera: Psyllidae). *J. ent. Soc. sth. Afr.* **36,** 191–202

Moulins, M. (1968). Les sensilles de l'organe hypopharyngien de *Blabera craniifer* Burm. (Insecta, Dictyoptera). *J. Ultrastruct Res.* **21,** 474–513

Moulins, M. (1969). Étude anatomique de l'hypopharynx de *Forficula auricularia* L. (Insecte, Dermaptère): Téguments, musculature, organes sensoriels et innervations. Interprétation morphologique. *Zool. Jb. Anat.* **86,** 1–27

Moulins, M. (1971). Ultrastructure et physiologie des organes épipharyngiens et hypopharyngiens (chimiorécepteurs cibariaux) de *Blabera craniifer* Burm. (Insecte, Dictyoptère). *Z. vergl. Physiol.* **73,** 139–166

Mound, L. A. and Halsey, S. H. (1978). *Whitefly of the World.* British Museum (Natural History), London

Mozley, S. C. (1971). Maxillary and premental patterns in Chironominae and Orthocladiinae (Diptera: Chironomidae). *Can. Ent.* **103,** 298–305

Mulkern, G. B., Anderson, J. F. and Brusven, M. A. (1962). Biology and ecology of North Dakota grasshoppers. I. Food habits and preferences of grasshoppers associated with alfalfa fields. *Res. Rep. N. Dak. agric. Exp. Sta.* **7,** 26 pp.

Mulkern, G. B., Toczek, D. R. and Brusven, M. A. (1964). Biology and ecology of North Dakota grasshoppers. II. Food habits and preference of grasshoppers associated with the sand hills prairie. *Res. Rep. N. Dak. agric. Exp. Sta.* **11,** 59 pp.

Mustaparta, H. (1973). Olfactory sensilla on the antennae of the pine weevil, *Hylobius abietis. Z. Zellforsch.* **144,** 559–571

Mustaparta, H. (1975). Responses of single olfactory cells in the pine weevil *Hylobius abietis* L. (Col: Curculionidae). *J. comp. Physiol.* **97,** 271–290

Mustaparta, H. (1979). Chemoreception in bark beetles of the genus *Ips*: synergism, inhibition and discrimination of enantiomers. In: *Chemical Ecology: Odour Communication in Animals* (F. J. Ritter, ed.). Elsevier/North Holland Biomedical Press, Amsterdam

Mustaparta. H., Angst, M. E. and Lanier, G. H. (1977). Responses of single receptor cells in the pine engraver beetle, *Ips pini* (Say) (Coleoptera: Scolytidae) to its aggregation pheromone, ipsdienol, and the aggregation inhibitor, ipsenol. *J. comp. Physiol.* **121,** 343–347

Mustaparta, H., Angst, M. E. and Lanier, G. N. (1979). Specialization of olfactory cells to insect and host-produced volatiles in the bark beetle *Ips pini* (Say). *J. Chem. Ecol.* **5,** 109–123

Myers, J. (1968). The structure of the antennae of the Florida queen butterfly, *Danaus gilippus berenice* (Cramer). *J. Morph.* **125,** 315–328

Nault, L. R. and Styer, W. E. (1972). Effects of sinigrin on host selection by aphids. *Ent. exp. Appl.* **15,** 423–437

Nault, L. R., Edwards, L. J. and Styer, W. E. (1973). Aphid alarm pheromones: secretion and reception. *Envir. Ent.* **2,** 101–105

Nayar, F. K. and Fraenkel, G. (1963). The chemical basis of the host selection in the Mexican bean beetle, *Epilachna varivestis* (Coleoptera: Coccinellidae). *Ann. ent. Soc. Am.* **56,** 174–178

Nielsen, J. K. (1978). Host plant selection of monophagous and oligophagous flea beetles feeding on crucifers. *Ent. exp. Appl.* **24,** 562–569

Nielsen, J. K., Larsen, L. M. and Sørensen, H. (1979). Host plant selection of the horseradish flea beetle *Phyllotreta armoraciae* (Coleoptera: Chrysomelidae): identification of the flavanol glycosides stimulating feeding in combination with glucosinolates. *Ent. exp. Appl.* **26,** 40–48

Nishino, C. and Takayanagi, H. (1979). Electroantennogram responses from parts of antennae of the American cockroach. *Appl. Ent. Zool.* **14,** 326–332

O'Connell, R. J. (1972). Responses of olfactory receptors to the sex attractant, its synergist and inhibitor in the red-banded leaf roller, *Argyrotaenia velutinana.* In: *Olfaction and Taste IV* (D. Schneider, ed.). Wissenschaftliche GmbH, Stuttgart

O'Connell, R. J. (1975). Olfactory receptor responses to sex pheromone components in the redbanded leafroller moth. *J. gen. Physiol.* **65,** 179–205

Otte, D. and Joern, A. (1977). On feeding patterns in desert grasshoppers and the evolution of specialized diets. *Proc. Acad. nat. Sci. Philad.* **128,** 89–126

Otter, C. J. den (1977). Single sensillum responses in the male moth *Adoxophyes orana,* (F. v. R.) to female sex pheromone components and their geometrical isomers. *J. comp. Physiol.* **121,** 205–222

Otter, C. J. den, Schuil, H. A. and Oosten, A. S.-van, (1978). Reception of host-plant odours and female sex pheromone in *Adoxophyes orana* (Lepidoptera: Tortricidae): electrophysiology and morphology. *Ent. exp. Appl.* **24,** 570–578

Otter, C. J. den, Behan, M. and Maes, F. W. (1980). Single cell responses in female *Pieris brassicae* (Lepidoptera: Pieridae) to plant volatiles and conspecific egg odours. *J. Ins. Physiol.* **26,** 465–472

Owen, W. B. (1971). Taste receptors of the mosquito *Anopheles atroparvus* van Thiel. *J. med. Ent.* **8,** 491–494

Owen, W. B., Larsen, J. R. and Pappas, L. G. (1974). Functional units in the labellar chemosensory hairs of the mosquito *Culiseta inornata* (Williston). *J. exp. Zool.* **188,** 235–248

Pao, B. and Knight, K. L. (1970). Morphology of the fourth stage larval mouthparts of *Aedes* (*Aedimorphus*) *vexans* (Diptera: Culicidae). *J. Ga. ent. Soc.* **5,** 115–131

Pappas, L. G. and Larsen, J. R. (1976). Gustatory hairs on the mosquito, *Culiseta inornata. J. exp. Zool.* **196,** 351–360

Peck, S. B. (1977). An unusual sense receptor in internal antennal vesicles of *Ptomophagus* (Coleoptera: Leiodidae). *Can. Ent.* **109,** 81–86

Peregrine, D. J. (1972). Fine structure of sensilla basiconica on the labium of the cotton stainer, *Dysdercus fasciatus* (Signoret) (Heteroptera: Pyrrhocoridae). *Int J. Insect Morphol. & Embryol.* **1,** 241–251

Perkins, B. D. (1973). Potential for water hyacinth management with biological agents. *Proc. Tall Timbers Conf. ecol. Anim. Control Habitat Mgmnt.* No. 4, 1972, 53–64

Pers, J. N. C. van der and Otter, C. J. den, (1978). Single cell response from olfactory receptors of small ermine moth to sex-attractants. *J. Ins. Physiol.* **24,** 337–343

Persoons, C. J. and Ritter, F. J. (1979). Pheromones of cockroaches. In: *Chemical Ecology: Odour Communication in Animals* (F. J. Ritter, ed.). Elsevier/North Holland, Amsterdam

Peters, W. (1963). Die Sinnesorgane an den Labellen von *Calliphora erythrocephala* Mg. (Diptera). *Z. Morph. Ökol. Tiere* **55,** 259–320

Petryszak, A. (1975). The sensory peripheric nervous system of the *Periplaneta americana* (L.) (*Blattoidea*). I. Mouthparts. *Zesz. nauk. Uniw. Jagiellońsk.* **21,** 41–79

Petryszak, A. (1977). The sense organs of the mouthparts in *Libellula depressa* L. and *L. quadrimaculata* L. (Odonata). *Acta. biol. cracov.* (*Zool.*) **20,** 81–100

Pettersson, J. (1971). An aphid sex attractant II. Histological, ethological and comparative studies. *Ent. scand.* **2,** 81–93.

Plachter, H. (1979). Zur Kenntnis der Prämaginalstadien der Pilzmücken (Diptera, Mycetophiloidea) Teil II. Eidonomie der Larven. *Zool. Jb. Anat.* **101,** 271–392

Popham, E. J. (1979). The micro-structure of the maxillary and labial papillae of *Forficula auricularia* (Dermaptera). *J. Zool., London* **188,** 353–355

Powell, J. A. (1980). Evolution of larval food preferences in microlepidoptera. *Ann. Rev. Ent.* **25,** 133–159

Priesner, E. (1980). Sensory encoding of pheromone signals and related stimuli in male moths. In: *Insect Neurobiology and Pesticide Action* (*Neurotox* 79). Society of Chemical Industry, London

Pritchard, G. (1965). Sense organs on the labium of *Aeshna interrupta lineata* Walker (Odonata: Anisoptera). *Can. J. Zool.* **43,** 333–336

Pritchard, G. and Scholefield, P. (1978). Observations on the food, feeding behaviour, and associated sense organs of *Grylloblatta campodeiformis* (Grylloblattodea). *Can. Ent.* **110,** 205–212

Rees, C. J. C. (1969). Chemoreceptor specificity associated with choice of feeding site by the beetle, *Chrysolina brunsvicensis* on its foodplant, *Hypericum hirsutum*. *Ent. exp. Appl.* **12,** 565–583

Rice, M. J. (1973). Cibarial sense organs of the blowfly, *Calliphora erythrocephala* (Meigen) (Diptera: Calliphoridae). *Int. J. Insect Morphol. & Embryol.* **2,** 109–116

Rice, M. J. (1976). Contact chemoreceptors on the ovipositor of *Lucilia cuprina* (Wied.), the Australian sheep blowfly. *Aust. J. Zool.* **24,** 353–360

Rice, M. J. and McRae, T. M. (1976). Contact chemoreceptors on the ovipositor of *Locusta migratoria* L. *J. Aust. ent. Soc.* **15,** 364

Rice, M. J., Galun, R. and Margalit, J. (1973a). Mouthpart sensilla of the tsetse fly and their function II: Labial sensilla. *Ann. trop. Med. Parasit.* **67,** 101–107

Rice, M. J., Galun, R. and Margalit, J. (1973b). Mouthpart sensilla of the tsetse fly and their function III. Labrocibarial sensilla. *Ann. trop. Med. Parasit.* **67,** 109–116

Richerson, J. V., Borden, J. H. and Hollingdale, J. (1972). Morphology of a unique sensillum placodeum on the antennae of *Coeloides brunneri* (Hymenoptera: Braconidae). *Can. J. Zool.* **50,** 909–913

Richter, S. (1962). Unmittelbares der Sinneszellen cuticularer Sinnesorgane mit der Aussenwelt. Eine Licht- und Elektronenmikroskopische Untersuchung der Chemorezeptorischen Antennensinnesorgane der *Calliphora*-larven. *Z. Morph. Ökol. Tiere* **52,** 171–196

Riegert, P. W. (1960). The humidity reactions of *Melanoplus bivittatus* (Say) (Orthoptera, Acrididae): antennal sensilla and hygro-reception. *Can. Ent.* **92,** 561–570

Röhler, E. (1906). Beiträge zur Kenntnis der Sinnesorgane der Insekten. *Zool. Jb.* (*Anat.*) **22,** 225–288

Rohr, W. (1978). *Untersuchungen zur postembryonalen Entwicklung von Acheta domestica* (*Insecta: Saltatoria: Ensifera: Gryllidae*). Diplomarbeit, Zoologisches Institut, Braunschweig

Rościszewska, M. and Fudalewicz-Niemczyk, W. (1974). The peripheral nervous system of the larva of *Gryllus domesticus* L. (Orthoptera). Part II. Mouthparts. *Acta. biol. cracov.* (*Zool.*) **17,** 19–39

Rowell, H. F. (1978). Food plant specificity in neotropical rain-forest acridids. *Ent. exp. Appl.* **24,** 651–652

Rowley, W. A. and Cornford, M. (1972). Scanning electron microscopy of the pit of the maxillary palp of selected species of *Culicoides. Can. J. Zool.* **50,** 1207–1210

Rudinsky, J. A. and Ryker, L. C. (1977). Olfactory and auditory signals mediating behaviour patterns of bark beetles. *Colloques. int. Cent. natn. Rech. scient.* **265,** 195–209

Ruth, E. (1976). Elektrophysiologie der Sensilla Chaetica auf den Antennen von *Periplaneta americana. J. comp. Physiol. A* **105,** 55–64

Ryan, M. F. and Behan, M. (1973a). The sensory receptors of *Tribolium* larvae. *Physiol. Zool.* **46,** 238–244

Ryan, M. F. and Behan, M. (1973b). Cephalic sensory receptors of the cabbage-root fly larva, *Erioischia brassicae* (B.) (Diptera: Anthomyiidae). *Int. J. Insect Morphol. & Embryol.* **2,** 83–86

Ryan, M. F. and Behan, M. (1973c). The sensory receptors of the carrot fly larva, *Psila rosae* (F.) (Diptera, Psilidae). *Bull. ent. Res.* **62,** 545–548

Salama, H. S. (1966). The function of mosquito taste receptors. *J. Ins. Physiol.* **12,** 1051–1060

Sanes, J. R. and Hildebrand, J. G. (1976). Structure and development of antennae in a moth, *Manduca sexta. Devl. Biol.* **51,** 282–299

Sass, H. (1976). Zur nervösen Codierung von Geruchsreizen bei *Periplaneta americana. J. comp. Physiol.* **107,** 49–65

Sass, H. (1978). Olfactory receptors on the antenna of *Periplaneta*: response constellations that encode food odors. *J. comp. Physiol.* **128,** 227–233

Sass, H. (1980). Physiological and morphological identification of antennal olfactory receptors of insects. In: *Olfaction and Taste VII* (H. van der Starre, ed.). Information Retrieval Ltd., London

Schafer, R. (1971). Antennal sense organs of the cockroach, *Leucophaea maderae. J. Morph.* **134,** 91–104

Schafer, R. (1973). Postembryonic development in the antenna of the cockroach, *Leucophaea maderae*: growth, regeneration, and the development of the adult pattern of sense organs. *J. exp. Zool.* **183,** 353–364

Schafer, R. (1977). The nature and development of sex attractant specificity in cockroaches of the genus *Periplaneta* IV. Electrophysiological study of attractant specificity and its determination by juvenile hormone. *J. exp. Zool.* **199,** 189–208

Schafer, R. and Sanchez, T. V. (1973). Antennal sensory system of the cockroach, *Periplaneta americana*: postembryonic development and morphology of the sense organs. *J. comp. Neur.* **149,** 335–354

Schafer, R. and Sanchez, T. V. (1976). The nature and development of sex attractant specificity in cockroaches of the genus *Periplaneta* I. Sexual dimorphism in the distribution of antennal sense organs in five species. *J. Morph.* **149,** 139–158

Schaller, D. (1978). Antennal sensory system of *Periplaneta americana* L. Distribution and frequency of morphologic types of sensilla and their sex-specific changes during postembryonic development. *Cell Tiss. Res.* **191,** 121–139

Schneider, D. and Kaissling, K.-E. (1957). Der Bau der Antenne des Seidenspinners *Bombyx mori* L. II. Sensillen, cuticulare Bildungen und innerer Bau. *Zool. Jb.* (*Anat.*) **76,** 223–250

Schneider, D. and Kaissling, K.-E. (1957a). Der Bau der Antenne des Seidenspinners *Bombyx mori* L. I. Architektur und Bewegungsapparat der Antenne sowie Struktur der Cuticula. *Zool. Jb.* (*Anat.*) **75,** 287–310

Schneider, D. and Seibt, U. (1969). Sex pheromone of the queen butterfly: electro-antennogram responses. *Science, N.Y.* **164,** 1173–1174

Schneider, D., Lacher, V. and Kaissling, K.-E. (1964). Die Reaktionsweise und das Reaktionsspektrum von Riechzellen bei *Antheraea pernyi* (Lepidoptera, Saturniidae). *Z. vergl. Physiol.* **48,** 632–662

Schneider, D., Kafka, W. A., Beroza, M. and Bierl, B. A. (1977). Odor receptor responses of male gypsy and nun moths (Lepidoptera, Lymantriidae) to disparlure and its analogues. *J. comp. Physiol.* **113,** 1–15

Schoonhoven, L. M. (1967). Some cold receptors in larvae of three Lepidoptera species. *J. Ins. Physiol.* **13,** 821–826

Schoonhoven, L. M. (1969). Gustation and foodplant selection in some lepidopterous larvae. *Ent. exp. Appl.* **12,** 555–564

Schoonhoven, L. M. (1972). Secondary plant substances and insects. In: *Structural and Functional Aspects of Phytochemistry* (V. C. Runeckless and T. C. Tso, eds). *Recent Adv. Phytochem.* **4,** 197–224

Schoonhoven, L. M. (1973). Plant recognition by lepidopterous larvae. *Symp. Roy. Entomol. Soc. London* **6,** 87–99

Schoonhoven, L. M. (1974). Studies on the shootborer *Hipsipyla grandella* (Zeller) (Lep., Pyralidae). XXIII. Electroantennograms (EAG) as a tool in the analysis of insect attractants. *Turrialba* **24,** 24–28

Schoonhoven, L. M. and Henstra, S. (1973). Morphology of some rostrum receptors in *Dysdercus* spp. *Neth. J. Zool.* **22,** 343–346

Scott, D. A. and Zacharuk, R. Y. (1971). Fine structure of the antennal sensory appendix in the larva of *Ctenicera destructor* (Brown) (Elateridae: Coleoptera). *Can. J. Zool.* **49,** 199–210

Selzer, R. (1979). Morphological and physiological identification of food odour specific neurones in the deutocerebrum of *Periplaneta americana. J. comp. Physiol.* **134,** 159–163

Shambaugh, G. F., Frazier, J. L., Castell, A. E. M. and Coons, L. B. (1978). Antennal sensilla of seventeen aphid species (Homoptera: Aphidinae). *Int. J. Insect Morphol. & Embryol.* **7,** 389–404

Singleton-Smith, J., Chang, F. J. and Philogène, B. J. R. (1978). Morphological differences between nymphal instars and descriptions of the antennal sensory structures of the nymphs and adults of *Psylla pyricola* Foerster (Homoptera: Psyllidae). *Can. J. Zool.* **56,** 1576–1584

Sinha, A. K. and Krishna, S. S. (1970). Further studies on the feeding behaviour of *Aulacophora foveicollis* on cucurbitacin. *J. econ. Ent.* **63,** 333–334

Sinitsyna, Y. Y. (1971). Bioelectric activity of the contact chemoreceptive sensilla of *Aedes aegypti* L. (Diptera, Culicidae) and food reactions to two-component solutions. *Ent. Rev.* **50,** 151–155

Slansky, F. (1974). Relationships of larval food-plants and voltinism patterns in temperate butterflies. *Psyche, Camb., Mass.* **81,** 243–253

Slifer, E. H. (1962). Sensory hairs with permeable tips on the tarsi of yellow-fever mosquito, *Aedes aegypti. Ann. ent. Soc. Am.* **55,** 531–535

Slifer, E. H. (1967). Sense organs on the antennal flagella of earwigs (Dermaptera) with special reference to those of *Forficula auricularia. J. Morph.* **122,** 63–80

Slifer, E. H. (1968a). Sense organs on the antennal flagellum of a giant cockroach, *Gromphadorhina portentosa*, and a comparison with those of several other species (Dictyoptera, Blattaria). *J. Morph.* **126,** 19–30

Slifer, E. H. (1968b). Sense organs on the antennal flagellum of a praying mantis, *Tenodera angustipennis*, and of two related species (Mantodea). *J. Morph.* **124,** 105–116

Slifer, E. H. (1969). Sense organs on the antenna of a parasitic wasp *Nasonia vitripennis* (Hymenoptera, Pteromalidae). *Biol. Bull.* **136,** 253–263

Slifer, E. H. (1970). The structure of arthropod chemoreceptors. *Ann. Rev. Ent.* **15,** 121–142

Slifer, E. H. (1974). Structures on the antennal flagellum of a katydid, *Neoconocephalus ensiger* (Orthoptera, Tettigoniidae). *J. Morph.* **143,** 435–444

Slifer, E. H. (1976a). Sense organs on antennal flagellum of *Grylloblatta campodeiformis* E. M. Walker (Orthoptera: Grylloblattodea). *Ent. News* **87,** 275–279

Slifer, E. H. (1976b). Sense organs on the antennal flagellum of a bird louse (Mallophaga). *J. N.Y. ent. Soc.* **84,** 159–165

Slifer, E. H. (1979a). Sense organs on the antennal flagellum of the mimosa webworm, *Homadaula anisocentra* (Lepidoptera, Glyphipterygidae). *J. Morph.* **162,** 163–174

Slifer, E. H. (1979b). Sense organs on the antennal flagella of four species of *Chrysopa* (Neuroptera: Chrysopidae). *Ann. ent. Soc. Am.* **72,** 529–531

Slifer, E. H. (1980). Chemoreceptors and other sense organs on the antennal flagellum of the flea, *Ctenocephalides canis* (Siphonaptera: Pulicidae). *Ann. ent. Soc. Am.* **73,** 418–419

Slifer, E. H. and Sekhon, S. S. (1961). Fine structure of the sense organs on the antennal flagellum of the honey bee, *Apis mellifera* Linnaeus. *J. Morph.* **109,** 351–381

Slifer, E. H. and Sekhon, S. S. (1963). Sense organs on the antennal flagellum of the small milkweed bug, *Lygaeus kalmii* Stal (Hemiptera, Lygaeidae). *J. Morph.* **112,** 165–193

Slifer, E. H. and Sekhon, S. S. (1964). Fine structure of the sense organs on the antennal flagellum of a flesh fly, *Sarcophaga argyrostoma* R.-D. (Diptera, Sarcophagidae). *J. Morph.* **114,** 185–208

Slifer, E. H. and Sekhon, S. S. (1970). Sense organs of a Thysanuran, *Ctenolepisma lineata pilifera*, with special reference to those on the antennal flagellum (Thysanura, Lepismatidae). *J. Morph.* **132,** 1–26

Slifer, E. H. and Sekhon, S. S. (1971). Circumfila and other sense organs on the antenna of the sorghum midge (Diptera, Cecidomyiidae). *J. Morph.* **133,** 281–302

Slifer, E. H. and Sekhon, S. S. (1972). Sense organs on the antennal flagella of damselflies and dragonflies (Odonata). *Int. J. Insect Morphol. & Embryol.* **1,** 289–300

Slifer, E. H. and Sekhon, S. S. (1973). Sense organs on the antennal flagellum of two species of Embioptera (Insecta). *J. Morph.* **139,** 211–226

Slifer, E. H. and Sekhon, S. S. (1974). Sense organs on the antennae of two species of thrips (Thysanoptera, Insecta). *J. Morph.* **143,** 445–456

Slifer, E. H. and Sekhon, S. S. (1978a). Sense organs on the antennae of two species of Collembola (Insecta). *J. Morph.* **157,** 1–20

Slifer, E. H. and Sekhon, S. S. (1978b). Structures on the antennal flagellum of *Zorotypus hubbardi* (Insecta, Zoraptera). *Notul. Nat.* no. 453, 8 pp.

Slifer, E. H. and Sekhon, S. S. (1980). Sense organs on the antennal flagellum of the human louse, *Pediculus humanus* (Anoplura). *J. Morph.* **164,** 161–166

Slifer, E. H., Prestage, J. J. and Beams, H. W. (1959). The chemoreceptors and other sense organs on the antennal flagellum of the grasshopper (Orthoptera; Acrididae). *J. Morph.* **105,** 145–191

Slifer, E. H., Sekhon, S. S. and Lees, A. D. (1964). The sense organs on the antennal flagellum of aphids (Homoptera), with special reference to the plate organs. *Q. J. microsc. Sci.* **105,** 21–29

Smith, C. M., Frazier, J. L., Coons, L. B. and Knight, W. E. (1976). Antennal sensilla of the clover head weevil *Hypera meles* (F.) (Coleoptera: Curculionidae). *Int. J. Insect Morphol. & Embryol.* **5,** 349–355

Städler, E. (1978). Chemoreception of host plant chemicals by ovipositing females of *Delia (Hylemya) brassicae. Ent. exp. Appl.* **24,** 711–720

Städler, E. and Hanson, F. E. (1975). Olfactory capabilities of the "gustatory" chemoreceptors of the tobacco hornworm larva. *J. comp. Physiol.* **104,** 97–102

Städler, E. and Seabrook, W. D. (1975). Chemoreceptors on the proboscis of the female eastern spruce budworm: electrophysiological study. *Ent. exp. Appl.* **18,** 153–160

Städler, E., Städler-Steinbrüchel, M. and Seabrook, W. D. (1974). Chemoreceptors on the proboscis of the female eastern spruce budworm. *Mitt. schweiz. ent. Ges.* **47,** 63–68

Starre, H. van der (1972). Tarsal taste discrimination in the blowfly, *Calliphora vicina* Robineau-Desvoidy. *Neth. J. Zool.* **22,** 227–282

Steinbrecht, R. A. (1970). Zur Morphometrie der Antenne des Seidenspinners, *Bombyx mori* L.: Zahl und Verteilung der Riechsensillen (Insecta, Lepidoptera). *Z. Morph. Tiere.* **68,** 93–126

Steinbrecht, R. A. (1973). Die Feinbau olfaktorischer Sensillen des Seidenspinners (Insecta, Lepidoptera). *Z. Zellforsch.* **139,** 533–565

Steinbrecht, R. A. and Müller, B. (1976). Fine structure of the antennal receptors of the bed bug, *Cimex lectularius* L. *Tissue & Cell* **8,** 615–636

Steward, C. C. and Atwood, C. E. (1963). The sensory organs of the mosquito antenna. *Can. J. Zool.* **41,** 577–594

Strausfeld, N. J. (1976). *Atlas of an Insect Brain.* Springer Verlag, Berlin

Strenzke, K. (1960). Metamorphose und Verwandtschaftsbeziehungen der Gattung *Clunio* Hal. (Dipt.). *Soumal. eläin–ja kasvit. Seur, van. eläin. Julk.* **22,** 1–30

Sublette, J. E. and Sublette, M. F. (1973). The morphology of *Glyptotendipes barbipes* (Staeger) (Diptera, Chironomidae). *Studies in Natural Science, Portales, New Mexico.* **1(6),** 1–82

Sutcliffe, J. F. and McIver, S. B. (1976). External morphology of sensilla on the legs of selected black fly species (Diptera: Simuliidae). *Can. J. Zool.* **54,** 1779–1787

Sutcliffe, J. F. and Mitchell, B. K. (1980). Structure of galeal sensory complex in adults of the red turnip beetle, *Entomoscelis americana* Brown (Coleoptera, Chrysomelidae). *Zoomorphology* **96,** 63–76

Takahashi, R. (1971). Monophagy and polyphagy among phytophagous insects: an evolutionary consideration. *Kontyû* **12,** 130–139

Tamaki, Y. (1977). Complexity, diversity, and specificity of behavior-modifying chemicals in Lepidoptera and Diptera. In: *Chemical Control of Insect Behavior: Theory and Application* (H. H. Shorey and J. J. McKelvey, eds). Wiley, New York

Thomas, J. G. (1965). The abdomen of the female desert locust (*Schistocerca gregaria* Forskål) with special reference to the sense organs. *Anti-Locust Bull.* no. 42, 20 pp.

Thomas, J. G. (1966). The sense organs on the mouthparts of the desert locust (*Schistocerca gregaria*). *J. Zool., London* **148,** 420–448

Thorsteinson, A.J. (1958). The chemotactic influence of plant constituents on feeding by phytophagous insects. *Ent. exp. Appl.* **1,** 23–27

Tjallingii, W. F. (1978). Mechanoreceptors of the aphid labium. *Ent. exp. Appl.* **24,** 731–737

Toh, Y. (1977). Fine structure of antennal sense organs of the male cockroach, *Periplaneta americana*. *J. Ultrastruct. Res.* **60,** 373–394

Tominaga, Y., Kabuta, H. and Kuwabara, M. (1969). The fine structure of the interpseudotracheal papilla of a fleshfly. *Annotnes zool. jap.* **42,** 91–104

Uchida, K. (1979). Cibarial sensilla and pharyngeal valves in *Aedes albopictus* (Skuse) and *Culex pipiens pallens* Coquillett (Diptera: Culicidae). *Int. J. Insect Morphol. & Embryol.* **8,** 159–167

Uvarov, B. (1977). *Grasshoppers and Locusts*, Vol. 2. Centre for Overseas Pest Research, London

Vale, G. A. (1980). Field studies of the responses of tsetse flies (Glossinidae) and other Diptera to carbon dioxide, acetone and other chemicals. *Bull. ent. Res.* **70,** 563–570

Vareschi, E. (1971). Duftunterscheidung bei der Honigbiene–Einzelzell-Ableitungen und Verhaltensreaktionen. *Z. vergl. Physiol.* **75,** 143–173

Viscuso, R. (1974). Studio comparato della lamina epifaringea di vari insetti Ortotteri. *Redia* **55,** 129–141

Viscuso, R., Longo, G. and Sottile, L. (1978). Studio comparato della lamina epifaringea degli Ortotteri e di altri ordini affini *Proc. XI Cong. Naz. Ital. Ent.* 95–103

Visser, J. H. and Avé, D. A. (1978). General green leaf volatiles in the olfactory orientation of the Colorado beetle, *Leptinotarsa decemlineata. Ent. exp. Appl.* **25,** 738–749

Voegelé, J., Cals-Usciati, J., Pihan, J.-P. and Daumal, J. (1975). Structure de l'antenne femelle des trichogrammes. *Entomophaga* **20,** 161–169

Waldorf, E. S. (1976). Antennal amputations in *Sinella curviseta* (Collembola: Entomobryidae). *Ann. ent. Soc. Am.* **69,** 841–842

Waldow, U. (1970). Elektrophysiologische Untersuchungen an Feuchte-, Trocken- und Kalterezeptoren auf der Antenne der Wanderheuschrecke *Locusta. Z. vergl. Physiol.* **69,** 249–283

Waldow, U. (1973). Elektrophysiologie eines neuren Aasgeruchrezeptors und seine Bedeutung für das Verhalten des Totengräbers (*Necrophorus*). *J. comp. Physiol.* **83,** 415–424

Waldow, U. (1975). Multimodale Neurone im Deutocerebrum von *Periplaneta americana. J. comp. Physiol.* **101,** 329–341

Waldow, U. (1977). CNS units in cockroach (*Periplaneta americana*): Specificity of response to pheromones and other odor stimuli. *J. comp. Physiol.* **116,** 1–17

Wall, C. (1978). Morphology and histology of the antenna of *Cydia nigricana* (F.) (Lepidoptera: Tortricidae). *Int. J. Insect Morphol. & Embryol.* **7,** 237–250

Wallis, D. I. (1962). The sense organs on the ovipositor of the blowfly, *Phormia regina* Meigen. *J. Ins. Physiol.* **8,** 453–467

Wang, C.-H. and Huber, F. (1976). Morphological study of the aphid antennae of *Aphis nerri* Boyer (Homoptera: Aphididae). I. Flagellar sensilla. *Bull Inst. Zool. Acad. sin.* **15,** 47–56

Ward, L. K. (1973). Thysanoptera occurring in flowers of a chalk grassland. *Entomologist* **106,** 97–113

Washio, H. and Nishino, C. (1976). Electroantennogram responses to the sex pheromone and other odours in the American cockroach. *J. Ins. Physiol.* **22,** 735–741

Washio, H., Nishino, C. and Bowers, W. S. (1976). Antennal receptor response to sex pheromone mimics in the American cockroach. *Nature, London* **262,** 487–489

Weide, W. (1960). Einige Bemerkungen über die antennalen Sensillen sowie über das Fühlerwachstum der Stabheuschrecke *Carausius* (*Dixippus*) *morosus* Br. (Insecta: Phasmida). *Wiss. Z. Martin-Luther-Univ. Halle-Wittenb.* **9,** 247–250

Wensler, R. J. (1977). The fine structure of distal receptors on the labium of the aphid, *Brevicoryne brassicae* L. (Homoptera). *Cell Tiss. Res.* **181,** 409–422

Wensler, R. J. and Filshie, B. K. (1969). Gustatory sense organs in the food canal of aphids. *J. Morph.* **129,** 473–492

Whitehead, A. T. (1978). Electrophysiological response of honey bee labial palp contact chemoreceptors to sugars and electrolytes. *Physiol. Ent.* **3,** 241–248

Whitehead, A. T. (1981). Ultrastructure of sensilla of the female mountain pine beetle, *Dendroctonus ponderosae* Hopkins (Coleoptera: Scolytidae). *Int. J. Insect Morphol. & Embryol.* **10,** 19–28

Whitehead, A. T. and Larsen, J. R. (1976a). Ultrastructure of the contact chemoreceptors of *Apis mellifera* L. (Hymenoptera: Apidae). *Int. J. Insect Morphol. & Embryol.* **5,** 301–315

Whitehead, A. T. and Larsen, J. R. (1976b). Electrophysiological responses of galeal contact chemoreceptors of *Apis mellifera* to selected sugars and electrolytes. *J. Ins. Physiol.* **22,** 1609–1616

Wieczorek, H. (1976). The glycoside receptor of the larvae of *Mamestra brassicae* L. (Lepidoptera, Noctuidae). *J. comp. Physiol.* **106,** 153–176

Wieczorek, H. (1978). Biochemical and behavioural studies of sugar reception in the cockroach. *J. comp. Physiol.* **A 124,** 353–356

Wigglesworth, V. B. (1959). The histology of the nervous system of an insect, *Rhodnius prolixus* (Hemiptera) I. The peripheral nervous system. *Q. J. microsc. Sci.* **100,** 285–298

Wilczek, M. (1967). The distribution and neuroanatomy of the labellar sense organs of the blowfly *Phormia regina* Meigen. *J. Morph.* **122,** 175–202

Winstanley, C. and Blaney, W. M. (1978). Chemosensory systems of locusts in relation to feeding. *Ent. exp. Appl.* **24,** 750–758

Wolk, G. M. van der (1978). The typology and topography of the tarsal chemoreceptors of the blow flies *Calliphora vicina* Robineau – Desvoidy and *Phormia terranovae* Robineau – Desvoidy and the housefly *Musca domestica* L. *J. Morph.* **157,** 201–209

Yamada, M. (1971). A search for odour encoding in the olfactory lobe. *J. Physiol.* **214,** 127–143

Yamada, Y., Ishikawa, Y., Ikeshoji, T. and Matsumoto, Y. (1981). Cephalic sensory organs of the onion fly larva, *Hylemya antiqua* Meigen (Diptera: Anthomyiidae) responsible for host-plant finding. *Appl. Ent. Zool.* **16,** 121–128

Yin, W.-Y. and Li, Y.-P. (1980). Scanning electron microscopy of antennal sensilla of the pink bollworm, *Pectiniphora gossypiella. Acta. ent. sin.* **23,** 123–129

Yokohari, F. (1978). Hygroreceptor mechanism in the antenna of the cockroach *Periplaneta*. *J. comp. Physiol.* **124,** 52–60

Zacharuk, R. Y. (1962). Sense organs of the head of larvae of some Elateridae (Coleoptera): Their distribution, structure and innervation. *J. Morph.* **111,** 1–22

Zacharuk, R. Y. (1980). Ultrastructure and function of insect chemosensilla. *Ann. Rev. Ent.* **25,** 27–47

Zacharuk, R. Y. and Blue, S. G. (1971a). Ultrastructure of the peg and hair sensilla on the antenna of larval *Aedes aegypti* (L.). *J. Morph.* **135,** 433–456

Zacharuk, R. Y. and Blue, S. G. (1971b). Ultrastructure of a chordotonal and a sinusoidal peg organ in the antenna of larval *Aedes aegypti* (L). *Can. J. Zool.* **49,** 1223–1229

Zacharuk, R. Y., Yin, L. R.-S. and Blue, S. G. (1971). Fine structure of the antenna and its sensory cone in larvae of *Aedes aegypti* (L.). *J. Morph.* **135,** 273–297

Zwarzin, A. (1912). Histologische Studien über Insekten. II. Das sensible Nervensystem der Aeschnalarven. *Z. wiss. Zool.* **100,** 245–286

Subject Index

Cumulative List of Authors

Numbers in bold face indicate the volume numbers of the series

Cumulative List of Chapter Titles

Numbers in bold face indicate the volume number of the series